Variations on Media Thinking

Posthumanities
Cary Wolfe, Series Editor

52 VARIATIONS ON MEDIA THINKING
    Siegfried Zielinski

51 AESTHESIS AND PERCEPTRONIUM: ON THE ENTANGLEMENT OF SENSATION, COGNITION, AND MATTER
    Alexander Wilson

50 ANTHROPOCENE POETICS: DEEP TIME, SACRIFICE ZONES, AND EXTINCTION
    David Farrier

49 METAPHYSICAL EXPERIMENTS: PHYSICS AND THE INVENTION OF THE UNIVERSE
    Bjørn Ekeberg

48 DIALOGUES ON THE HUMAN APE
    Laurent Dubreuil and Sue Savage-Rumbaugh

47 ELEMENTS OF A PHILOSOPHY OF TECHNOLOGY: ON THE EVOLUTIONARY HISTORY OF CULTURE
    Ernst Kapp

46 BIOLOGY IN THE GRID: GRAPHIC DESIGN AND THE ENVISIONING OF LIFE
    Phillip Thurtle

45 NEUROTECHNOLOGY AND THE END OF FINITUDE
    Michael Haworth

44 LIFE: A MODERN INVENTION
    Davide Tarizzo

43 BIOAESTHETICS: MAKING SENSE OF LIFE IN SCIENCE AND THE ARTS
    Carsten Strathausen

42 CREATURELY LOVE: HOW DESIRE MAKES US MORE AND LESS THAN HUMAN
    Dominic Pettman

41 MATTERS OF CARE: SPECULATIVE ETHICS IN MORE THAN HUMAN WORLDS
    Maria Puig de la Bellacasa

(continued on page 427)

Variations on Media Thinking

Siegfried Zielinski

posthumanities 52

UNIVERSITY OF MINNESOTA PRESS
Minneapolis • London

Chapter 17 was first published in English as "Designing & Revealing—Some Aspects of a Genealogy of Projection," in *Light, Image, Imagination: The Spectrum beyond Reality and Illusion*, edited by Martha Blassnigg (Amsterdam: Amsterdam University Press, 2013).

Copyright 2019 by the Regents of the University of Minnesota

All rights reserved. No part of this publication may be reproduced, stored in a retrieval system, or transmitted, in any form or by any means, electronic, mechanical, photocopying, recording, or otherwise, without the prior written permission of the publisher.

Published by the University of Minnesota Press
111 Third Avenue South, Suite 290
Minneapolis, MN 55401-2520
http://www.upress.umn.edu

Printed in the United States of America on acid-free paper

The University of Minnesota is an equal-opportunity educator and employer.

27 26 25 24 23 22 21 20 19   10 9 8 7 6 5 4 3 2 1

Library of Congress Cataloging-in-Publication Data
Names: Zielinski, Siegfried, author.
Title: Variations on media thinking / Siegfried Zielinski.
Description: Minneapolis : University of Minnesota Press, [2019] | Series: Posthumanities ; 52 |
Includes bibliographical references and index. |
Identifiers: LCCN 2018058229 (print) | ISBN 978-1-5179-0707-5 (hc) | ISBN 978-1-5179-0708-2 (pb)
Subjects: LCSH: Mass media—Historiography. | Mass media—Philosophy.
Classification: LCC P91 .Z539513 2019 (print) | DDC 302.23/0722—dc23
LC record available at https://lccn.loc.gov/2018058229

For Maja

Contents

Acknowledgments ix

Introduction. Generators of Surprise:
Diverse Media Thinking xi

I. Provocations

1. History as Entertainment and Provocation:
   The TV Series *Holocaust* in West Germany 3

2. Media Archaeology: Searching for Different Orders
   of Envisioning 35

3. Seven Items on the Net 43

4. Toward a Dramaturgy of Differences 49

5. From Territories to Intervals: Some Preliminary Thoughts
   on the Economy of Time/the Time 57

6. On the Difficulty to Think Twofold in One
   with Nils Röller 77

7. The Art of Design: (Manifesto) On the State of Affairs
   and Their Agility 91

8. "Too Many Images!—We Have to React": Theses toward
   an Apparatical Prosthesis for Seeing—in the Context of
   Godard's *Histoire(s) du cinéma* 99

II. Particular Archaeologies

9. The Audiovisual Time Machine: Concluding Theses
   on the Cultural Technique of the Video Recorder.......... 127

10. War and Media: Marginalia of a Genealogy, in Legends
    and Images 147

11. *Theologi electrici*: A Few Passages 169

12. Historic Modes of the Audiovisual Apparatus 187

13. "To All!" The Struggle of the German Workers Radio Movement, 1918–1933 207

14. Urban Music Box, Urban Hearing: Avraamov's *Symphony of Sirens* in Baku and Moscow, 1922–1923—A Media-Archaeological Miniature.......... 261

15. How One Sees 269

16. Lüology, Techno-souls, Artificial Paradises: Fragments of an An-archaeology of Sound Arts..................... 305

17. Designing and Revealing: Some Aspects of a Genealogy of Projection.. 339

18. Allah's Automata: Where Ancient Oriental Learning Intersects with Early Modern Europe. A Media-Archaeological Miniature by Way of Introduction 377

Publication History 403

Index... 407

Acknowledgments

Many thanks to Cary Wolfe for this opportunity to make a selection of my essays and shorter texts available to English-speaking readers. His crucial engagement and the confidence of his publisher, Doug Armato, have made possible and sustained this project. Aaron Jaffe, through his hospitality, also greatly contributed to its realization. Many thanks as well to Geoffrey Winthrop-Young and Lauren K. Wolfe, who, despite serious time constraints, undertook the translation of the original German texts; the oldest of these was written forty-four years ago, and the language must have been somewhat imposing. Mário Gomes did an excellent job proofreading the final texts. Stefanie Rau contributed patience, care, and her unique sensibility in creating the prototypes for the production of each text. Gabriel Levin supported the publication of the book in a wonderful and precise way. I thank David Fideler for his great work on the layout and design. I am also grateful to the institutions that supported this project in various ways, especially the University of Arts and Design and the Center for Art and Media in Karlsruhe. Thanks to the publishers who generously permitted the reprinting of individual essays. I owe very special thanks, as always, to Daniel Irrgang, who had an active hand in every phase of the project, including indexing, and whose extraordinary editorial, conceptual, and organizational abilities have been essential to the book's completion.

Translated by Lauren K. Wolfe

Introduction

Generators of Surprise
Diverse Media Thinking

Past centuries have provided us with plenty of those who prophesize and plenty of those who warn against the conquest of the last refuges of the *anthropos* by instruments and technical systems. Catholic mathematician Johann Zahn (1641–1707) believed the artificial eye *(oculus artificialis)* wielded such enormous power that the optical apparatus—a robust telescope with a projection chamber attached—could even extract impure spots from the supposedly pure and divine Sun, that it could, in other words, outwit astrophysical reality. Hegelian philosopher Ernst Kapp (1808–96) was urging as early as 1877 that culture itself be reconceived from a technological perspective as *organ projection* and that the structure of language was so intimately bound up with the nature of the state that the development of electronic communications networks and the kinematic concept of *disciplinary full-closure* represented the becoming-apparatus of the actual late nineteenth-century state. Friedrich Nietzsche (1844–1900) insisted toward the end of his life that the more his own psychophysical powers of handwriting deteriorated, the more his typewriter would become coauthor of his texts. His pencil was smarter than he was, Albert Einstein (1859–1955) is purported to have joked. Bertolt Brecht (1898–1956) knew already in the 1920s that art without technology was sheer absurdity. Walter Benjamin (1892–1940) assumed—if somewhat more serenely—that the typewriter would alienate the pen-holding hand of the litterateur only "once the precision of typographical forms were immediately assimilated into the conception of his books ... and the innervations of the commanding fingers had replaced the familiar hand."[1] Catholic iconoclast Marshall McLuhan (1911–80) wanted us to seek the agent of our sensibility and our understanding in the medium and nowhere else, though this first pop star in the global market of media thinking largely left open what he actually meant by this mysterious portent, the *medium*. Friedrich Knilli (1930–), who was raised among the cutting and sewing machines of his uncle's garment factory in the Austrian city of Graz and later studied mechanical engineering, came to understand the powerful materiality of the medial through the vibrating membranes of loudspeakers in Austria's and Germany's early radio studios and then developed from this his own psychophysical and aesthetic concept of the *total sound play (totales Schallspiel)*. That was about the same time that Jacques Lacan (1901–81) began insisting emphatically that even the unconscious was structured like language. It was also at about this time that structure itself began to win the upper hand over the Subject in disciplines ranging from ethnology to history and literature and even early theories of cinema. No longer were we Subjects but *Projects* that in the ideal circumstance projected worlds of our own—as Vilém Flusser (1920–91) consistently emphasized in his own unique way. In an equally eschatological gesture, Friedrich Kittler (1943–2011) claimed, on the basis of his "technical a priori," that all that was expressed and all that our eyes and ears received as symbolic material was first and foremost technology and that it always would remain technology, even at the vanishing point of its development.

Each of these diverse concerns, from prophetic and cautionary voices alike, has gained acceptance in various ways. What we refer to as our world is no longer thinkable without the medial. Mathematicians and physicists, medievalists, philologists of all kinds, theologians, philosophers, biologists, and art critics all know that they must deal with media—or at least with materials that are contingent on media—when they trawl through the containers, archives, and contemporaneous utterances that have been produced in their respective fields, in their endeavor to understand and to impart. All of them equally must learn to read, interpret, and calculate medial surfaces and materialities as well as the metaphysical messages intimately linked with these—messages that articulate and transport symbolic bodies and their networks.

The urgent question is, do these practices of *expanded hermeneutics* which technical objects and medial circumstances require of us demand in turn a separate academic discipline of their own? Must the heterogeneous, prismatic, aggregate phenomena of techno-aesthetic media be accommodated under a distinct and specialized system with delimited epistemic objectives and specific methods? Can we actually neatly organize the constant transgression of limits that the thinking of medial circumstances requires? Or must such an endeavor necessarily break down into paradox, as it did for instance in Pier Paolo Pasolini's (1922–75) world of poetry and politics? Or again—with reference to a field adjacent to media thinking—has it ultimately proved to be good for the arts and the infinite diversity of our ways of seeing that we in the German-speaking humanities have established a unique compartment for them, one that now threatens to turn into the massive and highly ordered curio cabinet of a hegemonic regime of *visual studies (Bildwissenschaften)*, in which the whole of our knowledge of what can be seen in objective form is filed, sorted, and archived?

Perhaps it is too soon to answer such questions with any certainty. It is, however, not too soon to pose them resolutely. For the tendency toward the establishment of media theory (in Germany, this has even been immoderately propagated as "media science" [*Medienwissenschaft*]) as its own discipline with its own laws, hierarchies, canons, power structures, conceptualities, and clearly defined origins, is quite strong. The disastrous consequence of such a circumscription, for instruction and increasingly also for research, is that the loops of self-reflectivity embarked upon by both apprenticed and established media experts assume ever more audacious forms and contents. Students now earnestly believe that the only legitimate content of media can be nothing but other media, and they write and act accordingly. Ever since critical thinking was cast out from the humanities, the medial has been confirmed and celebrated—and decelerated—as *the* communicative potential: for control and correction but also for culture. The *difference engine* has become an engine of management and design, even an engine of careers. To posit that nothing anymore can exist and thrive independently of mediation-machines tends to inflate those very machines into all-powerful, self-sufficient centrifuges positioned right at the center of what journalists still blithely refer to as society, where they whirl away, organizing their own academic circles around themselves.

Interdiscursivity implemented in experimental practice requires prismatic ways of seeing, judicious but always moveable viewpoints, artful variants as well as the development of elegant, multiperspectival narratives. Freestyle thinking—as I attempted to formulate that of Vilém Flusser[2]—without banisters to offer provisional support can become, in the long run, the movement of a prescriptive regulating-machine.

Elaborated media thinking needs in its immediate vicinity the depths and gravity of other modes of thought that are not oriented toward medial phenomena, with which it may periodically connect, by which it may be stimulated, urged on, occasionally reined in, and reminded of its place. The study of medial sensations and structures is not an end in itself, or else it would devolve into the very paradox in the void which Baudrillard never tired of criticizing. Ultimately, technical means of communication only serve to make encounter impossible. It is the imaginary that saves us in the ongoing acid test between the real and the symbolic; yet it punishes us at the same time, given its semblant character. It was Jacques Lacan, borrowing an exceptional media concept from Lucretius,[3] who so admirably formulated this in a number of variants.

Let's dwell for a moment, then, with this ancient thinker who has been so tremendously important in my own intellectual passages through medial phenomena. The *clinamen* is "the smallest deviation possible" that may take place "we know not when, we know not where," as Louis Althusser puts it, citing *De rerum naturae,* that incomparable natural-historical poem written by Lucretius in the last century prior to the Common Era. The clinamen causes an atom to "swerve" from its vertical plunge into the void, where "there occurs *an encounter* between one atom and another, and this *event* becomes *advent* on condition of the parallelism of the atoms, for it is this parallelism which, violated on just one occasion, induces the gigantic pile-up and collision-interlocking of an infinite number of atoms, from which a world is born"—a world, in other words, as an aggregate of atoms that is created through a chain reaction set off by the first swerve and the first encounter.[4] Althusser, along with ancient Greek natural philosopher Epicurus, was convinced that the origin of any world, thus any reality and any meaning, is due to a deviation, that deviation and not reason is the cause of the origin of the world.[5]

Considered from the perspective of deep time, my own media research has at its core been powerfully shaped by those thinkers, poets, and naturalists known to the histories of science as Atomists. Before Socrates, and of course beyond the great dividers, Plato and Aristotle, the Atomists conceive of the world fundamentally as turmoil, a ceaselessly streaming exchange of the smallest particles, energies, and signals, a world that does not yet require such severings as that between subject and object, active and passive, matter and mind, between the receiver on one side and the sender on the other. Anaxagoras, Anaximander, Democritus, Empedocles, Epicurus,

Lucretius, and others thought the world as perpetually colliding objects, as the billiard-reality of interobjectivity, two and a half thousand years before this concept again acquired effective power under the banner of things becoming independent, interobjective; they thought chaos, its complex regularities and its incalculabilities; they thought the world as comprising porous objects that articulate themselves and thus reveal themselves to our perception, just as much as we in turn are realized for them, become ecstatic for them and step out of ourselves. It was Martin Heidegger who rediscovered this world in the twentieth century and ontologically fundamentalized it with unnecessary severity. And for French philosophers of becoming and of energetic dialogue, too—from Gilles Deleuze to Félix Guattari and, in a different form, from Alain Badiou via Jean-Luc Nancy to Jacques Rancière, with whom I had the pleasure of teaching on the same faculty for a number of years[6]—this world full of motion and events is the only thinkable one, or better, the only attractive one with respect to a basic idea: that the world which is known to us has only a single raison d'être, which consists in the fact that it is changeable and that it is constantly changing.

Michel Foucault was a master of the kind of writing that makes us operatively conscious of where what we call our civilization comes from, why and how we have evolved into powerful beings; and he managed to pose these questions in such a way that we are able to critically examine what we call history even as we write it. Deriving it from an antihistorical concept developed by Nietzsche, Foucault designates this process as *genealogy*. It enables us to understand developments as labyrinthine, as movements associated with digressions and impasses, and it assumes a many-eyed seeing and a many-tongued writing.

After Nietzsche, Foucault expended enormous energy in an attempt to uncover how a scattered world is assembled to produce a specific world. In so doing, he increasingly substituted genealogical tactics for the meta-methods of archaeology, with their utopian and teleological promises. Genealogy "does not seek out and describe the 'things' that phenomenology holds to be the world, but rather delineates the *manner* in which the 'things' are 'made' into 'facts'"[7]—a variant of early ecological thinking that seeks to represent relation simultaneous with substance. Nietzsche had himself already proposed a methodological principle by means of which just such a genealogical representation could be realized. "Main proposition: no regressive hypotheses! . . . And as many individual observations as possible." "Task: to see things as they are! Means: to look on them from a hundred eyes, from many persons."[8]

As media researchers who think materiologically, we opt—in the event that we must choose—for the particulars we can experience and not the all-binding general which can only be thought. This I learned first and most profoundly from Friedrich Knilli, the machine builder. This choice we make is deeply connected with respect for the artifact, for the technical, biological, and cultural other, with respect for that which is not identical with us.

We do not need a new ontology, neither subject nor object oriented, to play together, critically and productively, with the things, facts, and circumstances, the words and concepts, that have to do with media or that are constituted and produced through media. Just as Nietzsche wished to provoke the dull, entrenched, and encrusted mentality of parochial philosophers and historians at the end of the nineteenth century, and like the clique surrounding Foucault also attempted to do in the last half of the twentieth, now at the start of the twenty-first century, we need another *move into the open*.[9] This is the title of Berlin-based philosopher Dietmar Kamper's invitation to unabashed exchange among architects, artists, musicians, philosophers, and media specialists—an invitation to a debate that need not lead to resolutions but to an intellectual adventure, that may not necessarily rule out academic chairmanships but that ultimately may not need them.

I have in my arsenal of language no better phrase for describing what the project toward a genealogy of media thinking is really about. The avant-garde is nothing but the reinterpretation of bygone presents, and genealogy proves above all to be an operation with a lofty aim, namely, to reopen the windows and doors onto that nervous, heterotopic space of possibilities that the thinking of media once occupied, and organize passages through the boredom that has taken root there, and recall the *gardens*[10] that poachers from the most disparate disciplines and schools of thought have in passing laid out there and cultivated.

In his book *Experimentalsysteme und die epistemischen Dinge* (Experimental systems and epistemic things, 2001), Hans-Jörg Rheinberger, a biologist, philosopher, and historian of science, uses the term *Überraschungsgenerator*, or "generators of surprise," to describe the most important function of the experiment in scientific laboratory activity. The term originates in the work of molecular biologist Mahlon Hoagland (1980) and characterizes the epistemic goal of a *cultura experimentalis* as I would like to see it. It is a great privilege to be able to write and publish. Use of this privilege is justified only if the works that we bring into the open, to the public sphere, are at least approximately capable of calling forth or recalling the peculiar quality contained in the moment of surprise. To consider texts as generators of surprise is likely a reliable formulation for the astonishment that one must never unlearn, neither in the sciences nor in media thinking.

I am extremely grateful to the University of Minnesota Press, and to Cary Wolfe in particular, for this opportunity to have created the present compilation from a diversity of my own writings. Thanks are due with respect not only to the latitude I've had in assembling these thematically heterogeneous genealogies and catalyzers of thought but a German-speaking author also may not take for granted that texts he wrote more than four decades ago will be disseminated within contemporary Anglo-Saxon discourse. Yet, when such a thing does occur, one hopes it will enable a reader to reconstruct a considerable stretch along the evolutionary arc of media thinking.

As early as 1965, Kurt W. Marek, using the anagrammatic pseudonym Ceram, published almost simultaneously in both Germany and the United States his explicit *Archaeology of the Cinema*. Around this same time, a number of art-historical and culture-historical texts were also operating with a distinctly archaeological gesture, such as Jurgis Baltrušaitis's fantastical writings on anamorphic art and the mirror[11] or Gustav René Hocke's superb 1957 work on mannerism in European art, *Die Welt als Labyrinth* (The world as labyrinth).[12] But archaeology first emerges as a notable thematic and methodological paradigm in the humanities as a discourse effect of the work of historian, sociologist, and philosopher Michel Foucault. *The Birth of the Clinic: An Archaeology of Medical Perception* (1963), *The Order of Things: An Archaeology of Human Sciences* (1966), and *The Archaeology of Knowledge* (1969) led a variety of disciplines, some with notable hesitation, to conduct analyses of historical phenomena that sought to interweave aspects of the political, the cultural, the technical, and the social—to conduct, in other words, *interdiscursive* analyses. At the Technical University of Berlin, where I studied, new research projects were being articulated on topics as diverse as the history of female labor (as in Karin Hausen's 1978 social history of the sewing machine), the intellectual and social history of mathematics, and the history of computing machines (as in Herbert Mertens's and Hartmut

Petzold's early studies). The periodical *Wechselwirkung* (Interaction), founded in 1979 in Berlin, provided a unique platform for this particularly active interdiscursivity between science and humanities. Such diverse archaeologies and genealogies evolved as academic attempts to intervene on the often encrusted systems of knowledge and organization in the established disciplines and to aggravate and alter these by means of critical, transdisciplinary reflection.

My first media critical publications emerged from just such a milieu, as did my early writings on the history of medial attractions like the Arbeiter-Radio-Bewegung (Workers' Radio Movement) of the interwar period. In today's terms, historicized in relation to hegemonic media apparatuses, one might deem this as one of the first hacker movements, vested in an aura similar to that which surrounded the self-styled Guerrilla Television of the electronic avant-garde of the late 1960s and early 1970s.[13] The epic gesture of intervening action that we had learned to extrapolate above all from Bertolt Brecht and his radio heuristic, but also from the hopeful potential in the writings of Walter Benjamin, was just as important to us as work on the utopian possibilities we saw in a collectivity in which, as a rule, there would be no exclusions and no hegemonic hierarchies in the exchanges among its members. Jürgen Habermas was as much on our minds as were the protagonists of critical theory, returned from their exile in the United States.

The opposite of this—the concentration of all power in relations of communication through ideology, stupefaction, hate, and envy—we knew too well from fascism. We were also able to observe, again and at close range, how the language of propaganda functioned in other manifestations in East Berlin. Critical engagement with one particular television event—a U.S. import called "Holocaust," as it appears here in one essay—was a central component of a comprehensive teaching and research project on the media of the Nazi machinery of murder and stupefaction which kept us busy for more than five years (1978–82) and from which three books resulted.

In my own development, I underwent an interim phase in this meta-methodological shift away from the study of media history toward the (an)archaeology and genealogy of media. As is so often recounted in the biographies of professional intellectuals, it was in the course of working on my dissertation that a whole new range of diverse research possibilities opened up for me. While still very much shaped by the philological tradition of the humanities faculty at my alma mater, and driven by a powerful curiosity about the world of apparatuses, physics, electrotechnology, and machine construction, I devoted myself, in a manuscript of more than six hundred pages, to a single artifact that I attempted to read and comprehend at the intersection of media-materiological, technical, temporal-philosophical, economic, and cultural perspectives. Today, I would describe a methodological endeavor like this as *expanded hermeneutics,* coupled with a particular philology: an almost exact philology of almost precise things.[14] The object of my investigation was an apparatus that could record audio and visual signals on an electromagnetically laminated tape so as to immediately reproduce them: the video recorder. The artifact and the technical system in which it was integrated fascinated me as an early *audiovisual time machine.* With this apparatus, it was not only possible to submit filmic and televisual programs to an analytical reading just like one would read a book but one could also intervene in their temporal structures and change them. Deeply influenced by the materialist variants of the Birmingham School of Cultural Studies, on one hand, and by German systems theorists and historians of technology like Günther Ropohl, on the other, in the closing chapter of my history of the video recorder (*Zur Geschichte des Videorecorders,* 1985) I termed this ensemble of machines and potential for action *Kulturtechnik*—or "cultural technique."

My own explicit (an)archaeologies of media began in close proximity with experimental practice and expanded hermeneutics. In the early 1990s, the University of Salzburg in Austria provided me with the wonderful opportunity to teach and research as a professor of *audiovisions,* which was the original title of a book I published in 1989. This context resulted in 1991 in the project we called "One Hundred—20 Short Films on the Archaeology of Audiovision," which was our contribution to a celebration of the first one hundred years of cinema history. In tandem with this, I prepared essays in the form of theses for the Austrian magazine *Eikon,* one of which is presented here for the first time in American English.

As these early media-archaeological miniatures began to appear, with the usual interval between completion and publication, the new, large-format technological project of establishing comprehensive telematic communications via the internet and the World Wide Web was well under way. Our writing was racing with the machines, as they rotated faster and faster. During the time I spent as founding rector of the Academy of Media Arts in Cologne (1993–2000), I began to publish more and more essays on the arts and artists that were engaging with these new circumstances, flanked by short manifestos or proclamations, provocations.

In the extreme rush of networked bustle, which also incorporated critique, I began to discover, in an indispensable countermovement, ever more of that dimension of the medial that I would go on to tamper with intensively for a good twenty years to come, much to my great intellectual pleasure: the deep time of the nexus of art, science, and technology. In *Variantology*, I came up with a new thinking and playing field, one in which I was able to investigate this exhilarating context as a unique poetics of relations.

The concept of variantology is a neologism that is ill suited to the purposes of standardization. There is clearly a paradox contained in it, one we are familiar with from other semantic iterations like Georges Bataille's "heterology" or the "heterotopias" of Michel Foucault. These too are indebted to a logic of diversity, of multiplicity. Contrary, divergent, mutually conflicting, or even mutually repellent phenomena that as a rule evade unification are gathered under a provisional roof in such a way that they may nevertheless drift apart again as needed. Variantology has to do with compounds or mixtures of a kind whose unmixing always remains within the realm of imagination. The invocation of the *logos* in the concept serves less to produce a closed systematic relationship than it does to perpetually irritate the concept's inventor and those who engage themselves with it. The international conferences on variantology that took place in Cologne, Berlin, and Naples between 2004 and 2008 were simply an invitation to a kind of collaboration in which there were no contracts to sign or programs to subscribe to; they represented an offer of hospitality that obliged the guests to nothing but an increased presence of mind in their actual physical presence.[15]

In contrast to the heterogeneous, with its heavy inflections of ontology and biology, the *variant* is more interesting, in methodological and epistemological respects, as a mode of lightness and movement. As such, the variant is equally at home in experimental science as it is in diverse artistic practices,[16] most forcefully in music. Variation, versioning, digression—in playing and interpretation—are an obvious part of the vocabulary as well as the everyday practice of composers and interpreters alike. In a narrower sense, the variant designates a modulation, say, from minor to major tonal series, brought about by a change in the interval.

The semantic field that I am trying to open by means of this concept has a primarily positive connotation. To be different, to diverge, to shift, to alternate, are themselves alternative translations for the Latin verb *variare*. Its connotation topples over into the negative only when used by the speaking subject as a means of exclusion and discrimination—which the word itself does not actually abide. To vary something that is present is an alternative to its destruction, an aesthetic strategy that played a remarkably sustaining role in the diverse avant-gardes of the twentieth century, in politics as well as art. And, of course, an attractive medial format also inheres in the concept, a format one relates to as one would to a sensation. Long before the cinema, the variety show was experimenting with combining diverse stage practices into a colorful whole that would come together only in the time of a given performance.

The heterogeneity of variantological research among various concepts of modernity, between the Occident and the Orient and among the multiplicity of forms of European culture, expresses itself in this volume in a series of individual genealogies: of seeing and of visual perception, of sound and of musical mood, of the electric theologians and of "Allah's automata." These are the earliest texts from this collection. Research into variantology is ongoing.

In a flat temporal dimension[17]—which is by now rather alien to the activities of those who think deep time—the evolution of concepts of media thinking has been under way for hardly more than a century. It has only been since the end of the Second World War, in other words, about seven decades, that scientific, theoretical, philosophical, semiological, and philological engagements with and through media have been articulated and processed as a distinct discursive field of their own—albeit ever more unmistakably and increasingly louder.

I have attempted a thought experiment in operationally grouping past and present media researchers and protagonists by generation—not least in order to temporally locate my own position in the context of this still fledgling genealogy of our field of intellectual energies. I have started from the presumption that we are presently well into the seventh generation of *explicit* media thinkers.[18] Given the accelerated development of the inter-discursive field in the second half of the twentieth century, I decided to scale the shift in generations following decade markers. The generational groupings are not determined by the age of the thinker but rather on the basis of important differences each one has individually contributed to this heterogeneous field of knowledge. I have paid special attention to intelligible discourse effects that have been observed in Europe and that have also had an intelligible impact in Germany, for instance.

Early thinkers through the end of World War II: Theodor W. Adorno, Rudolf Arnheim, W. Ross Ashby, André Bazin, Walter Benjamin, Henri Bergson, Bertolt Brecht, Karl Bühler, Claude Cahun, Ernst Cassirer, Germain Dulac, Sergei Eisenstein, Gisèle Freund, René Fülöp-Miller, Aleksei Gastev, Siegfried Giedion, Fritz Heider, Max Horkheimer, Harold Innis, Ernst Kapp, Siegfried Kracauer, Lev Kuleshov, Harold Lasswell, Kazimir Malevich, Filippo T. Marinetti, Solomon Nikritin, John von Neumann, Charles S. Peirce, Luigi Russolo, Ferdinand de Saussure, Hermann Scherchen, Claude Shannon, Wilbur Schramm, Alan Turing, Dziga Vertov, Paul Watzlawick, Hermann Weyl, Fritz Winckel...

First mid- and postwar generation (explicitly active since the 1940s and 1950s): Günther Anders, Peter Bächlin, Roland Barthes, Max Bense, John Berger, Maya Deren, Jean-Luc Godard, Richard Hoggart, Danièle Huillet, E. Katz/J. G. Blumler, Harry Kramer, Marshall McLuhan, Werner Meyer-Eppler, Abraham Moles, Raymond Queneau, Gilbert Simondon, Hans Heinz Stuckenschmidt, Wolf Vostell, Roman Wajdowicz, the Whitney Brothers, Norbert Wiener, Iannis Xenakis...

Second generation (explicitly active since the 1960s): Dieter Baacke, Nanni Balestrini, Gianfranco Baruchello, Konrad Bayer, Gilbert Cohen-Séat, Guy Debord, Umberto Eco, Vilém Flusser (in Brazil), Otto F. Gmelin, Jürgen Habermas, Helmut Heißenbüttel, Walter Höllerer, Friedrich Knilli, Ferdinand Kriwet, Gerhard Maletzke, Denis McQuail, Christian Metz, Franz Mon, Frieder Nake, Georg Nees, Ted Nelson, Nam June Paik, Pier Paolo Pasolini, Wolfgang Ramsbott, Jasia Reichardt, Gerhard Rühm, Marc Vernet, Paul Virilio, Peter Weibel, Oswald Wiener, Raymond Williams...

Third generation (explicitly active since the 1970s): Jean-Louis Baudry, Hans Belting, René Berger, Gábor Bódy, Jean-Louis Comolli, Gilles Deleuze, Mary Ann Doane, Franz Dröge, Hermann Klaus Ehmer, Thomas Elsaesser, Hans Magnus Enzensberger, VALIE EXPORT, Friede Grafe, Félix Guattari, Hans Ulrich Gumbrecht, Stuart Hall, Stephen Heath, Knut Hickethier, Horst Holzer, Stuart Hood, Eberhard Knödler-Bunte, Gerhard Lischka, Laura Mulvey, Friederike Pezold, Marcelin Pleynet, Hans Posner, Erwin Reiss, Michel Serres, Kristin Thompson, Sven Windahl, Peter Wollen...

Fourth generation (explicitly active since the 1980s): Jean Baudrillard, Peter Bexte, Teresa de Lauretis, Anne-Marie Duguet, Vilém Flusser (in Europe),

Dietmar Kamper, Friedrich Kittler, Sybille Krämer, Arthur and Marilouise Kroker, Werner Künzel, Pierre Lévy, Jean-François Lyotard, Joachim Paech, Miklòs Peternák, Hartmut Petzold, Hans-Ulrich Reck, Irit Rogoff, Avital Ronell, Florian Rötzer, Allucquére Rosanne Stone, Georg Christoph Tholen, Gerburg Treusch-Dieter, Christina von Braun, Michael Wetzel, Hartmut Winkler, Siegfried Zielinski...

Fifth generation (explicitly active since the 1990s): Marie-Luise Angerer, Peter Berz, Manuel Castells, Régis Debray, Manuel DeLanda, Bernhard Dotzler, Timothy Druckrey, Lorenz Engell, Wolfgang Ernst, Matthew Fuller, Ulrike Gabriel, Miriam Hansen, Donna Haraway, N. Katherine Hayles, Hans-Christian von Herrmann, Erkki Huhtamo, Brenda Laurel, Thomas Y. Levin, Geert Lovink, Lev Manovich, Dieter Mersch, Brian Massumi, Alla Mitrofanova, Claus Pias, Nils Röller, Henning Schmidgen, Bernhard Siegert, Andrey Smirnov...

Sixth generation (explicitly active in the 2000s and beyond): Arianna Borrelli, Knut Ebeling, Alexander Galloway, Mark B. N. Hansen, Erich Hörl, Ute Holl, Yuk Hui, David Link, Mara Mills, Jussi Parikka, Matteo Pasquinelli, Patricia Pisters, Gao Shiming, Hito Steyerl, Frederik Stjernfelt, Eugene Thacker, Tiqqun, Joanna Zylinska, et al.

Translated by Lauren K. Wolfe

Notes

1. Walter Benjamin, "Lehrmittel. Prinzipien der Wälzer oder die Kunst, dicke Bücher zu machen" [Teaching aids: The principles of tomes, or the art of making thick books]. *Gesammelte Werke* 4, no. 1 (1991): 105. Translation LKW.
2. See Vilém Flusser, *Flusseriana: An Intellectual Toolbox*, ed. Siegfried Zielinski, Peter Weibel, and Daniel Irrgang (Minneapolis: University of Minnesota Press/Univocal, 2015), 17. The book is a trilingual publication (English, German, Portuguese).
3. Nam si abest quod ames, praesto simulacra tamen sunt (For if what you love is distant, its images are present). Titus Lucretius Carus, *De rerum naturae (On the Nature of Things)*.
4. Louis Althusser, "The Underground Current of the Materialism of the Encounter," in *Philosophy of the Encounter: Later Writings, 1978–87*, ed. François Matheron and Oliver Corpet, trans. G. M. Goshgarian (London: Verso, 2006).
5. Althusser.
6. I am referring here to the European Graduate School (EGS) in Saas-Fee, Switzerland.
7. Tracy B. Strong, *Friedrich Nietzsche and the Politics of Transfiguration* (Urbana: University of Illinois Press, 2000), 54.
8. Friedrich Nietzsche, *Kritische Studienausgabe* (Berlin: de Gruyter, 1980), 170. Translation LKW.
9. *Umzug ins Offene* is the title of a volume edited by by Tom Fecht and Dietmar Kamper: *Umzug ins Offene: Vier Versuche über den Raum* [Move into the open: Four experiments about space] (Berlin: Springer, 2000).
10. "They see an entanglement of spaces that emerges as it would were we somewhere like a cinema auditorium. But the oldest example of a heterotopia may well be the garden." Michel Foucault, *Les hétérotopies/Le corps utopique* [Heterotopias/The Utopian Body] (Paris: INA, 2004).
11. See Jurgis Baltrušaitis, *Anamorphoses, ou magie artifielle des effets merveilleux* (Paris: Olivier Perrin, 1955), English translation *Anamorphic Art*, trans. W. J. Strachan (New York: Henry N. Abrams, 1977); and Baltrušaitis, *Le miroir: Essai sur une légende scientifique: révelations, science-fiction et fallacies* [The mirror: Essay on a scientific legend—Revelations, science fiction and fallacies] (Paris: Éditions du Seuil, 1978).
12. Gustav René Hocke, *Die Welt als Labyrinth: Manier und Manie in der europäischen Kunst—Beiträge zur Ikonographie und Formgeschichte der europäischen Kunst von 1520 bis 1650 und der Gegenwart* [The world as labyrinth: Manner and mania in European art. Contributions to the iconography and

formal history of European art from 1520 to 1650 and the present] (Reinbek, Germany: Rowohlt, 1957).

13 See the part history, part instruction manual by Michael Schamberg and Raindance Corporation, *Guerrilla Television* (New York: Holt, Rinehart, and Winston, 1971).

14 There is a chapter dedicated to this in my book *[... After the Media]: News from the Slow-Fading Twentieth Century* (Minneapolis: University of Minnesota Press/Univocal, 2013), 173ff.

15 This resulted in the five volumes of *Variantology* that I was able to publish through Walther König in Cologne between 2005 and 2011, in collaboration with a rotating pool of editors and on the basis of a worldwide economy of friendship with the contributing authors.

16 For a powerful contemporary example in the visual arts, see Allen Ruppersberg, *One of Many: Origins and Variants* (Cologne: Walther König, 2005).

17 We have elsewhere dealt extensively with the dimensions of deep time. See, e.g., *Deep Time of the Media* (Cambridge, Mass.: MIT Press, 2006); the German original was published in 2002.

18 For the distinction between explicit and implicit media thinkers, see Zielinski, *[... After the Media]*, esp. chapter 3, 173ff. Implicit media thinkers are not contained in the list.

I
Provocations

1 History as Entertainment and Provocation

The TV Series Holocaust *in West Germany*

A good thirty years following the violent dictatorship of the German fascists, there came a massive intervention in a postwar culture still grappling with its recent past, of all things in the form of a commercial television miniseries made in the United States. The moving saga of the Jewish Weiss family opened up what had thus far been a rather elite discourse to broader segments of the population. In collaboration with my then professor Friedrich Knilli, I spent the years between 1978 and 1982 developing a vast research project on this medial and political event. Knilli and I published several books together, and I concluded my studies by making a documentary film called Responses to "Holocaust" in Western Germany, *which I then took on tour in the United States, holding screenings in university towns and Jewish communities. During that time, I wrote the following essay for* New German Critique.

This essay was originally published in 1980 and slightly revised in 1986.

∷∷

The screening of the NBC series *Holocaust* by West German television's Third Channel in January 1979 is now history-political and media history. In the meantime, the four-part series based on Gerald Green's novel was rerun in November 1982, and since then, German audiences have seen hundreds of hours of programs thematicizing fascism and its crimes of mass murder against the Jews. These encompassed the most widely different forms of presentation and dramaturgical concepts: eyewitness accounts, documentaries, specials, feature films, and even series whose construction referred back to *Holocaust*.

In the meantime, a host of cinema films have been made by West German film "authors," each of whom attempted to get a grasp on the events of recent history. In Mario Offenberg's *Alptraum als Lebenslauf* (1982), Georgia T. recalls the years of her internment in Ravensbrück death camp. Axel Engstfeld documents the role of the Nazi judiciary in *Von Richtern und anderen*

Sympathisanten (1982) and demonstrates its continuity to a certain degree in the development of the Federal Republic. These are just two exceptional examples from the nonfiction genre. Peter Lilienthal, Rainer Werner Fassbinder, Michael Verhoeven, and Ottokar Runze, among others, have all made impressive feature films. *Holocaust* has passed into rock music and advertising and can be found on the covers of millions of publications in vastly different constellations of concepts and meanings. The most abhorrent phenomenon, "concentration camp pornography," has come into being or was developed further, for example, as a subject listing on the video film market. In the meantime, the fiftieth of Hitler's seizure of power has been commemorated at the most diverse levels of cultural and political activity—and not least by a veritable marathon of TV programs on the subject. From October 1982 to January 1983, Channels 1 and 2—ARD and ZDF—have devoted 150 hours of viewing time to this subject alone.

These past five years of works on fascism, both within and outside the medium of television, represent in their effects a context that, naturally, was received only in part by the West German public. When dealing in the following pages with the media event *Holocaust* and some dimensions of its effects, it must be remembered that it is but one element in a process and can only be lifted out of this process for the purposes of analysis. In reality, it has long since fused with other factors.

Holocaust was unique in the history of television in that the series immediately became the object of socioempirical investigations and cultural discourses. In this respect, the Federal Republic was no exception, and many surveys were undertaken, primarily to analyze aspects of the program's effects. My own work on *Holocaust* was done mainly within the framework of a teaching and research project at the Technical University of Berlin, the findings of which are documented in numerous essays and two book publications.[1] I mention our works at this stage to avoid repeated references to sources and material contained therein within the text of this essay and also because collective work inevitably means that one's own conclusions have absorbed those of coworkers with whom one has collaborated in the process of research.

"Catharsis of a Nation" was the somewhat arrogant term applied by *Der Spiegel* to the *Holocaust* telecast in 1979. Others dubbed it a "didactic play" (Lehrstück), whereas some who did not bank on any deeper effects called the visibly hefty emotional reaction of the TV audience "a flash in the pan"

FIGURE 1.1. The Berlin research project on the reception of the television series *Holocaust* resulted in a one-thousand-page report, which was finally published as a book by Elefanten Press, publisher for education and academic studies (Berlin, 1982). The title translates as *"Holocaust" for Entertainment: Anatomy of an International Bestseller.*

(the German word used was *Strohfeuer*—a sarcastic choice of metaphor considering the program's subject). Indeed, there were many variations on these attempts to arrive at a definition, and all had one thing in common: the intent of reducing the significance of this multimedia event to a single category and to bestow on it a clear pedagogical, ideological, or political meaning. The thing simply had to be filed away somewhere, and appropriate labels were readily at hand.

Being concerned with analysis of the mass media, we, too, felt the need to classify the *Holocaust* media event and to define it for purposes of coming to terms with the most atrocious period of Germany's history and, in particular, its reflection in the mass media. This resulted in an analysis of the NBC series and the body of published criticism on it as well as a scrutiny of the many and varied dimensions of its effects. From this it became clear that a single, unequivocal interpretation would fall short of the main-faceted reality. We availed ourselves of the category of contradiction to track down *Holocaust* and its functions. At first glance, this might appear to be an evasion, but its sophistic character disappears the moment a closer look is taken at the dimensions of cause and effect in the process of the *Holocaust* event. This television film was enacted and produced both with the object of entertaining and with the historical and moral purpose of rousing its viewers; it also displayed this ambivalence in its function for West German audiences.

Holocaust was a diversion—it made people excited and then calmed them down again—but it also represented a provocation: it triggered discussions and manifest activities. It raised the awareness of the German public regarding fascist anti-Semitism and its ghastly effects. Thus it also possessed a dimension of enlightenment. With the category of contradiction, we can move toward a definition of the media event that comprises both its reproductive and productive functions. In this sense, *Holocaust* had attributes of a dialectic of entertainment. But let us first try to take this odd conceptual construction apart and substantiate it.

The Framework of *Holocaust* in the Federal Republic of Germany

"Anti-Semitism ... is something like the ontology of advertising."[2] What Theodor Adorno expressed by using this metaphor in 1964 had been formulated in various ways in the philosophical fragments gathered in *Dialectic of Enlightenment* together with Max Horkheimer twenty years previously. One of the key "elements of anti-Semitism"[3] that appears again and again is the structural connection between anti-Semitic thinking and behavior and the mechanisms of the culture industry—cultural commodity production, "mass production," "ticket mentality," "ticket thinking," the "lusting" after "property, acquisition," the substitution of judgment with "exchange," the "stereotype," the "cliché." For the two exponents of critical theory, these are all attributes and categories that are ideologically co-constituents of anti-Semitism.

By stressing the structural relationship of anti-Semitic stereotypes with the "invariables of advertising psychology,"[4] Adorno's aim in 1964 was quite evidently the strategy of enlightenment that he wanted to see implemented as the basis "in the struggle against anti-Semitism today"—"only emphatic enlightenment with the whole truth is of any use; strictly foregoing anything which resembles advertising."[5]

In this respect, the TV series *Holocaust* is by no means abstemious. On the contrary, it is a truly outstanding product of the North American—that is, the capitalist culture and industry—and as such relies fundamentally on advertising psychology. As a generalization, one can say that the show's structure exhibits the characteristics of mass production in a particularly pointed and condensed form: in the construction of the plot as a family drama, in the characters' profiles, in the suspense curves of sequences and the separate installments, in the scenes and settings, in the use of music, and so on. The lowest common denominator is the organizing principle of the mass audience. Following this line of thought to its logical conclusion means that in its accumulation of clichés, the TV series bases itself on attributes of anti-Semitic thinking—it possesses anti-Semitic elements. For even without the interspersed advertising spots—which is how the series was presented in the Federal Republic and also, for example, in England—its affinity with commodity aesthetics, its construction according to the dramaturgy of advertising, is evident. Admittedly, its subject is lethal anti-Semitism itself. *Holocaust* wants to denounce and condemn anti-Semitism, to show the suffering of its victims and render them comprehensible to the audience, to touch the emotions of the guilty, the uninvolved, the successive generations, and to arouse identification with the victims. In consequence, it assails advertising as an accumulation of clichés with its very own strategies. *Holocaust* is not enlightenment but *anti-advertising*.

But isn't that a tautology? In the age of advanced mass communication, where conditions prevail dominated by Coca-Cola, Disco, Wrigley's, and Levi's—in short, the cultural hegemony of commodity aesthetics—what else can enlightenment be other than animadverting? Doesn't one have to submit to the conditions of commodity exchange if one wants to convey anything to the army of consumers? In the face of these conditions, one can withdraw one's desire to instruct and refuse to have any part of them. But wouldn't this mean that one ends up talking to oneself? It is clear at least that it would not be possible to reach those whose everyday experience is dominated by the outward manifestations of commodity aesthetics—those for whom commodity circulation is the actual form of communication in work and leisure. If we take the Adorno–Horkheimer thesis as a starting point, does the quality of the commodity really play a role here? Is a distinction between the TV offerings "Nazi persecution of the Jews" and "Kojak" really significant at all? These are some of the questions—deliberately overstated, of course—that particularly preoccupied certain critical intellectuals before *Holocaust* was telecast, but only marginally afterward, because the reaction of the country's inhabitants was so overwhelming.

The program was discussed and debated and aroused mourning for a good week in firms; in schools; around the dinner table; in pubs, living rooms, and churches; and even within the hallowed halls of the German Parliament.

The Adorno quotation cited earlier occurred to me immediately after seeing the original American version of *Holocaust*. A few months later, when official rejection of the series was at a pitch here, I thought increasingly of something another writer had said, one who, like Adorno, had fled to the United States from the Nazis: "The tenet—a work of art is a commodity—would be a tautological assertion if it were not for the fact that there is, nevertheless, something more of function in it, something which constitutes its main value."[6] With Brecht's insistence on the character of art—including industrially reproduced art—as a vehicle of use-value, and while keeping the *Dialectic of Enlightenment* at the backs of our minds, it may perhaps be possible to approach the phenomenon of *Holocaust* in the Federal Republic.

There were times one gained the impression from the discussions around and about *Holocaust* that the presentation of Nazi atrocities on the screen had only just begun with this TV series. Of course, the fact of the matter is that it was the first American TV series on this subject to penetrate German living rooms. The tradition of audiovisual expositions on the Third Reich and its crimes can be traced back to the period immediately after the liberation of Germany from fascism. The East German DEFA's first film, shot amid the ruins of Berlin, is concerned not only with the victims but also very much with the guilty perpetrators. *Die Mörder sind unter uns (Murderers among Us)* by Wolfgang Staudte was premiered in October 1946. The female protagonist is a young Jewess, a survivor of the death camps who tries to settle down in bombed-out Berlin. It took a little longer in the Western Allied occupied zones. The first feature film to thematicize Nazi crimes against the Jews—among other subjects—had its premiere in June 1947—*In jenen Tagen (In Those Days)*, directed by Helmut Käutner.[7]

In the former Soviet occupied zone—later the German Democratic Republic—the aesthetic treatment of fascism in films evolved into a distinctive subject that runs continuously through the GDR's film and television history. Some of the most important productions are *Ehe im Schatten* (1947), *Affaire Blum* (1948), *Rotation* (1949), *Der Rat der Götter* (1950), *Stärker als die Nacht* (1954), *Der Hauptmann von Köln* (1956), *Sterne* (1959), *Professor Mamlock* (1961), *Nackt unter Wölfen* (1963), *Die Abenteuer des Werner Holt* (1965), *Ich war neunzehn* (1968), *Trotz alledem!* (1972), *Jakob der Lügner* (1975), and *Mama, ich lebe* (1977). This represents only a selection of the feature films made on the subject. In addition, a large number of documentaries, miniseries (like the five-part *Krupp und Krause*), and TV plays have been produced. In accordance with the GDR's attitude toward fascism, this tradition of antifascism signifies neither a foremost nor an exclusive concern with anti-Semitism. Obviously, some of the productions concentrate on this subject—like *Ehe im Schatten, Affaire Blum, Sterne,* and *Jakob der Lügner*—have all, in the meantime, practically become classics of this genre. In the postwar West Zones, now the Federal Republic, many documentaries and compilations were produced. The most well known of these, Erwin Leiser's *Mein Kampf* (1960) and *Deutschland, erwache!* (1968), attempt antifascist instruction through the critical interpretation of original audiovisual documents of the Third Reich. In general, though, fictional or dramatic treatments of German fascism remain conspicuous by their absence from the cinematic screen.[8] Of course I am not referring here to the more or less apologetic war films, which enjoyed a veritable boom in the 1950s and early 1960s. Even the modest start that was made with feature films like *In jenen Tagen, Morituri, Zwischen Gestern und Morgen,* and *Lang ist der Weg* was not continued. For a long time, *Rosen für den Staatsanwalt* (1959), which deals with the uninterrupted careers of Nazi judges in the Federal Republic, made by Wolfgang Staudte after he switched from the East German DEFA to Western production companies, remained a notable exception. The first full-length cinematic production that centers on Nazi anti-Semitism belongs to the "post-Holocaust" era—*David* by Peter Lilienthal. The Berlin Film Festival, which highly acclaimed Lilienthal's film, took place in February 1979, shortly after the telecast of *Holocaust*. The tradition of the Holocaust's treatment on television is a slightly different story. Particularly in the 1960s, when the young Federal Republic was going through its first serious identity crisis, there were a number of dramatizations with an antifascist slant, for example, Rolf Hädrich's TV play *Der Schlaf der Gerechten* (1962), Egon Monk's *Ein Tag. Bericht aus einem deutschen Konzentrationslager. Januar 1939* (1965), Dieter Meichsner's *Wie ein Hirschberger Dänisch lernte* (1968), Aleksander Ford's *Sie sind frei, Dr. Korczak* (1973, as a coproduction with Israel), or *Aus einem deutschen Leben* (1977), directed by Theodor Kotulla. Otherwise, West German TV producers, program directors, commercial film distributors, and the institutions responsible for educational film distribution depended on foreign films, but only to a limited extent on those of the DEFA, mainly films from Poland, Yugoslavia, Czechoslovakia, the Soviet Union, and the United States. *The Diary of Anne Frank* (1959), directed by George Stevens, must be mentioned here.

The recalcitrance on the part of most of the older generation of film directors regarding dramatic film treatments of fascism and its crimes against the Jews is not difficult to explain because they spent crucial years making their careers in the Nazi propaganda and dream factories. They were the ones to profit from the "elimination" of their Jewish colleagues from the film business or, at the least, shut their eyes to it. Apparently, though, the younger German cineasts have also had great difficulty in approaching the subject. Where attempts have been made, these have taken the form of trauma, as in the Fassbinder film *Ein Jahr mit dreizehn Monden*. The figure of the former inmate of Bergen-Belsen, Anton Seitz, a social climber who rises from brothel manager to formidable property agent in Frankfurt, is a clumsy attempt at psychologizing a concentration camp survivor. Apart from such exceptions, the Nazi genocide, its manifestations and consequences, is taboo.

This situation changed—to begin with, in a negative sense—just before the advent of *Holocaust*. In the wake of the considerable boost given by Joachim C. Fest and Christian Herrendoefer's film about the career of a dictator, the West German media was veritably swamped by a "Hitler wave." Fascism even became an entertainment attraction on talk and quiz shows or was used as a trapping in adventure series like *Es muss nicht immer Kaviar sein,* which was based on a best seller by the highly successful writer J. M. Simmel.[9] A boom in Nazi relics was noticeable at the flea markets of West Germany and Berlin. Prices for original editions of Hitler's *Mein Kampf* skyrocketed. Even the Stars of David that the Jews were forced to wear by the fascists were being hawked for hundreds of West German marks—provided they were the "real thing," of course.

This vogue was in full swing in April 1978 when news about the telecast of *Holocaust* in the United States began to spread. Here of all places, where the market for Nazi nostalgia was flourishing, where they had no compunction about producing Super 8mm fascist porn with swastikas on G-strings and gasps of "Heil Hitler!" between copulations, there was an outcry. Indignation about the commercialization of the Nazi genocide was the main tenor of the critics. Elie Wiesel's evaluation of the television *Holocaust* as a trivialization of "an ontological event" that resulted in "a soap opera" was adopted by most of the large daily newspapers and broadcasting authorities. This quotation must have appeared hundreds of times or was simply paraphrased.

Precisely because afterward nobody wanted to talk about the initial indignant outburst, it is important to recall those first reactions of published opinion before the *Holocaust* telecast: *Der Spiegel,* which, incidentally, was full of praise for the event later on, was worried about "the German diplomats in the USA, for 'Holocaust' will provoke a new wave of anti-German feeling."[10] The *Frankfurter Allgemeine Zeitung* asked, on the verge of desperation, whether "horror is only comprehensible through banalities?" and retaliated with, "Mistrust of the undemonstrative power of authentic documents is the catch of this miniseries."[11] This in turn supported the imploring argument made by the chief moderator of the news show *Tagesthemen,* Claus Stephan, in a telecast that went on the air after the last installment was shown in the United States: "Not a fraction of the horror of a gas chamber can be conveyed in literal form, with the help of actors: quite the contrary. This artificial, supposedly artistic portrayal obliterates the ghastly truth. Any and every black-and-white photograph of a gas chamber taken after the end of Nazi rule has more effect: silent horror in the face of the murders committed." The *Rheinischer Merkur* moralized, "The most lenient objection to 'Holocaust' is that respect has been sacrificed here to play-acting: respect for reality and respect for the victims."[12] After an initially favorable review, Axel Springer's *Die Welt*, which after all can hardly polemicize much against its own market strategies, was soon complaining that genocide was being portrayed as a "Teutonic phenomenon" and demanding, in the good company of the conservative international law specialist Alfred de Zayas, a revision of American school textbooks because the TV guides in the United States that were distributed when *Holocaust* was telecast did not even take into account "the expulsion of Germans from Eastern Europe."[13]

However, these opinions changed. The hypocritical indignation over the marketing of Germany's past by the Americans was superseded by a political insight: the confrontation was inevitable. And this was due in no small measure to the fact that the Federal Republic was about to celebrate its thirtieth birthday and that international opinion had been following the "Hitler wave" very attentively. The change of heart was further encouraged by the positive reception of the series in other countries. The United Kingdom's BBC was the first to show *Holocaust* in Europe. Audience response was not high, but the response in the press was considerable. Belgium followed suit and then—significantly—Israel, in September 1978. This was the external signal that prompted the final decision in favor of telecasting *Holocaust*. Heinz Hübner, the program director of Westdeutscher Rundfunk, the broadcasting authority that had purchased the series, stated, "The film is a political issue, and, if it can be shown in the country [Israel] of those directly affected by the Holocaust, then it is not asking too much of the Germans if it is shown here too—to those involved in those events and those who came after."[14]

At the end of October 1978, the telecast dates set for *Holocaust* (January 1979) were publicized, and then the public institutions began to prepare almost feverishly for the coming event. Institutions involved in political education commissioned TV guides. The Federal Center for Political Education devised a nationwide representative survey on the subject. Teacher's unions produced teaching aids. Magazines planned and published special issues. The television authorities themselves produced two comprehensive documentaries designed to prepare the viewing public for the coming film.

These activities reached a first peak on the fortieth anniversary of the so-called Night of the Broken Glass (Reichskristallnacht). For the very first time—at least to my knowledge—a broad discussion of Nazi war crimes against German and European Jews took place. It was the first time that something akin to deep concern and the attempt to transmit this became apparent in the mass media. Pictures of burning synagogues filled the inside pages of the large daily newspapers or the front covers of illustrated weeklies. The provincial papers from Bavaria to the Frisian Islands published information, in most cases scrupulously collected and compiled, on what happened to the Jewish citizens in that particular region during the years 1938 to 1945. Again and again, the coming event *Holocaust* was the point of reference.

It was in the wake of this publicity, of this media marathon for an anniversary of grief, that the cautious relegation of the series to the Third Channel was seen to be absurd. Long before the telecast, the *Holocaust* week had become a social event that no institution having any interest in the workings of West German minds could afford to ignore. The TV magazines of the largest publishing houses, Springer, Bauer, and Gruner & Jahr, did the rest. A week before the telecast, they had nearly all the appropriate headlines on the cover of their publications and numerous articles inside: "'Holocaust' on All Third Channels: The Drama of the Jews!" *(Fernsehwoche)*, "The Series That Moved the World" *(Gong)*, "For America It Was a Spectacle—and for Us?" *(Funk Uhr)*, "The Most Controversial TV Series of All Times Is Coming!" *(TV Hören und Sehen)*, and the like. Just because of the different TV magazines vying with one another and using superlatives, news of *Holocaust* must have reached over 80 percent of the West German adult population. Then there were the introductory television programs. The Arbeitsgemeinschaft der Rundfunkanstalten Deutschlands (ARD) made an attempt to prepare viewers for *Holocaust* with two of its own documentaries on the history and background of Nazi anti-Semitism, which were produced especially for this occasion. Political parties, Jewish communities, and Christian church organizations encouraged people to watch the program. Last but not least, neo-Nazi activities (directed at the "6 million—the hoax of the century") contributed to the wide publicity—anti-Semitic leaflets, street-corner agitation, and, naturally, the most spectacular variation: bomb attacks on Channel 1 transmitters during the second *Holocaust* introductory documentary. Only against this background of very varied, extremely contradictory circumstances is it possible to account for the fact that almost a quarter of all West German TV sets were tuned in to the first installment of the miniseries in January 1979. What followed during the week *Holocaust* was telecast is unique in the history of Federal Republic television. Although considerably fewer than half of all TV households on average tuned in to the Third Channel,[15] virtually nothing else was talked about at places of socialization in our society: in schools, universities, factories, offices, and living rooms. There was always someone present who had seen *Holocaust* and who provoked discussions and arguments—about fascism, its origins, the persecution of the Jews, guilt, and responsibility. In short, not only what had been seen on TV was discussed but also issues of social and historical relevance that had greater ramifications.[16] The American TV series about the families Weiss and Dorf succeeded in bringing about a phenomenon where documentaries, curricula, stage plays, historical novels, poems, and feature films before it had failed. For many days, words like *Nazi crimes, concentration camps, anti-Semitism, Auschwitz, neo-Nazism,* and, finally, the term *Holocaust* itself pervaded and satiated the daily lives of most West Germans. What had formerly been the province of a few educated circles and certain political groupings became a public event of the first order.

A Tentative Evaluation of the *Holocaust* Reception in West Germany

Since the telecast of *Holocaust,* the question of its effect has been asked time and again and has remained unanswered just as often. Indeed, it would be difficult to offer a satisfactory and theoretically consistent answer to two incongruous premises: on one hand, it is assumed that a heterogeneous product like the television *Holocaust*—with regard to its ideological and aesthetic structure—which addresses itself to such a diverse audience, could produce a single specific effect; on the other, it presumes the series itself is the cause of possible or observed effects. Here two simple findings can be advanced as an objection, yielded by several years of study of this national and international television event. The series produced as many dimensions of effect as it assembled different audience groups in front of the TV sets; the effects of *Holocaust* were manifold and qualitatively different—different both on an international level but also within a given country like the Federal Republic. Furthermore, the series was part of a complex process of public discussion that functioned as a framework and itself interacted with the series: the four episodes of *Holocaust* can only be isolated from this context by an act of force.

For example, our analysis of the viewers' mail to the television authorities after the first telecast of *Holocaust* in 1979 (around nine thousand letters, telegrams, telexes, and postcards) found that the TV studio discussions

HISTORY AS ENTERTAINMENT AND PROVOCATION | 15

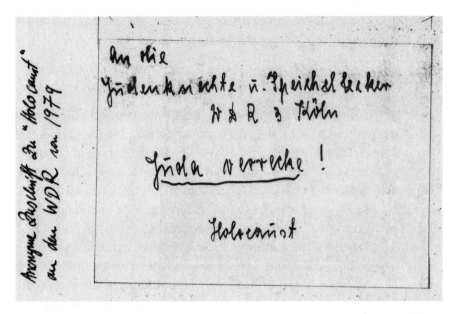

FIGURE 1.2. Postcard 8308 written to the WDR, the broadcast station that screened the *Holocaust* series in 1979 for German prime-time television (ARD). The postcard was one of around nine thousand that the research group at the Technical University Berlin analyzed. From Friedrich Knilli and Siegfried Zielinski, eds., *Betrifft "Holocaust"—Zuschauer schreiben an den WDR* (Berlin: Spiess, 1983), 361.

after each installment played a far more important role with regard to the articulation of specific audience needs than did just the series alone.¹⁷ This finding was confirmed when the series was rerun in November 1982—during prime time, on the first channel, but without the mobilizing journalism of 1978–79. The resonance—or lack of it, to be precise—could in no way be compared to that of the first telecast, which proved that this kind of backup is of crucial importance for the dimensions of qualitative effect of such a TV event.

For these reasons, the effect of the bare series *Holocaust* will not be my concern here. My interest lies in certain dimensions of the *Holocaust* event's discourse effect. Among them is the question as to what functions the media event centering on *Holocaust* had for specific social groups and areas, and whether it has helped us in the Federal Republic along the difficult road of comprehending fascism and its crime of genocide against the Jews—with the object of preventing it from happening again. The targets of scrutiny are thus behavioral modes, attitudes, and activities that could be observed in the wake of *Holocaust* and that referred explicitly to it. The selection presented may afford a glimmer of the greater whole to become more visible.

Public Agitation, Consternation, and Disavowal of Guilt

One read and heard everywhere that *Holocaust*'s most outstanding feat was the "emotional shock" it produced. I must admit that this category means very little to me. Not only is it difficult to verify or substantiate but my main problem with it is that it is too exclusively located in the emotional makeup of the individual, access to which is very difficult and usually well-nigh impossible.[18] Also, "shock" implies "without consequence or paralysis," and *Holocaust* certainly was not without consequences.

Already before the telecast and immediately after, individual and social activities were discernible that were bound to have at least a limited influence. The broadcasting authorities received thousands of telephone calls and letters; newspapers, journals, and public institutions likewise received sacks of mail; intense discussions took place in both the public and private spheres—these are all signs of active thought and refection processes not usually associated with everyday television consumption. For the viewers themselves, these activities had the function of making public what was occupying their thoughts. All are courses of action presupposing a high degree of excitement and concern while revealing nothing of the actual quality of this reaction. When, in January 1980, the topical TV magazine *Report* showed a program about the crimes Germans committed against Soviet soldiers during the last war, a comparable storm of indignation was unleashed. That very evening, the television studios received more than two thousand telephone calls, nearly all of which were outraged protests at this supposed "defamation of the Germans."

It is important to note that the agitation over *Holocaust* is certainly not deducible from elements of the series alone, such as the presentation of brutal physical violence in the scenes depicting the gas chambers, mass murder, torture, and rape, nor from the appeal to the viewer's pity and the identification proffered with victims and murderers. The very way the whole *Holocaust* event was constructed demanded public agitation: the controversy over the decision to telecast it or not, the extensive coverage in the TV magazines, the concentration of the other mass media on the event, the compression of the experience into the space of a week, and the splitting of society into two camps for and against *Holocaust*. The nonviewers did not have much to say during that week because practically nothing else was discussed.

This visceral reaction is symptomatic of the gaping hole that must have existed in the minds and emotions of the viewers. *Holocaust*'s impact sheds by no means a favorable light on the Federal Republic. On the contrary, it demonstrates the thirty-year-old failure of the mighty cultural industry to transmit the experience of fascism and its capital crimes to the mass of the inhabitants of this country—moreover, to transmit them as complex, in a way that does not relegate the problems to an affair of intellectuals and does not bypass the emotions.

On the other hand, the agitation stirred up has been subject to much overinterpretation. It was attributed with a potential for drawing general conclusions about the development of politicohistorical consciousness, that is, the assumption of long-term effects. These were hasty conclusions if society as a whole is being considered. Undoubtedly, the most prominent political event coinciding with the reaction to *Holocaust* was the debate about the statute of limitations for Nazi crimes. Results of the survey carried out by the Federal Center for Political Education demonstrate that the number of those opposing limitations increased immediately after the telecast (although 39 percent is still way below half of the total population), but then fourteen weeks later, the advocates of retaining limitations (that is, in favor of Nazi crimes no longer being prosecuted) had a relative majority of 41 percent.[19] This development took place *after* Parliament had already voted in favor of abolishing limitations for all cases of murder.

But the agitation caused by *Holocaust* had another dimension that pervaded private and public discussion then as now: the articulation of an overwhelming desire to be finished morally with the subject, in every sense of the word. This desire is shared in different ways by very disparate social groups, and it is attributable to extremely antithetical motives. For example, a large proportion of our parents' generation feels personally accused when reminded of the inhuman crimes of their former leaders and would dearly like to free themselves once and for all from the moral pressure weighing on them. This applies similarly to a section of the postwar generations to whom categories like "historic guilt" appear to mean less and less the younger they are. The two poles these generations represent came together in a striking manner with regard to their reactions to the invasion of Lebanon by Israel and specifically to the case of the massacres of Palestinians at Sabra and Chatila.

At last it was time to adjust the balance. At last—or so it seemed—the one side was relieved of their burden and the other had found an opponent, who appeared to them to equal the Nazis' terror: in their eyes, the victims had become offenders themselves. And they came so hastily to the historically and morally false conclusion that the Holocaust is divisible; the blame could be laid on the Jews for another Auschwitz—Sabra and Chatila. Even if this attitude did not come over so crassly in public,[20] it illustrated very clearly how far we still were, three years after the first telecast and in the wake of 1982's rerun, from understanding the history of the Nazi genocide.

The Mobilization of New Forms of (Self-)Consciousness

Whereas for three decades in postwar West Germany, the connection in any positive way to German fascism had been prohibited, it became very apparent that *Holocaust* had had a "liberating" function on this taboo. The growth of a new self-consciousness among those who had "gone along" with fascism, who had "taken part," is still for me today one of the strangest effects of the event. Accepted authorities and powerful figures who determine public opinion, such as the editors of *Der Spiegel* and *Stern*, Rudolf Augstein and Henri Nannen, made loud and perfidiously "honest" confessions (and presumably expected admiration for it): "Did I really know nothing at all about it? Yes and no. We all knew Dachau. Dachau stood for incarceration and Nazi brutality—for all concentration camps. I came home from the war on the Eastern front knowing nothing about gas chambers or the systematic murdering. War had made me insensitive. I suddenly realized that all the time my one concern had been my own and my family's fate. I had lost sight of the fate of the Jews."[21] "I, at any rate, knew that in Germany's name defenseless people were being exterminated in the way vermin are extinguished. And unashamed, I wore the uniform of an officer in the German air force. Yes, I knew about it and was too cowardly to resist. Was there anything I could have done at that point in time? That is not the question."[22]

In a kind of way, it became rather socially decorous to have "been around then" and to publish your sincere or just recently concocted struggles with your conscience. It was not the resistance fighters or the victims who were now put forward as an example but instead the pitiable followers. Obviously one of the reasons lay in the massive positive echo this met with in the general public. For many, this was a common decisive experience. Long before television's *Holocaust,* Elie Wiesel had anticipated this turn of public opinion. In a justifiable overstatement, he alludes to the relationship between public opinion and the victims and survivors of the concentration camps: "It will not be long before they are felt to be unwelcome intruders. The spotlights are now on the murderers. These are presented in film, they are analyzed in detail, they are endowed with characteristic human traits. At first they are viewed objectively, then with sympathy."[23] It is a worthwhile exercise to examine the dramatic construction of the figure of Erich Dorf, the Nazi protagonist of *Holocaust,* in the light of this statement.

This phenomenon was also characteristic—albeit in a diminished form—at the level of representative politics. It was not a politician like Willy Brandt, who emigrated because of the Nazis, was active in the resistance movement, and is reviled even today as a "traitor to the people" because of it (and not only by the Nazi press), who threw himself into the moral fray but it was his party comrade and federal chancellor, Helmut Schmidt. As a former Wehrmacht soldier at the front, he shares his biography with the majority of German men now over fifty and thus provides the exemplary model for identification. The most penetrating instance of this took place about seven months after the telecast of *Holocaust.* Against the venerable backdrop of the Berlin Reichstag, Schmidt braved the harmless questions of a few American, English, and French journalists. This live television broadcast completely and utterly vindicated all those present, on- or offscreen. The TV flock of the Federal Republic, which had gathered together for the occasion, listened sympathetically to their chancellor. He told of the scruples and prickings of conscience that he had because of the contradiction between the demands of social duty and knowing why he was required to obey. No one contradicted, delved deeper, or attempted a provocation, neither during the program nor in the press reports that followed. Is it any wonder, then, that criticism of the top politician and West German figurehead, Karl Carstens, formerly in Nazi service, has become so quiet as to be inaudible? The opposite is true. Sympathy with him is going up all the time. After all, he is in office now, and respect is due to that much authority, isn't it? A good illustration gives the conservative *Frankfurter Allgemeine Zeitung*'s little résumé of December 1979: "Before the election of the present Federal President, there was some legitimate criticism of a candidate as to whether he was fit for what is, after all, a political office, and there were also the jarring war cries of some of his political opponents. Today, after the first six months in office, neither the office nor its new holder has sustained any lasting injury. Carstens is becoming acquainted with his new duties."[24] In a society that has elevated the career to the highest of ethical values, it cannot and must not be reprehensible to subordinate one's duty to resist to the exigencies of one's career.

But *Holocaust* awakened another, positive variety of new self-awareness in those social minorities who were also victims of the Nazi regime of violence. They are still suffering in the present day because of their identity—Gypsies, homosexuals, the politically persecuted. For the first time in the history of the Federal Republic, their voices have been heard loudly and frequently in public, demanding recognition and ideological and material redress. Perhaps this protest was provoked by the fact that the TV series hardly touched on their persecution and suffering. They have started to insist on their rights and to publicize their stories—with increasing success.

Removing the Taboos and Playing Down History

The adaptation of the historical Holocaust according to the principles of commodity aesthetics of commercial television production made it possible for the first time to give an international mass audience access to aspects of that horrible reality. After all, more than a quarter of a billion people in thirty-two countries watched *The Story of the Family Weiss* the first time it was screened. In West Germany, the audience numbered approximately 12 million. In an international comparison audience size, however, the Federal Republic was only in sixth place, and in the ratings list of all countries that screened *Holocaust*, we occupied only seventeenth place. Nevertheless, for many, especially younger viewers, it was their first confrontation with a part of the history of the European Jews and their tormenters, because this part of German history was one that had hitherto been rigorously excluded from public debate in our institutions of education. The ceremonies of the annual Week of Fraternity or commemorations of the Kristallnacht (1938); dramas like Peter Weiss's *The Investigation* or Rolf Hochhuth's *The Deputy*; the shamefully few trials of the former concentration camp butchers of Maidanek, Dachau, and Auschwitz—these were registered by only a fragment of the general public, part of the cultural and political elite. But because of the TV series, the word *Holocaust* passed into the common language and replaced to a certain extent terms borrowed from the bureaucratic euphemisms of the Nazis, like *Final Solution*. *Holocaust* became a metaphor for inhumanity.

But this process of lifting taboos was also two-sided, and the negative one has begun to predominate. What Elie Wiesel anticipated after the miniseries's premiere in the United States—that this "ontological event" would be dragged down to the level of everyday normality—has begun to come about in the past years in the most vehement fashion and accompanied by the most incredible phenomena.

In the public consciousness, one Holocaust has become many. The TV series unleashed a veritable Holocaust inflation. The metaphor was drained of historical content and since then has been applied to each and every event manifesting inhuman dimensions of violence: the "Nuclear Holocaust" (although this is one example where the ontological analogy is least controversial), the "Holocaust in Cambodia," the "Armenian Holocaust," the "Palestinian Holocaust," and so on, through to the marketing of the term by the cultural industry—rock music promoters billed a gig of several bands as a "Heavy Metal Holocaust," and the final, most degenerate use is the film title *Zombie Holocaust*, which is sold on the European home video cassette market by CBS Records, among others, a "nasty" video of the cannibalism genre.

One would not feel outraged at these faux pas—to put it mildly—of the culture industry and a few businessmen who think themselves particularly smart were it not for the fact that there are parallels in the public discussions in society. The spectacular remarks of Heiner Geissler, a present minister and general secretary of the Christian Democratic Union (CDU), placing the blame for Auschwitz on those who died there,[25] or the recent characterization of the Nazi pogroms as "still human" compared to the social system of the Soviet Union, also by a CDU politician,[26] were evaluated by public opinion as individual blunders and which—and this is part of the contradiction—unleashed a furor of protest, not only from victims. These examples stand at the same time for the loss of adequate consciousness vis-à-vis the unique individuality of the historic Holocaust, a loss that threatens to become greater as time passes and the warning voices of witnesses are silenced by death. This loss of political and moral culture has already had consequences for everyday life in West Germany. People have become quick to make comparisons with the Nazi era. A few broken windows at young squatters' demonstrations are put on a par with the terrors of the Nazis' Night of Broken Glass; members of the Green Party are compared directly with the Nazis; members of the peace movement exhibited a complete lack of historical and moral sensibility by holding nocturnal *Mahnwachen* (warning vigils) on the doorsteps of members of Parliament before the Bundestag's vote on the deployment of new nuclear weapons; the harmless—and justified—occupation of an alternative West Berlin magazine's offices by women's groups was termed a *Rollkommando*. The television *Holocaust* has contributed to fascism being dragged into the arena of everyday conflicts, and we must be very careful indeed lest in this way it become everyday and banal.

Neo-Nazis: The Double Edge of Increased Publicity

Onlookers in other countries may have gained the impression from the event *Holocaust* in West Germany that the neo-Nazi movement here has expanded and spread. This is not so, but it has not diminished either. What did happen was that it acquired a quality it did not possess two years ago: the movement commands a great deal of public attention and tries to get more in an increasingly militant way. This is one of the most obviously contradictory functions that *Holocaust* helped provoke. Whereas the general public and the mass media have become more sensitive to what is happening on the extreme right, the neo-Nazis have become more greedy for publicity and attempt to create it by staging spectacular provocations. Only very few of these are ever brought before a court of law. Nonetheless, the Federal Crime Department's list of offenses by neo-Nazis exposed in 1979, the year *Holocaust* was first telecast, is menacing: "Fifteen hundred and eighty cases of seizing and impounding literature, 3,400 of distribution of newspapers, journals, posters, and leaflets, 447 of daubings on walls, 24 of desecration of graveyards, 22 of bodily injury, 18 of seizure of weapons, 5 of arson, 3 of planted explosives, 3 of threatened assassination."[27] And this is only part of the extensive catalog.

Even more brutal is the fact that neo-Nazi militancy has succeeded in intimidating sections of the populace. The following is taken from a report of the German news agency, dpd, on one of the most active neo-Nazi groups, the Wiking Jugend, in December 1979, and the last sentence sounds only too familiar: "According to security service information, about 30 young people are members of the 'Wiking Jugend' in Berlin, and they wear uniform-like clothes. This clothing is mainly black with high boots. Since the beginning of this year, they have increasingly engaged in violent activities, according to police sources. Thus several fights have been picked by the group. A series of searches of group members' houses resulted in the seizure of warning

FIGURE 1.3. Writing on the wall: one of many graffiti in Berlin Kreuzberg, 1978. Photograph by the author.

shot and gas weapons as well as truncheons. The insecurity caused in the general public by this group is being taken 'very seriously' by the police. *However, inquiries were being impeded because persons threatened and molested by members of the group often only gave evidence to the police very reluctantly"* (emphasis added).

All this belongs to the immediate context of the *Holocaust* event. Some of the actions by neo-Nazis in West Berlin even referred directly to it as when a discussion was interrupted in the Haus der Kirche by a "storm troop" in spring 1979. The double edge of the publicity accorded to such actions showed later that the organized right wing was becoming increasingly militant, not even shrinking from murder. Twelve people died and 213 were severely injured in a bomb explosion at the Munich Oktoberfest in 1980 because of a bomb planted by an associate of the Hoffmann Group. "The neo-Nazis have been underestimated" was the unanimous opinion of the press. Those who had given this very warning at the time of the *Holocaust* event were either not taken seriously or branded as out-of-touch left-wingers.

In recent years, young people have been showing increasing affinity with the extreme right. It is they who are affected most by social declassification and isolation. The tightly organized nucleus of the neo-Nazis has successfully recruited new members from various youth subcultures that are particularly vulnerable to fascist ideology, like the skinhead soccer rowdies. Since the Federal Republic, too, has begun to feel the impact of capitalism's present structural crisis, the image of their foe has changed—or at least outwardly. The immigrants and particularly the Turkish "guest workers" are the Jews of today. They know full well that this choice meets with broad agreement on the part of the silent majority.

In the fight against the new and old Nazis, it is of crucial importance that another section of youth come to realize what dangers their future may hold. Using their own means of expression, they will certainly achieve more than the most well-meaning institutionalized pedagogy. One outstanding example of many is the "Holocaust" song of the postpunk band Tank of Danzig from their LP *Not Trendy* (1982):

> You read it in a book
> You saw it on my TV screen
> To you it's a nightmare
> To some it's a dream, a dream
>
> Remember Belsen, remember Auschwitz
> They try to say they didn't exist
> Don't let them put this country in chains
> Don't let six million die in vain, in vain
>
> They hide their real character
> Behind the German flag
> You'd better watch out
> They are hiding for a comeback, a comeback
>
> Belsen
> Auschwitz
> Dachau.

Activation of the Cultural Sector

In the politicocultural sphere, the functional event of *Holocaust* as enlightenment was fruitful, the most visible effects being in the educational institutions. School students, who had never heard anything about the Nazi genocide, pressed their teachers to deal with the subject in class. Other teachers used the provocative occasion to thematicize a subject that was long overdue. The effects of this were not limited to the immediate teaching situations. More visits to concentration camps were arranged. The desire to talk to survivors of the Holocaust or members of the resistance was voiced. Particularly, the concept of antifascism attained a new status. Working groups of school students constituted for the purpose organized a special antifascist "day" or "week," often in the face of resistance by the school bureaucracy and individual teachers. At these events, photo documentations of Nazi atrocities, among other things, were compiled and exhibited; plays were staged; discussions were held; and other activities were arranged.

This movement in the schools, which is continuing, is also not free of contradiction. On the forty-first anniversary of the Kristallnacht (which, like the following examples, was hardly mentioned in the press), there were individual cases of stiff opposition from school students to the subject being handled in class. Many of my friends are teachers, and they reported cases of massive protest among "their" school pupils at the past being "dragged up" yet again. Although remarks like "It was the Jews' own fault" and "Too few were gassed anyway" were rare, it is ominous that their classmates did not put up any opposition; they reacted with uncertainty and disinterest.

A particular occurrence at this time serves to illustrate official reaction to the school students' informative activities. At a West Berlin trade school, the school administration organized the first Anticommunist Week—as a necessary counterweight to the antifascist activities, said the director. In the future, it will be held annually as Antitotalitarian Week. In this way, the original opponent is exchanged for the desired opponent in the present—totalitarian theory in practice.

In the domain of the visual and performing arts, *Holocaust* led to the rediscovery of a subject—pictures from the concentration camps and everyday life under the Nazis. Socially committed galleries put together exhibitions with themes like *Children in the Concentration Camps* (Elefantenpress Galerie, West Berlin) or *Resistance: Not Conformism* (Cultural Office, Karlsruhe), which included works used in *Holocaust*. The most impressive is *Survival and Resistance,* a cooperative exhibition of the West German "German-Polish Society" and the State Museum of Oświęcim-Brzezinka (Auschwitz-Birkenau), which has been touring major cities for months and has met with a great and positive response.

But of far greater significance than the numerous activities emanating from the media event was their content—the way the series was utilized as a starting point. The story of the Weiss and Dorf families disclosed a meaningful way to confront the past in that it dealt primarily with the everyday experiences of the murderers and their victims. Moreover, this was accomplished without totally neglecting the level of high politics. In this way, *Holocaust* succeeded above all in giving the postwar generations an idea of how people lived under fascism. And afterward, this served as a useful point of departure. Since then, there have been many instances of the encouragement of "oral history," of people taking up the search for traces of everyday life under the Hitler regime. Many places in West Germany have thus acquired regular historical signposts. For example, in West Berlin, a youth organization, the Stadtjugendring, organizes tours of the city to "famous sites in the history of the Labor Movement, Fascism, and the Resistance." This is an attempt to remedy the fact that for many young people, the imposing architectural remains of power and persecution, such as the Reichstag, the Jewish Community Center in the Fasanenstraße where once the great synagogue stood, the elegant villa in Wannsee where fanatical German bureaucrats decreed the "Final Solution" for the Jews forty years ago, or the execution factory at Plötzensee, have become so much dead matter. In particular, the fiftieth anniversary of the Nazi seizure of power was used as an opportunity to delve into the history of whole districts of Berlin, such as Kreuzberg, Neukölln, and Wilmersdorf. The results were presented to the public at exhibitions, in brochures, and in films. Streets, sites, and names became known again, even though they are not in history textbooks.[28]

One of the most remarkable signs of a modification in cultural consciousness was the thematicization of the long-forgotten and repressed Yiddish culture. After a number of committed art cinemas—like the Kommunales Kino in Frankfurt or the Arsenal in West Berlin—had organized highly regarded festivals of Yiddish films, Zweites Deutsches Fernsehen (ZDF) risked offering such a program in 1983 to a mass TV audience. The genre's history was presented in an introductory film, *Das Jiddische Kino,* and in March and April, the second channel screened three of the most seminal works: Joseph Green's *Ein Brief an die Mutter* (1938), Maurice Schwartz's *Tewje, der Milchmann* (1939), and *Jidl mit der Fiedel* by Green/Przybylski (1936). This was undoubtedly the first authentic contact with the Yiddish language and culture for the majority of West German viewers.

And forty-four years after the infamous anti-Semitic treatment for the edification of concentration camp henchmen and the ideological mobilization of the populace, the most controversial and popular figure in recent Jewish German history made a reappearance: *Joseph Süss Oppenheimer* is the title of a new documentary drama by Gerd Angermann, which ZDF set in contrast to the fascist film interpretation *Jud Süss* by Veit Harlan.[29] Just how sensitive this taboo area is in the consciousness of public institutions is demonstrated by the vacillations over the projected screening date of this program. The original date set was January 24, 1984, but because of Chancellor Kohl's trip to Israel on the same day, the telecast was postponed until February 14. The ZDF press office explained that they did not wish to encumber the "rather sensitive visit of the Chancellor to Israel with even a hint of problems."[30] The widespread identification of Jews as Israelis, which is part of the ideological fundament for the reversal of the victims'/perpetrators' roles during the Lebanon conflict, finds its adequate representation here in the attitude of our state-run television authorities.

In West Germany, probably more than in other countries, the two central categories of content in television communication are opposing ones: information and entertainment. This point of view is shared not only by the producers of aesthetic commodities and the program directors but also by their critics. The mixing of information with entertaining strategies is forbidden, just as entertainers successfully defend themselves against having anything to do with society. Thus, on West German TV, dry political magazines or features coexist with mindless shows and quiz programs, unattractive news broadcasts with detective series or "pure" sport.

This strict division of labor received a provocative challenge from the *Holocaust* event. In the discussion on the pros and cons of telecasting the series, care was taken to point out that there was more to this American family story than adventure, crime, love, tears, and the fascination of horror. The fact that *Holocaust* was consigned to the Third Channel confirmed that the program directors were scared stiff of any wide popular effect it might have. At the time, this is symbolic of the contradiction within the American TV series itself: an important historical theme has been dismantled, cut up into snacks for easy consumption, which stimulates the famous tingling in the stomach; evokes sorrow, hatred, and identification; and, at the same time, throws up questions, triggers discussion, and activates the gray matter.

This describes a qualitatively new phenomenon in the development of the culture industry, actuated by the American entertainment factory, which is, anyway, the most developed in existence. Marketing is suddenly no longer one-dimensional but ambiguous. The purity of the charge of manipulation brought against everything this factory churns out does not hold good when confronted with a decisive slice of German history that has been turned into a commodity. Despite the commercial packaging of the Nazi mass murder, the subject loses little of its monstrous explosiveness and still provokes proportionate contention.

The *Holocaust* of television is not the only example of this phenomenon and its young tradition. Indicative of this trend, for example, are the TV series *Roots* and *King* and the cinema productions reflecting the Vietnam experience, such as *The Deer Hunter, Coming Home,* and the most spectacular, *Apocalypse Now*. It seems that the culture industry has discovered social contention and reflection as a medium of marketing the goods. It appropriates the social issue, but—and this takes place independently of its intention—by doing this, the issue itself becomes effective in a mass contact for the first time. The discussion is wrenched away from its confinement in intellectual circles and placed firmly in the area of everyday public debate.

All this could not fail to affect the way that West German television dealt with fascism and especially the mass murders against the Jews. In addition to the standard documentaries, a considerable number of new productions modeled themselves directly on the dramaturgical and aesthetic principles of the NBC series without actually trying to copy it. Both for TV and cinema, the most popular subject became the biographical film (or series) that very often used literary autobiography as a basis for the script.[31] These included *Die Welt in jenem Sommer* (January 1980), directed by Ilse Hoffmann and based on the autobiographical novel of the same name by Robert Muller, who has never returned to Germany from exile in England; *Kaiserhofstrasse 12* by Ann Ladiges and Rainer Wolffhardt, based on the book by Valentin Senger, telecast December 1980; *Die Kinder aus Nr. 67 oder: Heil Hitler, ich hätt' gern ein paar Pferdeäppel*, a feature film set in the milieu of Berlin's "Hinterhöfe" by Usch Bartelmess-Weller and Werner Meyer and which was first shown in 1980 in cinemas and then screened by ZDF in March 1982; *Ein Stück Himmel*, an eight-part miniseries based on Janina David's autobiography, directed by Franz Peter Wirth and telecast by ARD in April–June 1982; and Ottokar Runze's cinema film *Stern ohne Himmel*, in which film buffs can detect a sarcastic piece of casting by the director—one of the adult protagonists is

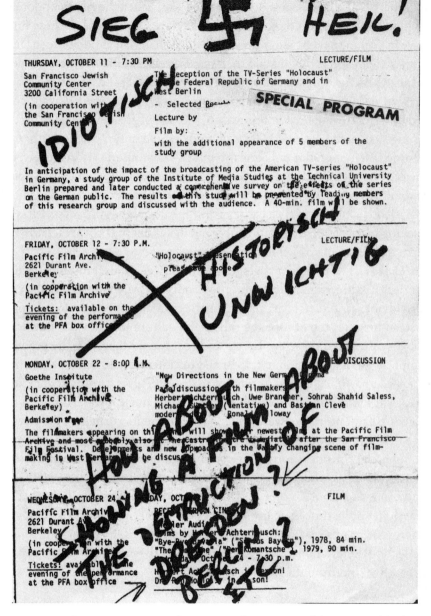

FIGURE 1.4. Comments on a poster that announced screenings and discussions of the film *Responses to HOLOCAUST in Western Germany,* written and directed by Siegfried Zielinski (1979), in San Francisco and Berkeley on October 11 and 12, 1979: "Idiotic," "historically unimportant." The film was shown at many locations in the United States and can be found in the archive of the Museum of Broadcasting, now the Paley Center for Media, New York City.

played by Malte Jäger, who, in 1940, played Karl Faber, the Aryan opponent of Jud Süss in the film directed by Veit Harlan. One aspect all these films have in common is the story perspective: all share the child's or teenager's view of the world of victims and their persecutors. Another common aspect is their thematic level: all concentrate on everyday life and on the death and survival of Jewish children and youngsters and the way their "Aryan" contemporaries treat them. For this reason, they are of great importance for the younger generations in West Germany and deserve similar promotion to that afforded *Holocaust*. But only one other film—a miniseries—has received a measure of that: *Blut und Ehre: Jugend unter Hitler (Blood and Honor: Youth under Hitler)*, which in the meantime, has also been screened in the United States and would certainly have been unthinkable without *Holocaust*. It is an exciting piece of TV entertainment about life under the brown dictatorship from the perspective of children and teenagers.

Holocaust as anti-advertising, caught between entertainment, which renders its subject harmless, and provocative—professional culture critics did not want to resolve this immanent contradiction in 1979 and opted instead for one side of it or the other. The culture industry has now long since synthesized these antitheses and recognized that with the growing significance of entertainment, the need for interpretation and orientation has also grown. At the beginning of 1984, *The Day After* was screened in West Germany, the CBS film about a fictitious nuclear war. Although the voices are not nearly so heated, the same arguments are being put forward in a discussion about the permissibility and ethical integrity of combining commercial considerations with a warning about the nuclear catastrophe, just as was the case five years ago. To take this product of the cultural industry and use it as a fruitful point of departure, to develop its effects further, as was done successfully in many instances with *Holocaust*, would certainly be energy spent in a meaningful way. This would mean the development of a dialectic of entertainment originating not from the products of mass culture themselves but from the way we utilize them.

Translated by Gloria Custance

Notes

1. See Friedrich Knilli and Siegfried Zielinski, eds., *Holocaust zur Unterhaltung: Anatomie eines internationalen Bestsellers* (Berlin: Elefanten Press, 1982); Knilli and Zielinski, eds., *Betrifft: "Holocaust": Zuschauer schreiben an den WDR. Projektbericht unter Mitarbeit von E. Gundelsheimer, H. Mass und F. Ostermann* (Berlin: Spiess, 1983). These publications also contain comprehensive bibliographies of international research literature.
2. Theodor W. Adorno, "Zur Bekämpfung des Antisemitismus heute," *Das Argument* 6 (1964): 29.
3. Max Horkheimer and T. W. Adorno, *Dialektik der Aufklärung* (Amsterdam, 1948; bootleg print), 199ff.
4. Adorno, "Bekämpfung."
5. Adorno.
6. Bertolt Brecht, *Gesammelte Werke*, vol. 18 (Frankfurt am Main, Germany: Suhrkamp, 1967), 201.
7. For a detailed analysis of these two films, see Siegfried Zielinski, "Faschismusbewältigungen im frühen deutschen Nachkriegsfilm," in *Sammlung 2: Jahrbuch für antifaschistische Literatur und Kunst*, ed. Uwe Naumann, 124–33 (Frankfurt am Main, Germany: Röderberg, 1979).
8. I have not included treatments of this subject in other media, that is, the stage plays by Max Frisch *(Andorra)*, Rolf Hochhuth *(Der Stellvertreter* and his latest *Die Juristen)*, and Peter Weiss *(Die Ermittlung)*. All are of importance for the cultural history of the Federal Republic, and a study of their relationships would be of great value.
9. Cf. my essay on "Fascism as 'Entertainment,'" *Konkret* 8 (1978): 42. For the Fest/Herrendoerfer Hitler film, see Jörg Berlin, *Was verschweigt Fest? Analysen und Dokumente zum Hitler-Film von J. C. Fest* (Cologne, Germany: Pahl-Rugenstein, 1978).
10. *Der Spiegel*, July 17, 1978.
11. *Frankfurter Allgemeine Zeitung*, April 20, 1978.
12. *Rheinischer Merkur*, April 28, 1978.
13. *Die Welt*, July 10, 1978.
14. Evangelischer Pressedienst, ed., *Holocaust: Der Mord an den Juden als Fernsehserie* (Frankfurt am Main, 1979).
15. By way of comparison, the eleven-part TV dramatization of the German classic *Buddenbrooks* by Thomas Mann, which was screened a few months later in 1979, had audience participation averaging 46 percent for the first seven installments. That is nearly 10 percent more than *Holocaust*'s average of 37 percent over four nights.
16. The survey carried out by the Federal Center for Political Education in Bonn paid particular attention to the conversations and discussions that went on

during and after the event. For a summary of the results, see Tilman Ernst, "Umfragen der Bundeszentrale für politische Bildung," *Bild der Wissenschaft* 6 (1979): 74–80.
17 Knilli and Zielinski, *Betrifft Holocaust*.
18 In November 1982, the ratings averaged 23 percent, that is, considerably lower than for the first telecast in 1979. Just how little was done in preparation for the rerun can be seen from the fact that part 1 reached a mere 16 percent, whereas the last installment did at least achieve 30 percent.
19 See Ernst, "Umfragen," and Ernst, "'Holocaust' und politische Bildung (Ergebnisse der dritten Befragungswelle)," *Media Perspektiven*, December 1979, 819–27.
20 To cite just one example, a passage from an interview with the French writer Jean Genet that appeared in the first issue of the magazine TIP (West Berlin) in 1984. Neither answer nor question is lacking in clarity: "TIP: Could one say that the Palestinians are the Jews of today? Genet: Yes. Today the Palestinians are the martyrs, and the Jews are the Nazis."
21 Rudolf Augstein in *Der Spiegel,* January 29, 1979, 20.
22 Henry Nannen in *Der Stern,* February 1, 1979, 5.
23 Elie Wiesel, "For Some Measure of Humility," *Sh'ma* 5/100 (October 1975). Cited in Bruno Bettelheim, "Holocaust: Überlegungen ein Menschenalter danach," *Der Monat,* December 1978, 15.
24 *Frankfurter Allgemeine Zeitung,* December 18, 1979.
25 Heiner Geissler said this in front of the Bundestag during one of the debates on the deployment of new American medium-range missiles (Pershing II) by way of a comparison to the present-day peace movement, which he made responsible for an eventual World War III.
26 It was the CDU member of parliament Anton Feyssen from Hildesheim who made this comparison in November 1983 during a public debate. For further details and the protest of Heinz Galinski, president of the Jewish Community in Berlin, see *Frankfurter Rundschau,* December 12, 1983.
27 *Die Tat,* December 7, 1979, 1.
28 These activities by galleries, youth centers, district cultural departments, artist's associations, and so on, were coordinated by the Berliner Kulturrat, constituted expressly for this purpose. It gave rise to considerable controversy, for the Berlin Senate then refused to supply the funds necessary for a majority of the various activities. See Berliner Kulturrat, ed., *1933 Zerstörung der Demokratie: Machtübergabe und Widerstand, Ausstellungen und Veranstaltungen, Programm 1983* (Berlin, 1983).
29 See Knilli et al., *Jud Süss: Filmprotokoll, Programmheft und Einzelanalysen* (Berlin: Spiess, 1983). For the controversy over the director of *Jud Süss* and the legal trials for "crimes against humanity" against him, see my book *Veit Harlan: Analysen und Materialien zur Auseinandersetzung mit einem*

Film-Regisseur des deutschen Faschismus (Frankfurt am Main, Germany: Fischer, 1981).
30 *Frankfurter Allgemeine Zeitung,* December 17, 1983, 24.
31 See S. Zielinski, "Aspekte des Faschismus als Kino und Fernseh-Sujet: Tendenzen zu Beginn der 80er Jahre," in *Sammlung 4, Jahrbuch für antifaschistische Literatur und Kunst,* ed. Uwe Naumann, 47–56 (Frankfurt am Main, Germany: Röderberg, 1981).

FIGURE 2.1. Dialectics of (un)visibility: In astrophysics, protuberances are the masses of burning hydrogen that flare up from the Sun's surface at a speed of approximately six miles per second and reach a height of up to thirty thousand miles. Seen through a telescope, at the edges of this extravagantly wasteful star dynamic, forms glow against the blackness of space: slender fountains, shapes reminiscent of plants. These phenomena can be observed especially well during a total eclipse of the Sun, when the Moon shuts out the light from the fiery ball. Wilhelm Zenker drew this sketch to record his observations of the Sun's eclipse in summer 1887. From Wilhelm Zenker, *Sichtbarkeit und Verlauf der totalen Sonnenfinsternis in Deutschland am 19.8.1887* (Berlin: Dümmler, 1887), appendix.

2   Media Archaeology

Searching for Different Orders of Envisioning

> *Michel Foucault's concept of archaeology was for me, at first, a guiding principle for branching out into another kind of historiography of the media. With his analysis of Bentham's Panopticon, Foucault established the panoramic view as the organizational apparatus of a disciplinary society. My own primary motivation was to discover aesthetically and politically different, agitating ways of envisioning that might open Foucault's concept—which is above all critical of power and in this sense also relatively constrained—to transgressive experiences. The corresponding method I developed was thus an an-archaeology of the arts and media, which later evolved under the neological concept of* variantology. *This chapter is an early text along this research pathway.*
>
> *This essay was originally published in 1994.*

⁂

> The old world
> Neared its end.
> The pleasure garden
> Of the young race
> Withered.
> And out
> Into free space
> Strove the mature
> No longer childlike people.
> The gods had vanished.
> Nature stood
> Lonely and lifeless,
> Deprived of soul by the strict number
> And the iron chain.
> Laws came into being.
> And as into dust and thin air
> The immeasurable bloom
> Of thousandfold life
> Vanished into words.
> —Novalis

Because arctic coldness threatens to envelop the dominant sciences and the hegemonic culture, fashionable searches are now under way to uncover archaeological material coated with the patina of history.

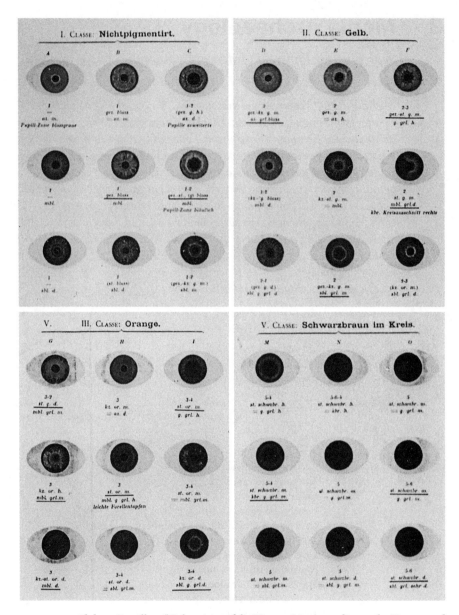

FIGURE 2.2. Alphons Bertillon, "Colorations of the Human Iris, According to the Texture and Quantity of the Orange-Yellow Pigment," 1870s. From Alphons Bertillon, *Das anthropometrische Signalement* (Bern: A. Siebert, 1895).

There is a desire for lush and abundant testimonies of the creation of texts and images, of imaginings, of the stimulating and frequently delirious thinking between word and icon. To be sure, neither my concept of archaeology nor the related work my students and I keep embarking on are free of such romantic leanings. But then again, why should they? My guiding motivations and ideas are rooted in a deep mistrust of history, or rather historiography, which I acquired by reading, among others, Michel Foucault; and they are fueled by notions of structuration that Gilles Deleuze and Félix Guattari never ceased to advocate, that is, open systems of energetic fields caught in highly complex exchanges of forces and shapes, in search of underground connections not yet exposed that, in turn, need conjunctions that do not constrain appearances but instead help unfold them. Fabrics. Rhizomes. Filigree root systems. Weeds.

Archaeology (of media or of audiovision) would, then, be a method of foregrounding the resistant local discursivities and the expressions and conceptualizations of technologically based imaginings and worldviews that are at work within our largely linear and chronologically constructed history. As I see it, this project has two aspects:

1. It is a matter of uncovering and interdiscursively confronting those energies and expressive potentials that hitherto have either been ignored by media history or suppressed by professional textual histories.
2. Furthermore, it is a matter of closely examining these exposed phenomena to determine the extent to which their uncovering and exhibition have served to smoothen, harmonize, or even remove their unruliness and complexity by subjecting them to historically alien concepts and perspectives.

The latter point is particularly relevant for my work on serial, sequential, or chronophotography. With its rigid focus on attempts to either approach or distance itself from the reproduction of living bodies, the cinema myth—that is, the notion of cinematography as the vanishing point of heterogeneous designs and technologies—has all but ostracized the specific contrariness of the actual diversity of concepts, artifacts, and technological systems. This is, no doubt, highly unsatisfactory; yet supported by an understanding of media committed to heterogeneities and connections between differentiated materialities, we are able to stage interventions. Our goal is to create a genealogy of a different, technologically based envisioning, such as we are pursuing now at the Cologne Academy of Arts and Media with our project "Speleology of the Other Gaze."[1]

The most difficult aspect is to translate into tactical reality what Deleuze and Guattari called "antigenealogical." How does the reconstruction of historical phenomena avoid the dangers of once again burying uncovered manifestations of the living underneath strict linear orders?

This conflict, which ultimately can only be solved by paradoxes, clearly speaks to a presence that my archaeological work is linked to and which for the sake of orientation I will delineate in four theses:

1. With history in our back and a not yet fully determined telematic in the shape of a menetekel ahead of us, we are confronted with two paradigmatic processes running in opposite directions. On one hand, we have the primarily industrial project of universalization; on the other, the potential or possible diversification of expressive practices, institutionalizations, organizations of meaning, and aesthetic configurations within the domain of the audiovisual.

This needs to be understood and—following Foucault's genealogical concept—tactically unfolded: we need to oppose the projected universalization of the codes and the medial machinery, which clearly have approached the limits of standardization and are on the verge of spilling over into program differentiations, with an insistence on and subsequent development of a multiplicity of aesthetic (and theoretical) practices. Both processes need to be brought into a relationship marked by both incompatible tension and playful exchange.

2. This, in turn, is directly connected to the former. There are two ways in which art, indeed aesthetic practice in general, can relate to—and develop new identities in—these processes of change:

Either it challenges and stages interventions targeting the most advanced technologies. That is to say, it constantly has to cross the boundary into science, into the domains of engineering and informatics, to prepare itself for its role as a special playground in the network of symbolic actions. But this means that it also has to engage contexts of meaning that are both accessible and adjacent to the symbolic: the fields of knowledge, discourse, syntax, the rational and relational, and so on. Here we make history, and history is being made.

Or—and I am very consciously using exaggerated formulations—art resolves to dedicate itself to life. In so doing, it may use and interpret technology; it may recruit it as a tool for the production of appearances. Essentially, its contexts of meaning are desire and death, and its main task is to equip their phantoms with signifiers of the imaginary. Cinema has been doing nothing else for the last hundred years. And yet—though it may be unfashionable to utter such sentiments in view of the dominance of the technoimaginary—I would not want it any other way. Or, to stick to photography in order to highlight the emerging principle, the abstraction and the high-grade (and highly legible) order of computer-generated images only hold attraction for me because I also have the opportunity to enjoy the pictures of, say, Joel Peter Witkin. The pathologies of the rational and the sexual, or whatever society applies these labels to, were already lumped together in the late nineteenth-century days of Richard von Krafft-Ebing.[2] The latter is the subtext of the former. In the history of the eye and the gaze, the Marquis de Sade and Georges Bataille are two sides of one medal.

3. There is a very tense relationship between photographic and extra-photographic reality, between two-dimensional plan, on one hand, and multidimensional process—or at least voluminous object—on the other. If we, albeit somewhat superficially, further pursue the two paradigms sketched above, we arrive at two possible ways of distinguishing how photographic acts approach bodies. They have diametrically opposed intentions. Or, phrased in a more rational vein, they are inscribed by different cognitive interests. The combination of subject, lens, and lighting turns the body into a sensation: the body in the image is a sensational body, and what is most interesting about this body is how its features are located on—and expressed by—its surface rather than how it smells, tastes, or feels. But as long as we are focused on the body as an extended object in space equipped with a specific consistency and density, and with an equally distinct inner (physical) composition and functionality, lens-equipped systems tend to be a hindrance. The subject becomes obsolete and anachronistic. The act of measuring may

contain its share of presumptiveness, but it is in the first instance not geared toward sensationalizing its object but understanding and changing it.
4. Finally, and in great brevity, the fourth thesis, which happens to be the paradox everything has been leading to: the distancing means of apparatus-based perception lend themselves to the dramatic *staging* of the voluminous, because in the hands of artists and designers, these very means can be used to transcend this type of perception. Be it at rest or in motion, the voluminous is rendered transparent by means of virtually excluding the lens from the apparatus-based envisioning process. This lends itself less to the staging of bodies than to their analysis, repair, and murder.

A brief contrast between the differing concepts of E. J. Muybridge and E.-J. Marey on the photographic reconstruction of movement on a two-dimensional surface may serve to clarify that the two possible poles of technology-based envisioning were, at rock bottom, already contained in the serial, sequential, and chronophotographic practices of late nineteenth-century photography. All this, however, has so far been marginalized by media histories; they have either obscured or blurred the differences by taking the pertinent photographs into custody, as it were, by trimming them in such a way that they all appeared to be pointing ahead to the next hegemonic medium—cinematography.

In essence, these two reconstructions of movement belonged to two different scopic regimes. Marey's visual sampling, his reduction of physical information down to kinetically relevant points, is linked to a tradition that seeks to enhance the *legibility* of the world. It leads to the "statistical image" (Abraham A. Moles) and reaches its tentative goal in the algorithmic computer image. Muybridge's concept, in turn, is informed by a mimetic tradition; his images are raw material for the imagination and deception as well as for interpretation and analysis.

Such strong dualisms, however, are not to be taken ontologically. Rather, they are useful contrasts that clarify the media-archaeological process and its present aesthetic and theorist practices. But once these conceptual divergences have been made clear, the stark contrast must be dissolved and turned into possible symbioses. After all, such exclusive and rigidly compartmentalized dimensions do not exist in real history; rather, they assume the shape of adjacent, overlapping, interlocking, and interactive fields.

However, highlighting the different orders of perception is of considerable significance to research and teaching in the theoretical-historical context of aesthetic practices. It can help us fathom and experimentally probe the perspectives that govern our engagement with media technologies, depending on the thematic fields in which we move and the varying intentions with which we position images and texts and organize tones.

But to be clear about this, *within* the arts, and *within* the study of the arts, it is not possible to conduct such probes and laboratories while avoiding any engagement with the elaborate media technology involved. That would entail nothing less than falling back behind early romantic conceptualizations of the utopian union of strict number anarchic life force.

> Gone was
> The all-powerful faith
> And the all-changing,
> All-uniting
> Celestial companion:
> Fantasy.
> Cold and sullen,
> A north wind
> Blew across the frozen plain
> And the marvelous home
> Vanished into the ether
> And the infinite expanse of heaven
> Filled up with
> Glowing worlds.
> —Novalis, *Hymns to the Night*, 1799

Translated by Geoffrey Winthrop-Young

Notes

1. The concept of speleology or the study of caves harkens back to Gustav René Hocke's studies of Mannerism. The project at the Cologne Academy of Arts consists of two parts: the first, historicotheoretical part is dedicated to exposing aspects of an other history of the gaze, or of the gaze of the other, ranging from Presocratic conceptualizations of the gaze and premodern notions of envisioning to the nineteenth century with its mad encyclopedias, or encyclopedias of madness. The second, laboratory-based part focuses on the development of aesthetic strategies for putting to the test contemporary artistic practices. Theoretical discourse and machine-based laboratory work are closely linked.
2. The reference is to Richard von Krafft-Ebing's groundbreaking *Psychopathia Sexualis* (first published in 1886).

3   Seven Items on the Net

> *This short, proclamatory text was precipitated by a profane and coolly formulated announcement on the internet on November 30, 1994, concerning the suicide of Guy Debord, the head of the French Situationists. More than most early web users, I wanted to clarify for myself what it was that the internet excluded and, above all, what it would become: a pitiless, pragmatic enunciatory nexus of the blatant and the self-evident—this side of secrecy and beyond any secret whatsoever.*
>
> *This essay was originally published at the beginning of 1995.*

⁂

1.

Now and again, unforeseen events burst into the telematic net. In a December issue (December 15, 1994) of the magazine *Fineart—Art & Technology Netnews,* Jeremy Grainger broke the AP news story via Fringeware that Guy-Ernst Debord had committed suicide. The report was terse: "He was 62. . . . Little known outside France, Debord denounced what he called 'the show-biz society' and declared that performing arts should be based on powerful emotions, passions, and sexual desire. His ideas were influential among theoreticians and essayists who achieved prominence in the May 1968 student-led cultural revolt that shook French society." That was it. That the cofounder of the Situationist International—who in *Society of the Spectacle* had diagnosed more than twenty years ago that all direct experience had given way to representation, who in the same book had attested that telecommunication "reunites the separate but reunites it as separate"—had died by his own hand did not affect the tidily arranged symbols on the net one pixel or their author. In 1952, at the age of twenty-three, Guy Debord made a film with a dialogue seemingly organized on random principles. The title was *Howlings in Favour of Sade.* At one point, the second voice says, "The perfection of suicide is in ambiguity." In the script, this is followed by a stage direction: "5 minutes' silence during which the screen remains dark."

2.

The way language is used on the net is most affirmative of life. As a principle, the language is positive, animated, apologetic, smart. It bristles with energy. It is an electronic fountain of youth. The computers, their technical designers, and the connections set up enable and facilitate and support (for example, nature). Programs lead and organize and select. Landscapes are created, as are populations or generations, that even develop dynamically and are at liberty to unfold in (self-)organization. The interfaces must be interactive and empathic (in the Aristotelian sense) or even biocybernetically interactive; that is, they have to organize something alive within the closed circuit. Their secret agents don't have trench coats with turned-up collars to hide their faces, they're not up to anything, as yet you will search in vain for them in the underground; they are tourist guides standing in the spotlights, inviting us to go surfing, leisurely. Many decades after their discovery by theoretical physics between the wars, the waves of possibilities in which quantum truths are now formulated exclude the violence of contexts/connections; they are not waves of pain or of ecstasy. "The linking of sensor data with parameters of user interaction permits meaningful correlations over and above various output modalities." In Chris Marker's *Sans Soleil* (1983), inspired by the music of Mussorgsky, we encounter a Japanese man who is always making lists of things, for example, of things that make the heart beat faster. I started to make a list of phenomena, phantoms, and modi that I miss on the net and in the columns of speech on the subject that are getting longer by the minute. Here are some of the favorite substantives:

> ambiguity
> anger
> attack
> collapse
> crime
> cruelty
> danger
> dark anguish of spaces
> daze
> death
> deviance

discomfort
discongruence
doubt
drive
ecstasy
eczema
evil
excess
hysteria
incest
interruption
irritant
lust
macrogenetosomia praecox
monster
neurosis
obsession
passion
pathology
risk
scream
seduction
uneasiness
yearning

3.

Although many differences existed between, for example, Artaud, Bataille, Duchamp, or Leiris, for all that the dissidents of the Surrealist movement had a common focal point from which they developed their relationship to the (intellectual and art) world, they disrupted their own marginal tributary as well as the larger mainstream because of their rejection of any kind of functionalized ethics, their resistance to one-dimensional rationality, their celebration of unrepressed pleasure, and their aesthetic development of desire as an existential mode. To them, it was of imperative significance that their thinking be far removed from any hierarchical structures and that their aesthetic practices be immanently and wildly heterogeneous juxtapositionings (philosophy and cultural critique took over these paradigms at a much later date, notably with the work of the duo Deleuze–Guattari). Particularly for characters of a passionate and tortured/suffering disposition, like Antonin Artaud, the focal point of artistic praxis was the nondispersable duality of experience and sensation (with a radicalness only comparable to Bataille's work in literature), which he confronted with the pure praxis of the concept; indeed, this also essentially shaped the work of Duchamp, for all his extravagances and craziness. On what does the hyperrealistic avant-garde orient itself? What orientation is it capable of elaborating and capturing for itself? The unconscious appears to have been consciously written to death after Freud and Lacan (who neglected to adhere to his own dictum that "there are problems one must decide to abandon without having found a solution") and, above all, after their innumerable adepts and interpreters. In the 1950s and 1960s, Activists, Situationists, and performance artists threw their own bodies into the fray, to the point of (self-)mutilation and (self-)immolation, against the discourse and the dispositives of power. So, will there now be a reorientation toward concepts, toward the natural and life sciences, toward the illusion of a continuity, a flow, a beautiful order in chaos? Or will there be the creation of new, artificial bodies in the form of bodies of knowledge and their mise-en-scène as aesthetically experienceable volumes in the tele-age, moving and ephemeral artifacts in antiquated space?

4.

The experimental work of the group Knowbotic Research suggests one possible avenue: their creations and workshop processes are factional, that is, they are extracted both from empirical data and from the realm of fiction, to which they always seem to want to return. In Circe's net, they strive to direct its visualization (knowledge and its organization) while at the same time hinting at a seduction, without which art as a sensibilizing terrain for the experience of the enigma is no/thing at all. To develop this character of the double-agent, the "knowbots" have been assigned a second mode of existence that can assume form outside of the net: in the event, in the one-off mise-en-scène of publicly accessible space, they become once again empirical bodies, sensations.

5.

The most complex mysticism praxis with the most complex language that I know of is the theoretical Kabbalah: "[A] technique for exercising reason or, instructions for use of the human intellect ... it is said, that angels gave the Kabbalah to Adam after being expelled from the Garden of Eden as a means whereby to return there."[1] The ten Sephiroth with their twenty-two connecting pathways constitute a sheer inexhaustible, network-like reservoir of associations, connections, punctuations; its construction principle is binary, and it is built of the basic tensions of theoretical reason *(chockmah)* and the power to concretize, to form *(binah)*. The only meaningful mode in which the Kabbalah can be read and revealed over and over again is that of interpretation. In this, the Kabbalah and art are akin.

Edmond Jabès's texts are philosophical poems. In a discussion with Marcel Cohen about the unreadable, he was asked what he meant by the "subversion" of a text, to which he replied by referring to the beginning of each and every subversion: disruption/interference. The paradox, that he himself operates with grammatically correct sentences and words that retain their connotative meanings, he resolves kabbalistically:

> I have not attempted to ruin the meaning of the sentence nor of the metaphor: on the contrary, I have tried to make them stronger. It is only in the continuity of the sentence that they destroy themselves, the image, the sentence, and its meaning when they are confronted with an image, a sentence, a meaning, that I consider to be just as strong. To attack the meaning by rebelling against the sentence does not mean that it is destroyed. On the contrary: it is preserved because a path to another meaning has been opened up. All this appears to me as though I were confronted by two opposing discourses that are equally persuasive. This results in the impossibility of privileging one over the other which, in turn, constantly defers the control of the meaning over the sentence. Perhaps the unthinkable is just simply the mutual suspension of two opposite and ultimate thoughts.[2]

There can be a key here to how aesthetic action within orders and structures might unfold, between Pentagon, academe, and the market, which afford only slim possibilities for temporary interference, the filigree weaving of labilities.

6.

On the net, there is no art of this kind (yet): it has had no time to develop a notion of the Other, the vanishing point of which would be Death. The model for Net Culture is life, and because there it has relinquished its unique existence, it easily and usually becomes a model. The algorithms used by the engineers and artists who are working more or less secretly on the orders of the Circe Telecom have been copied from the bio-logical, life form(ula)s translated into mathematics. Genetic algorithms are useful and fascinating because of their proximity to this life. They are bursting with strength and confidence. For art, it would be worthwhile to attempt to invent algorithms of (self-)squandering, of faltering, of ecstasy, and of (self-)destruction as an experiment. In full recognition and acceptance of the risk that perhaps there would not be much to see or hear, these would be transformed into sounds and images. In the universal shadow, in the dark halo, where the strong light bodies of knowledge of Knowbotic Research move but that also prevents them from dispersing, there is a presentiment of this secret.

7.

"When art becomes independent, represents its world in dazzling colors, a moment of life has grown old, and it cannot be rejuvenated with dazzling colors. It can only be evoked in remembrance. The greatness of art only begins to appear at the dusk of life."[3]

Notes

1. Katja Wolff, *Der kabbalistische Baum* (Munich: Knaur, 1989), 10.
2. Edmond Jabès, *Die Schrift der Wüste: Gedanken, Gespräche, Gedichte,* ed. Felix Philipp Ingold (Berlin: Merve, 1989), 69–70.
3. Guy Debord, *The Society of the Spectacle* (London: Rebel Press, 1992), 71.

4 Toward a Dramaturgy of Differences

In the mid-1990s, the Dutch V2_Institute for the Unstable Media invited six artists and thinkers to take part in an interesting experiment. In internet-mediated dialogue with one another, the authors were to develop positions on the question of the medial interface. The book that ultimately resulted from this experiment was similarly unusual: in the tradition of Raymond Queneau's Cent mille milliards de poèmes (1961), *fragments of individual authors' various texts were spliced together in the manner of an* ars combinatoria. *For the University of Minnesota Press, I've extracted my own contributions to that dialogue and here formulated them discursively.*

This text was originally published in 1997.

⁂

> For why, the senseless brands will sympathize
> The heavy accent of thy moving tongue,
> And in compassion weep the fire out;
> And some will mourn in ashes, some coal-black,
> For the deposing of a rightful king.
> —WILLIAM SHAKESPEARE, *Richard II,* act 5, scene 1

An interface separates things, or the concept would make no sense.
An interface connects things, or the concept would make no sense either.
An interface marks a difference.

Consciousness, which we use to form ideas and to express them, for instance, in speech, is an interface—the oldest one we know. This is where world/worlds/reality/realities are formulated. With the help of this interface, we try to comprehend what confronts us, in short: the other, in the broadest sense of the word, that which is not identical to us.

The interface determines the relation of the one to the other which is different and fundamentally unknown, and vice versa: through the interface, the other presents itself to the one, and it does this with respect to those aspects that are understandable.

In telematics, as in any form of communication which is based on technology, the interface separates and connects the worlds of acting subjects on one side and the worlds of working machines and programs on the other side. It separates as well as connects media-people and media-machines. It is the borderline where the medium takes its shape.

In following the ideas of psychoanalyst Lacan, I make a distinction in the media discussion between the real and reality/realities (whether this is a reality that can be subdivided into many different realities, or whether I'm working with a concept of multiple reality anyhow, is really of secondary importance to me; in whichever direction one looks for the answer to this, it will not solve the problem but merely shift it). In principle, the real cannot be spoken in words or be laid down or understood in other sign systems. It is the ultimately unutterable. In simpler words, it is the area of the mind that is furthest from the symbolic and closest to physical experience, for example, to desire and death. It is the unknown, which confronts us in the other and of which it is the nonconstructed part. Bataille might have called it the impossible—as distinguished from the possible/virtual.

Reality refers to those dimensions of world/worlds that we can formulate and understand and that have been (co-)constructed by us. At the end of the twentieth century, the realities we know have been strongly permeated and occupied by media. The internet, for instance, is an interdiscursive media-reality, built from and established by social, technical, cultural, and aesthetic realities, among others.

In dealing with virtuality or the virtual (I try to avoid marketing jargon by not using the term *virtual reality*), we are not dealing with a material world, a thing-world or a life-world, but with something immaterial, something that only exists as a program, as a symbol that we can perceive sensorially in the form of signs (*in abstracto*, figuratively, in image spaces, as icons, imaginary). The virtual now still has prosthesis-like extensions into the thing-world/life-world, in the form of data helmets, data spectacles or data gloves, the mouse, keyboard, and so on. But these are merely the outward surfaces for the program underneath. The virtual is appearance; it fulfills the function of consciousness. The virtual is the interface, a rather friendly English word that denies the dramatic aspect of separateness implied in the German *Schnittstelle*.

The big chance offered by extended telematics is to build a tension between the local and the global, between local identities and events, on one hand, and worldwide processes/structures, on the other hand: to respect the other in its otherness, to maintain and support its autonomy, and at the same time not to isolate it but to enable it to participate in the global exchange process. A permanent mixing (in the alchemist sense) without losing sight of the possibility of demixing. This will be the challenge for the next few years and decades.

Artists have a prominent task in creatively producing this tension. Ideally, they are the agents of the local. Locality is the context of their actions, because without locality, subjectivity cannot flourish. Artists therefore are especially predestined and suited to support and stimulate the local, and to create a sensitivity to its unfolding.

If I understand telematic art to be the art of connectionism, a form of *n*-dimensional connections, then its imaginative content would be the connection of the heterogeneous. To connect once again what is already globally/universally connected would be a tautology. My idea of a connectionistic International encompasses the permanent rebellion of the individual against the whole as well as the hope for innumerable defeats of the whole against the individual. Relationships are ongoing experiments—or graves. The idea of the internet as a mass grave is a horrible picture (of reality).

Current efforts in telecommunication and especially in the World Wide Web are aimed at hiding the differences between media-people, and media-machines and media-programs. The most important and dominating means in this hegemonic strategy is the illusion. Not in the sense of anything being at stake but more in the sense of a risk-free identification with the world of icons, symbols, and relations as it appears on the screen. This illusion is currently practiced in two ways: either through concepts of a primarily spatial orientation in the tradition of *ars memoriae* or through concepts that are primarily time oriented, as in the classical dramaturgy of Aristotle. The goal of this strategy, which is essentially a double strategy, is that the one (media-people) must have the illusion of being able to fully enter the other (e.g., the media-machines). This we call virtual reality or telepresence. The other must enter the one by means of the illusion and must be able to take on its identity. This is essentially the world of metaphors.

The experience of the virtual (which is not identical with its program nor therefore with its symbolic content) is a firsthand one, and therefore a subjective one. To put it quite bluntly, if this experience were to be organized collectively, that is, if the program of the experience would be forced upon us, it would mean a sort of digital fascism. In my view, collectiveness only has a place in the subjective experience of virtuality in the sense of being a utopian potential, a longing, an inkling of an impossible state. The possible (virtual) unfolds itself in a creatively/beautifully/aesthetically acceptable manner only in the tension between it and the impossible. All metaphors stemming from the linguistic phylum of "fluidity," streams, the ocean... originate in longing.

Metaphors are comparisons. As means of denotation (including aesthetic denotation and its potential meaning), metaphors hover between imagery, symbolism, and mystery. Metaphors arise from the need and power of mind as well as of emotion "not to be content with the simple, the usual, the plain, but to endeavor to search for something else, to linger with what is different, to unite what is twofold."[1]

Metaphors are constructed to enhance, to compress, to enrich, to uplift something; otherwise, they would only proliferate in the imagination of their constructor. This "something" is either spiritual or physical. Metaphors are constructed to ennoble the physical with the aid of the spiritual or to visualize the spiritual, make it profane and concrete (objectify it) by a comparison with the physical.

In this world of metaphors, the relation to life is remarkably central. Biology as a discipline has kept its leading role. The underlying idea is that life is continuous, flowing, forever moving (and that therefore it is also harmonious). The world of machines and programs has been constructed and calculated according to a plan. Everything in it is based on numbers and on the logical and systematic relations between numbers. In this sense, it is a coherent and consistent world, in all the complexity that playing with numbers allows. The world of living things is not based on such a reliable scheme. The main difference is that it is principally irreversible. Technological, social, and cultural systems are extremely discontinuous in both their origins and their current expanse. All metaphors that promise a free flow of information, that want to describe the ocean as a navigational grid—streams, genes, rhizomes, and the like—or that want us to experience communications structures as trees or root systems will fail in this respect.

Telematic networks connect technical artifacts and complex systems of artifacts with political, cultural, and aesthetic structures, and therefore they already connect the "twofold." The net itself is already a comparison, a trivial image, a shallow rendition of the complex communication process the word evokes with me. If we are primarily concerned with its connective potential, we should try to describe it as such: connectionism versus a naive fishermen's village idyll.

Not only in the present discussion on and about the net is this connection of complex physical units with immaterial units and structures once again being compared/combined with the living/life, or aspects thereof. This also contains the effort to uplift what is profane (the technical, the political) as well as the objectification/externalization of what is not or barely understandable, of what is structural (and therefore in essence spiritual). Both tendencies need to be criticized.

In telematic networks, nothing flows, let alone freely. The process of the net is discontinuous and determined locally as well as globally. In telematic networks, nothing evolves "genetically." They obey the laws of programs/rules, however complex they may be. Only to those for whom life itself is a program (or for whom the program is life) this equation makes sense. The filigree, nonhierarchical root systems are hidden, under the ground. What grows out of them is hierarchically structured, because as soon as it becomes visible, it is connected to the process of civilization. Telematic networks have already progressed to the stage of visibility.

The landscape metaphor is useful insofar as landscape, like architecture, is already of a conceptual/constructive nature. Landscape is a civilized, structured living space. But—to ask a trivial question—is it what we really want? If developed telecommunication phenomena and structures are potentially (and therefore virtually) dynamic, chaotic, nonlinear, and multidimensional, then what sense does it make to submit them to the same laws that created our landscape?

Of course metaphors have a practical use as well, for instance, for those who build programs and machines. Karlheinz Barck formulated this as follows in his study on poetry and imagination where he discusses Giovanni Battista Vico's concept of "Scienza Nuova": "Combining parts, constructing analogue relations, comparing and distinguishing between similarities of things—these characteristics define the methodology of the inventive work of engineers, and those are ... the hallmarks of any metaphorical language."[2]

In my view, what we need is a language (of text, images, sounds, and their connections) that does not conceal the technical and political/cultural character of artifacts, systems of artifacts, and the structures of extended telecommunication in the wider sense but that exposes it, evokes it, and refers to it when it is being used. Discontinuity, dynamics, switches, contacts, drivers, energies, interruptions, power, distribution—the potential in relations is as rich as the technical and political/cultural domain itself.

I insist on the dualism of media-people and media-machines or programs. Dualisms are a necessity, if we want to arrive at some clarification at all. Maybe they are a transitional stage, but I'm convinced that art that dramatizes the interface as a border between the one and the other is the only way to reach a quality of connection that distinguishes itself from a simple decision for the one or the other.

I'd like to make a case for an experiment with an interface

- that is not based on "virtual reality" but on contingencies, on potential single events instead of a homogeneous, calculated make-believe world
- that leaves open the possibility that it is a (technically/grammatically) constructed world that we are dealing with through media, and through which we are to gain access to the other
- that allows for an experimental relation of the user with the interface;
- that isn't so much a catharsis but more of an epic provocation
- that doesn't forget, however, that the world of communication is a world of sensations without which no one would bother to make contact with the others or the other

Why do we always assume we have to start everything from scratch and that we have to reinvent the world as a whole every day again?

Anyone who is theoretically or artistically interested in the interface should at least be familiar with the following literature: the work of S. J. Gould (especially *Punctuated Equilibria: The Tempo and Mode of Evolution Reconsidered*, 1977); *Tractatus Logico-Philosophicus* by Wittgenstein; the work of Edmond Jabès (especially *On Subversive Agency*); *Bio-Adapter* by Oswald Wiener, the radical subjective concept of an interface from the perspective of the individual as a collection of attractors; Brecht's *Kleines Organon für das Theater* and his *Me-ti—Buch der Wendungen*; W. F. Gutmann's *Organismus und Konstruktion I* (1987); Ilya Prigogine's *Vom Sein zum Werden: Zeit und Komplexität in Naturwissenschaft* (German edition, 1979); and the key essays on the heuristic of media apparatuses by dentist Jean-Louis Baudry.

Notes

1 G. W. F. Hegel, *Vorlesungen über die Ästhetik,* vol. 1, *Theorie-Werkausgabe* (Frankfurt am Main, Germany: Suhrkamp, 1977), 117–18.
2 Karlheinz Barck, *Poesie und Imagination: Studien zu ihrer Reflexionsgeschichte zwischen Aufklärung und Moderne* (Stuttgart, Germany: Metzler, 1993), 38.

5 From Territories to Intervals

Some Preliminary Thoughts on the Economy of Time/the Time

This text was originally composed as a lecture for a symposium held on the occasion of the Steirischer Herbst festival in Graz, Austria, that Peter Weibel organized in 1999–2000, where Harun Farocki and Bruno Latour also spoke. By the turn of the last century, at least in the most industrial societies, it had become quite clear that bitter territorial disputes were not going to cease. At the same time, it was already apparent that a certain hegemony over time itself and its rhythmicization would come to play an increasingly significant role. Neither the political nor the cultural elite—so it seemed to me at the time—were prepared for this. This text was also a preliminary study for the monograph I published in German in 2002 under the title Archäologie der Medien *and that appeared in English in 2006 as* Deep Time of the Media.

This essay was originally published in 2001.

⁂

Prelude

The decisive question posed by those who, after the First World War, laid claim to be fighting for a better life for the many, was "to whom does the world belong?" Bertolt Brecht had asked the same question in the second half of his title for the film *Kuhle Wampe* (1932), framed as an urgent question mark regarding territorial ownership (in the broadest sense: of factories, land, resources, machines, etc.). At the present time and for the decades to come, however, it appears to me that the decisive question is "to whom does the time belong?" From the beginning to the end of this century, a remarkable shift has taken place in the political and economic relationships of power in which the media are involved and where they have forced the pace: away from the power of disposal over *volumina* and territories and toward the power of disposal over time—not so much its expansion but rather its fine structuring and its form of intensity.

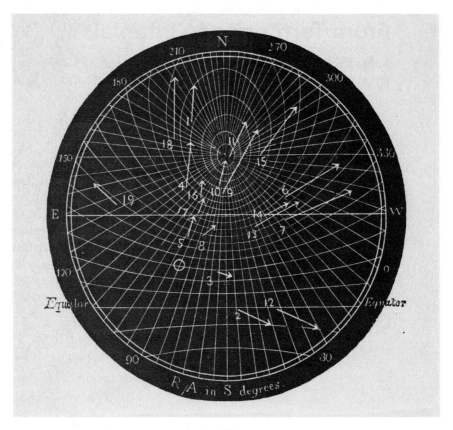

FIGURE 5.1. Cosmic space in a two-dimensional projection. To render graphically the movements of stars, a fixed point is taken (in this case, situated in the constellation of Orion) against which their trajectories are plotted. From Étienne-Jules Marey, *Le Mouvement* (Paris: G. Masson, 1894), 67.

Introduction

From Territories to Intervals was an attempt to sensibilize us for this question, which, as yet, still sounds unfamiliar. I tried out two different methodological approaches. The first was a short archaeological expedition where I traced the genesis of the question's formulation with the aid of two examples. The second was a conscious reference to Peter Weibel's rule of the "chronocracy" from the late 1980s, where I took up some of his ideas and spun them out in my own way in the form of a brief pamphlet. Here I shall only summarize the arguments of the first part (I spoke extempore and presented many visual examples, for example, the fantastic monologue of the time researcher from Alexander Kluge's *Der Angriff der Gegenwart auf die übrige Zeit* [*The Blind Director*, 1985]); the second is presented here in a more elaborated version.

My introduction in Graz was in the form of three preliminary remarks that referred to other contributions given at the symposium; however, as they are important for an understanding of the pamphlet, I include them here:

1. That which over ten years ago was firmly kept at the periphery of public attention and discriminated against by both politics and the (art) market has been moved to the center stage of power cravings and demands for legitimation during the last decade of this century: experiments undertaken by artists within the tension that exists between technology and information, on one side, and aesthetic expression and processes, on the other. Concepts such as interactivity, terms such as *communication* or *creativity*, which formerly had a rather casual meaning and, in some contexts, even an emancipatory or seditious one, became the leading concepts when the Western economies emerged into the so-called information society. They became the concepts of the controllers and the arrangers who then began to demand them as the paradigms of a new sociality. For the poets and the thinkers of the media, the issue is not so much to survive in this situation as to survive in dignity.
2. Stylized, neoexistentialistic creeds like that of Calvin Klein, "She is always, and never the same" (the advertising slogan for the most recent CK product, Contradiction, in his series of perfume-answers to the fundamental questions of life), need to be inverted in their sense: always the same, and never myself. The current media-tactical concepts, as well as critiques, of political economy still operate as a rule with the self as their center of gravity, the traditional concept of the subject. To take the shattered self as the point of departure, to understand it as a figment of the hegemonic imagination when it appears in an intact form while at the same time bringing an identity into play that is an oscillating, yet powerful opposite to the Other, might indicate one possible way out of this wretched situation.
3. It is worthwhile rediscovering Georges Bataille's notion of the "abolition of the economy," faced as we are with social exchange that is becoming progressively based on effective communication processes.

The attempt by the controllers and the arrangers to subject also the economy of signals and information to the productivity paradigm of the nineteenth century should be countered by an economy of extravagance for which art processes could be tested as models.

Archaeological Foray

Marey and Clearing Up a Misunderstanding

Until now, Étienne-Jules Marey's work has been interpreted by media and art people mainly in the context of film and photography. He is understood as a maker of images. This is a misunderstanding. I maintain that Marey, the doctor and physiologist, was not really interested in illusions, in superficial sensations, in the production of appearances. (I understand this thesis as roughly analogous to the widespread misconception that the computer is an image-machine. In my understanding, the computer is an artifact, a technical system for controlling other machines and automated processes, respectively, whether they are production, information, or war and military machines. In this sense, it is a medium.)

Proceeding on the assumption held by the most advanced experimental physiologists of the nineteenth century—all that we are able to perceive with our sensory organs are appearances—for Marey and his colleagues, it was a question of "eliminating the sensory appearances," as Wilhelm Wundt, among others, expressed it in his *Grundzüge der physiologischen Psychologie* (Essentials of physiological psychology). Their concern was to exclude interpretation and not to introduce it. (The illusion-machine of cinema, in which tradition Marey is repeatedly included, is first and foremost a machinery of interpretation.)

In this, Marey was following a central idea of Julien Offray de La Mettrie (whose book *L'homme machine*, published in 1748, should not be cited ad nauseam in the form of a few phrases, usually thirdhand, but rather people should actually read it): bodies or objects in motion may be exceedingly attractive to the senses (the reasons for this cannot be described exactly), but they also conform to mathematical laws. To expose the structures and quantitative relationships of these, to transform them into patterns and graphs that can be used for making calculations—this was the goal of the physiologists' quest.

To put this more pointedly, the experimental physiologist made no claim that the body he examined with measuring instruments was the whole body

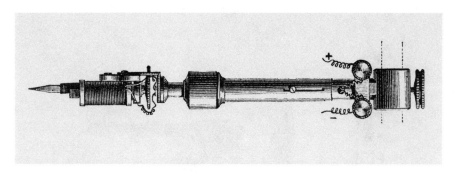

FIGURE 5.2. Electric chronograph (small model). From Marey, *Le Mouvement*, 139.

but was fully aware that his object of study represented only a special part of the body's reality. The chair that Marey held at the Collège de France from 1869 was titled "Histoire naturelle des corps organisés"—natural history of organized bodies. Marey's chronophotographs of animals and humans in motion are "clocks to be watched," as Roland Barthes once observed. In most contemporary reproductions of his pictures, the measuring instruments (to measure time and space) have been omitted. Marey deliberately included them as references for the bodies in motion on each individual recorded image. The object of this exercise was to register the movements of the body as a function in the form of a graphic impression or expression. From Carl Ludwig Anton's Kymographion of 1847 via the long chain of Marey's own graphic recording devices (Polygraph, Myograph, Sphymograph, etc.), experimental physiology pursued a path that led away from the bloody dissection (frequently, vivisection) of bodies and toward the bloodless anatomy of curves and readings. This was the beginning of a "second nature" that was aided and produced by instruments, which could be analyzed and reanalyzed at will. The aim was to develop a language for natural scientists, similar to the notation of musicologists—a kind of Esperanto in which colleagues from all over the world could communicate without there being any misunderstanding, a kind of universal language. The main task of the "station Marey," which was opened in 1902 near Paris, was to standardize and test physiological measuring instruments. At this point in time, cinema was on the way to its first industrial phase and developing into a "fantasy machine" (Fülöp-Miller). The fields of application that Marey was working on and which financed his equipment were sport (functional gymnastics), the military, and the workplace. These were not a later addition but were integrated into his R&D work.

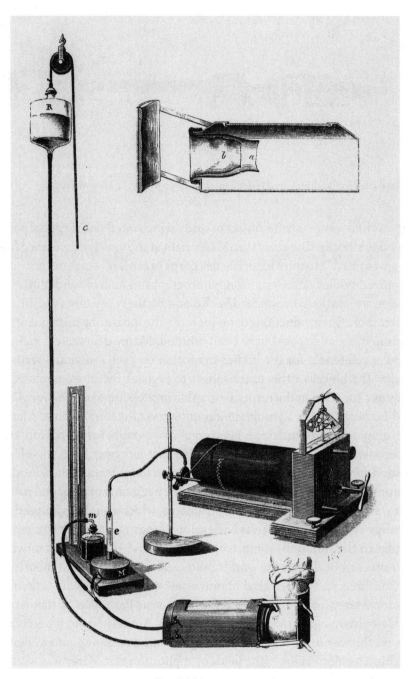

FIGURE 5.3. Apparatus ensemble of the experiment to inscribe the pressure in the vessels of the hand. From Marey, *Le Mouvement*, 614.

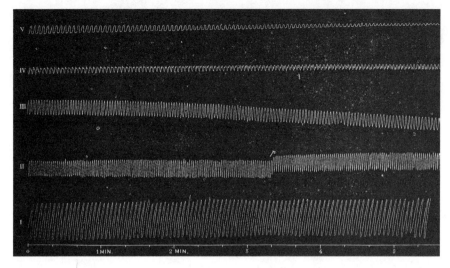

FIGURE 5.4. Various rhythms of the systoles of tortoises' hearts with the amplitude of each one. From Marey, *Le Mouvement*, 614.

Aleksei Gastev and the Proletarian Human Machine

We are well acquainted with the principles of scientific management that Frederick Winslow Taylor devised in the United States and, for example, the experimental work that Gilbreth carried out before the First World War. Not at all well known, however, is that in revolutionary Russia, Taylorism was combined in a radical way with the materialist and constructivist ideas of the avant-garde.

In 1914, Lenin was still at the stage where he wrote an essay with the title "The Taylor- System: The Enslavement of Man by Machines." His accusation read, "Recently, in America, the advocates of this system have instituted the following method: a small electric lamp is fastened to a worker's hand. The movements of the worker are photographed and the movements of the lamp analyzed. It is ascertained that certain movements are superfluous and the worker is forced to avoid these movements, that is, to work more intensively and not to waste even one second on resting.... Systematically, the cinematograph is being applied."

By 1920, the first Soviet Central Institute for research on work (Zentralny Institut Truda, ZIT) was founded in Moscow, and in the course of the decade, others followed at other locations in the Soviet Union. The frequently stated aim of these institutes was to structure the twenty-four hours of the

FIGURE 5.5. Runner equipped with special shoes to probe the pressure of the foot on the ground and carrying the device to register his paces. From Marey, *Le Mouvement*, 156.

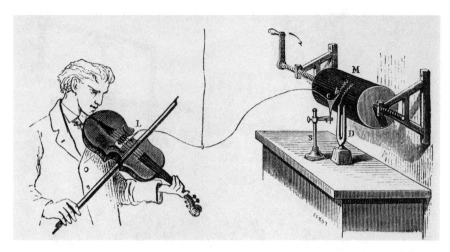

FIGURE 5.6. Transmission of a violin's vibrations via a brass wire to a registering stylus; method of Cornu and Mercadier. From Marey, *Le Mouvement*, 642.

day as effectively as possible (as it was not possible to stretch them), that is, to subject them rigorously and at all costs to the paradigm of productivity. The work of the All-Russian Association of the Time League also served this goal: they distributed timepieces to the public, who had to record the events of their daily lives with meticulous precision on so-called chronocards (including everyday actions from eating to personal hygiene). Then, with the aid of participating observers and experimental equipment, statistics were compiled on all kinds of work processes with the aim of achieving, through training, a gradual fusion of the workers with the machines. According to the ideas also of a section of the avant-garde who believed that the human organism was fundamentally a mechanical one, now powerfully enriched by the soul of electricity, the perfect symbiosis of labor and apparatus was sought.

Media apparatuses were the outstanding instruments, both in the work research institutes and at the Time League. At the institutes in Kazan, Moscow, and Petrograd, there existed large machine parks of the experimental apparatuses used by graphic physiology in the nineteenth and early twentieth centuries. Film and photographic studios were erected for recording the movements of metalworkers or tram drivers, for example. Camera teams invaded the factories and filmed people engaged in the production process. "It is necessary to overcome the limitations of time and space through organization," proclaimed one of the initiators and driving forces of this

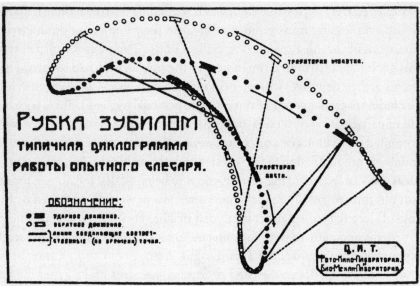

FIGURE 5.7. Russian Taylorism: a demonstration in the biomechanical laboratory of Aleksei Gastev. A metalworker is being observed at work with a sighting telescope and a coordinate frame. The time trajectory of the movement of the hammer is then transferred onto a diagram. From René Fülöp-Miller, *Geist und Gesicht des Bolschewismus* (Zurich, Switzerland: Amalthea, 1926).

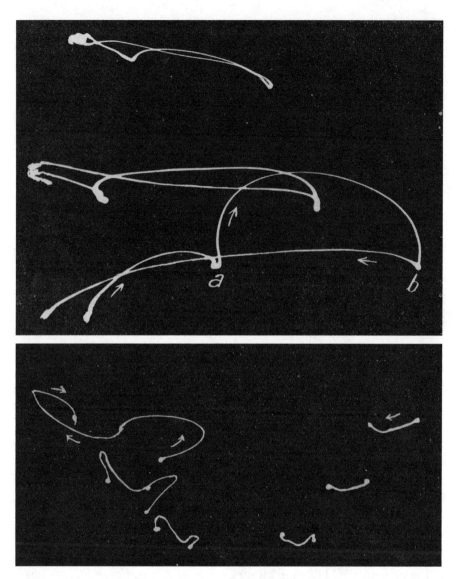

FIGURE 5.8. Light traces of workers' movements. Research work at the Central Institute of Labor, Moscow, 1920s.

movement, Aleksei Kapitonovich Gastev, at the beginning of the 1920s. In the preceding decade, he had belonged to the movement of the Russian Futurists, lived and worked for a time in Paris as a journalist, and became known for his rigorous development of a rigid economy of poetic language. The last series of poems that he wrote and published he called simply *Package of Orders* and numbered his lyric arrangements consecutively. They consisted of short, often one-word lines, reductions to the absolute bare essentials of communication, more like signals than messages ("Contact / Shunt / Stop," from Order 02). After Gastev had placed his powers of intensification exclusively at the service of the young Soviet economy, among other things, he invented a binary code of productive motion: using only the components "lift" and "press" with varying intervals and rhythms, it would be possible to (re)construct all the motions required in industrial production. Aleksei Gastev died in 1938 at the hands of Stalin's executioners, at about the same time as Vsevolod Meyerhold, whose theater concept of biomechanics was likewise a worldview radically translated into art.

The center of thought and theory of this movement was the Central Work Laboratory of the Institute for Brain Research at the Technical University in Petrograd, to which the father of acoustics, Ernst Florens Friedrich Chladni, had dedicated his *Entdeckungen zur Theorie des Klangs* (Discoveries relating to a theory of sound) of 1787 with deep respect. The director of this early neurobiological institute was the psychologist Vladimir Bekhterev. He died in mysterious circumstances in 1927, the day after he—in his capacity as physiologist—had examined Stalin's brain.

Research at the Petrograd institute focused on reflexology, studies on the ability to concentrate and hygiene in the workplace. Of particular importance were experiments with music structures, theories of the interval, of rhythm, and on the physiopsychological importance of music. For example, P. K. Anochin, a physiologist, published in 1921 the findings of his research on "The Influence of Major and Minor Chords on Activation and Inhibition of the Cerebral Cortex"; Bekhterev himself wrote an essay in an anthology on "Reflexology of Labour" with the title *Intellectual Work from a Reflexological Perspective and Measurement of the Ability to Concentrate,* where he investigated both the effect of Beethoven's Moonlight Sonata and of the overture to Gounod's opera *Faust* on intellectual work. A further project of particular interest for our context was the attempt to develop a system of musical notation for registering the movements of the human body, which was carried

FIGURE 5.9. Typist and Taylorism in Germany. From K. A. Tramm, *Psychotechnik und Taylor-System*, vol. 1, *Arbeitsuntersuchungen* (Berlin: Julius Springer, 1921).

out in cooperation with the Kazan institute of the scientific organization of labor; in 1928, the results were published by K. I. Sotonin in an essay.[1]

For the future, one of the most striking phenomena in connection with a time-based economy is the incredible concentration of working time in those sectors of the plebeian intelligentsia where the greatest potential for innovation is found. High performance in the brainwork of programming is comparable to the high performance of training the physis in sport. Both the one and the other demand young bodies. Just as the stars in the swimming pool or the gymnasium who carry off the medals get younger all the time, the programming arenas, for example, of George Lucas's Industrial Light and Magic or James Cameron's Digital Domain are populated by kids at the machines who probably bring their teddy bears with them. The period of time when these teens and tweens achieve their highest performance spans at most eight or ten years. After that—if they have earned and saved enough—they can or must take a very long surfing holiday in Hawai'i, like the top professionals in sport train for a second career or retire completely from the process of social reproduction.

For a Praxis of Kairos Poetry: *Against Psychopathia Medialis*

1. Karl Marx wrote for posterity. Thanks to his mania for scrupulously citing his sources, the remark of an anonymous contemporary was recorded in volume 26.3 of Marx and Engels's collected works (German edition), who, by succinctly summing up his own notion of economy, formulated what later became the touchstone of Marx's critique of established bourgeois economy: "A nation is only truly wealthy, if no interest is paid on capital; if the working day is six hours long instead of twelve. Wealth means to have time at one's disposal; nothing more, nothing less."

At a historical juncture where time has now been declared the most important resource for economy, technology, and art, we should not pay so much attention to how much or how little time we have. Rather, we should take heed of who or what has the power of disposal over our time and the time of others. The only efficacious remedy for bitter melancholy as the all-pervasive attitude toward the world is to assume, or reassume, the power of disposal over our own life's time. Only then is the future conceivable at all—as a permanent thing of impossibility.

The visual and audiovisual apparatuses that we work with are all time machines.

Their origins lie in the first founding era of the new media in the nineteenth century—they are prostheses, artificial limbs, for dealing satisfactorily with the impossible. Chronos—chronology's time is that kind of time which disposes of life by using it up. History. Chronology fits us into the order of things. Illness can be chronic, but never passion. Chronology cripples us because we are only mortal and we shall pass. Machines live longer. The computer scientist and engineer Danny Hillis, who has played a decisive role in constructing the massively parallel architecture of today's supercomputers, among other things, is currently working on the prototypes of a clock that will begin operation in the year 2001 and run for exactly ten thousand years.

The ancient Greeks understood only too well the dilemma we would get ourselves into with *chronocracy* (the term was coined by Peter Weibel) as the dominant time mode, and they introduced two further concepts of time: Aeon and Kairos. They were the antipodes of powerful Kronos who, ultimately, devoured his own children. We find the transcendental dimension of Aeon suspect—time that stretches far, far beyond our and the Earth's lifetime, which is supposedly "pure." The fastest way from zero to infinity, as Alfred Jarry once defined God. On the other hand, we value Aeon as a possibility of uniting time and life as a virtual power from which the vitality of life springs. By contrast, Kairos stands for the art of doing the right thing at the right moment; he is the god of the auspicious moment.

Yet only the interplay of all three conceptions of time can preserve us from alienation: to give the auspicious moment a charge from an Aeonian battery and/or challenge chronic time's power of disposal over life by applying both Aeon and Kairos tactics—to me, this appears to be a possibility whereby we may survive in dignity with and within the time machines.

2. In one of his early works, Jean-François Lyotard wrote, "Our culture singles out for special favor that which it places in the limelight—the only scene which it considers to be an event: the moment of exchange, the immediate, the sensational, the 'real' time, which for our culture, is the only time that is alive. This moment, in which accumulated 'dead' time is realized, one can call obscene."[2] Recently,

an American electronics firm presented a system which in the future will allow information to be invoiced per bit; it will be a matter of complete indifference whether the information is text, images, or sounds. The bit, as the smallest techno-unit, will become the new abstract currency, the basic coinage for an economy of text, image, and sound production, which will also include those art forms that are expressed and realized with and through the media. We should not tolerate either one of these kinds of expropriation of the so very valuable moment—neither that of the culture industry, with its obscene concentration of life's time in the staged and celebrated sensation, nor the installation of a universal measure of time and economy that assaults or runs over the capabilities of our own, proper faculty of perception.

The relationship of technical processes vis-à-vis time can be described thus: even the variables that influence the technical process from the outset (observation, inspection, control) are time dependent. In the course of the technical process, they undergo conversion. At the starting point of any and every machine–machine or human–machine system, we find time-dependent experiential variables. Such processes may also be termed dynamic. As intellectuals, or as artists, the very least we can do is to ensure that the conversion which takes place mid-process makes a sharp and qualitative distinction between the variables that are effective from the beginning and the end results. This means to work effectively at the interface, to dramatize it. Processed/designed time has to be able to give us back time that life has taken from us (this is one of Jean-Luc Godard's finest thoughts on cinema). Otherwise, it is time wasted, time lost. We should not allow ourselves, time and again, to fall short of the capabilities of machines.

3. In 1934, Max Horkheimer published his *Notes on Germany* under the pseudonym Heinrich Regius; he calls them *Twilight* in the title. In a small section with the heading "Time Is Money," he remarks, "Although one shrinks from falling into the generalizations of common platitudes, time is not money, but money is time, as well as health, happiness, love, intelligence, honor, peace. For it is a lie that if you have time, you also have money; with mere time you won't get any money but vice versa is certainly true."[3]

4. "We wander around in circles in the night and are consumed by fire"—this is how Guy Debord described his situation as a professional Situationist: the movement of roaming about, which was the title of the last film he made before his last will and testament. The first known timepieces, from ancient China, were rectangular metal reliefs structured like labyrinths. In the depressions, a slowly igniting powder was strewn and the burning of the powder showed the passage of time. Guy Debord offered his body and his imagination for measuring the time in which he lived. Yet what would be a viable alternative of action to the Situationist one, which consumes its own identity? Naturally, and theoretically: to be fire instead of burning powder. However, to take up this position is not an option for us, for we are (among other things) of the very stuff that time uses up (unless we want to play God). What we can do is to intervene in the rhythm and velocity of the burning, put in stops, and organize the intervals in between.

A superior time policy means to fight for the upper hand in time consumption and time use. It appears that one must be ready to face certain loss, both in the sense of self-consuming (Guy Debord) and of self-squandering (Georges Bataille). However, then loss is not a category of a fatal economy, if we succeed in making it an enrichment of others in a grand way. Otherwise, the act of consuming would be religious and of squandering, ideological. Both have had devastating effects in the recent past, for which Germany stands as an example.

5. If it is so that under the aegis of expanded interactivity at the interface of media-people and media-machines "creativity" has become a fundamental social competency, and although the traditional model of the artist in art is now a discontinued line, it appears to be becoming a general, central model for social action, then it is appropriate to work on complementary identities at the very least. The competencies in life that will be required increasingly of intellectuals and artists in the future are already tangible but as yet not translatable into concrete strategies and tactics: chaos pilots and Kairos poets; people who are not only capable of dealing with confusing arrangements but are also able to organize them; and those who catch the auspicious moment

(for example, in the cinema or on the web) and charge it with energy. Without a relationship to complexity, and without a relationship to time—both are inextricably bound up with each other—I cannot imagine that advanced praxis in art and thought is possible.

6. "In general we always seek—(in potlatch) or in actions or in contemplation—that shadow which per definition we are unable to grasp hold of; that we helplessly call poetry, profundity, or intimacy of passion. That we will be deceived is inevitable, should we attempt to grasp this shadow."[4] Under the new economy, the task will also be not to relinquish the attempt to express the inexpressible. With regard to how we handle the visual and audiovisual time machines, these can constitute a powerful unity: work on the living heterogeneity of the arts of image and sound (but without master- and slave-media, without accepting a new universal machine) and the sensitivity for the right moment, the auspicious moment for life, imagination, and the media. Moreover, this should also constitute the lowest common denominator for any contemporary or future academy—at least for those among them that rise to the challenge of intervening with art and artistic means in the current processes that are transforming society.

Conclusion: Brecht on His Hundredth

>Ich sitze am Straßenrand
>Der Fahrer wechselt das Rad.
>Ich bin nicht gern, wo ich herkomme
>Ich bin nicht gern, wo ich hinfahre.
>Warum sehe ich den Radwechsel
>Mit Ungeduld?[5]

>I sit at the roadside
>The driver is changing the tire.
>I didn't like where I have come from
>I don't like where I am going to
>Why I am impatient
>For the tire to be changed?

Translated by Gloria Custance

Notes

1 We continue to investigate this germ cell of the technoscene of the twentieth century. I suspect that it strongly influenced the artistic avant-garde in the Soviet Union. Dziga Vertov, for example, was a student there at the time that he was occupied with montages of sounds. The interesting texts we have found so far are all in Russian so that lengthy translation work is necessary. I would like to thank Maria and Boris Barth for their preliminary research in St. Petersburg in 1998.
2 François Lyotard, German translation in *Ästhetik und Kommunikation* 67/68 (1987): 40.
3 Heinrich Regius, *Dämmerung: Notizen in Deutschland* (Zurich, Switzerland: Oprecht & Helbling, 1934), 28.
4 Georges Bataille, *Die Aufhebung der Ökonomie—der verfemte Teil, Zweiter Teil*, 2nd extended ed. (Munich, Germany: Matthes & Seitz, 1985), 106.
5 Bertolt Brecht, "Buckower Elegien," in *Gesammelte Werke* (Frankfurt am Main, Germany: Suhrkamp, 1967), 1009.

6 On the Difficulty to Think Twofold in One

with Nils Röller

In 2000, an extraordinary group of physicists, neuroscientists, artists, mathematicians, and philosophers came together to celebrate the sixtieth birthday of Otto E. Rössler, chaos theorist and scientific polymath. The high point of this gathering was a video conference held with the Kewalo Basin Marine Mammal Laboratory in Hawai'i—an experimental lab conducting brain research on dolphins—during which Otto Rössler communicated with the dolphins via electronic screen, microphone, and loudspeaker. It was in this context that my then colleague Nils Röller and I formulated some foundational thoughts on the interface—or, more aptly in German, Schnittstelle. This essay was originally published in 2001.

⁂

I.

Usually, we academics speak in public about what we have already done. This is safe territory. Today, we want to talk about something that we would like to do because we think it is both meaningful and necessary. Therefore what we are going to put before you now must be understood as a rough draft: a preliminary sketch of a "short organon" for the interface.

To make it clear what this is about, a few prefatory remarks are necessary.

1. Art that is expressed via media is becoming more and more the art of the interface. Theory of media seems to become theory of the interface. This statement reflects in a very specific way and as a particular phenomenon what the computer is or can be. It is, first and foremost, a machine for measuring, monitoring, controlling, and connecting. It is not a machine for creating individual particular realities but one for guiding sequences, processes, time, and combinations of heterogeneous fields of knowledge or activities; of scattered and distributed actors, machines, and programs; of

combinations of different forms of expression and drawing materials: text, image, sound, haptic, and so on.

In short, we wish to treat the question of the medium as the classical question about the in-between. Computers—and, to an ever greater extent, computers in networks—are intermediate machines: machines and programs that operate in between various time processes and realities. It is quite possible that this will facilitate our approach to the question, which is of great concern to us from an artistic perspective, of subjectivity. It is easier to formulate this than the question of what happens in the cut, in the difference.

When German Romanticism was in full flower, Novalis complained that "everywhere we seek the absolute but find only things." Two hundred years later, if we turn this around, it makes sense as a question: why don't we seek the absolute in and between the particular things, in and between the singularities, by observing and at the same time respecting them?

2. A short etymological excursion: In engineering, *interface* denotes the connection between various machine processes and subprocesses for controlling and moving and for time processing, including the processes in which man–machine relations play an important role. The German term *Schnittstelle* clearly marks the place in between different realities; as separation, as a dramatic place where it is not possible to rest.

In English, *interface* is the meeting of two faces or surfaces: "the surface forming a common boundary of two bodies, spaces, or phases; the place at which independent and often unrelated systems meet and act on or communicate with each other."[1] An interface could be the "principal side presented by an object or the facade of a building." With the interface, we are referring to the active, striking, or working surface or edge of implements or tools like the sharp side of a knife (opposite the back). There we are again—*Schnittstellen* are dramatic places at which it is not possible to rest or stay.

3. What was and is the interface in the older forms of artistic praxis? For example, cinema: interface is here the apparatus in the widest sense of the term (*l'appareil de base*/basic apparatus in the sense of Jean-Louis Baudry), that is, everything in which the concept of the filmmaker/author is realized as a film for and inside of the cinema (shooting, cut, sound, montage, etc.). Analogous to cinema, it is in the theater: dramaturgy, movements and voices of actors, scenography, costumes, masquerade, music, light—all these and many more aesthetical qualities are coformulating the experience that we call drama or opera.

4. At the present time, the dominant trend is toward humanization of the interface as a special variant of the humanization of labor and the workplace. This expresses very clearly the human longing for standardization, for the universal. At issue is the "humanization of areas that are not yet humanized," with a view to appropriating them.[2] A reference to those who once took or still take the risk to explore the other side of the Moon: the importance of experiment (read through Klossowski's idiosyncratic critique of political economy, *Living Currency*): "In the meantime, the manufacture of appliances itself has become acquainted with temporary infertility. This is becoming ever more apparent because the accelerated rhythm of manufacture forces it incessantly to prevent inefficiency (in the products)—it has no other possibility except to resort to wastefulness/extravagance. The experiment, which is the condition preceding efficiency, presupposes wasteful errors. To test experimentally what can be manufactured with regard to profitable operation leads to elimination of the risk of an infertile product at the price of squandering material and human energy."[3]

Our experiment tonight is (at least for us) a risky one. We have taken a text that at the moment belongs to the forgotten ones, totally unfashionable, dusty, more than half a century old—Bertolt Brecht's *Kleines Organon für das Theater (Short Organon for the Theater)* from 1948—and we try to read through it the questions we have regarding the interface. The hypothesis in doing so is our suggestion that, nowadays, we urgently need something like an *Organon for the Interface*.

Brecht's short *Organon* was written as a wonderful polemic against the laziness and satiation of established classical theater. But it is much more than that. It was also written in a cultural situation, when industry started to take over the new media of Brecht's time—radio, cinema, and already television—and, moreover, in the same year that Alan Turing wrote his legendary text on intelligent machines:

> With my father I had already talked from continent to continent, but together with my son I saw the moving images of the explosion in Hiroshima. (*Organon,* end of item 16)[4]

Furthermore, it was written already to oppose the attempts by the new rulers in the East to use Brecht's theater as a service for their ideology. "Theatre has to stay something luxurious, which then means that one lives for luxury. Pleasures need defense less than anything else" (item 3).

So, let us explore some details. We focus on seven points.

1. Regarding the most important and general point, the Aristotelian concepts of empathy *(Einfühlung)* and purification versus the epic concepts of pleasurable learning and thought that intervenes (*eingreifendes Denken* in the sense of Brecht). Not constructed as a dualism, but a tension. No ontological contradiction between crying and counting or laughing and calculating (Brecht: on the relationship of reason/rational logic and feelings, particularly the first sections of the *Organon*).

The aim is not to replace entertainment but to implement a position (intervene) within it. At the beginning of the *Organon*, Brecht writes something rather strange: "Let us cause general dismay by revoking our decision to emigrate from the domain of the merely pleasant, and even more general dismay by announcing our decision to take up lodgings there. Let us treat the theater as a place of entertainment, as is proper in an aesthetic discussion, and try to discover which type of entertainment suits us best" (*Organon*, preamble).

This was a harsh critique against those adepts and critiques of Brecht, who wanted to put him completely outside of the process of theater by making him a god of the avant-garde, a fetishist of theory and dry scientific thinking—those people, as he writes elsewhere, who were applauding him by saying "yes, you are right, I agree, 2 + 2 = 5." The imperfectly understood *Lehrstück* (didactic drama). This—for us—involves a critique of the fetishization of the ASCII-file and the written algorithm as the worship of idols (Friedrich Kittler, but more his adepts).

2. Free interchange and transparency (the most difficult of all) toward the users on one side and to the machines and programs on the other. It must be made apparent that the dramatic construction is something that has been produced, something that is synthetic, artificial, and not the iconographic plastering over of the two sides (item 55).

3. The gesture of showing something and, at the same time, showing that you are showing something, Kronos and Kairos, the programming of two different kinds of time—or, better, to program time from two different points of view, the time that is portrayed and the point in time from which the first kind of time is programmed. Dual data processing, the date of the original event and the date of its actual reproduction. See also the example of the invention of non-Euclidean geometry in the second half of the nineteenth century by leading Russian mathematicians, artists, and charlatans.

The dynamization of an infinite number of lines running through one point opposite to a given line that do not intersect with the given one.

Brecht has tried to make this necessity of double data processing clear by developing the idea of an actress who is not behaving like a parrot or a monkey but who is able to play something that has happened from the perspective of actuality, as the auspicious moment. "Important is, what has become important. Such an alienation [*Verfremdung*] of a person as exactly this person and exactly this person now, is only possible if the following illusion is not created: the actress must be the figure, and the play should be the event" (item 51).

4. "We need a theater that not only enables feelings, insights, and impulses, which are allowed by the current historical field of human relationships on which the activities [*Handlungen*] take place, but we need a theater that uses and generates ideas and emotions, which play a role in the change of the field itself" (item 35).

We are trying to make a connection here with what Maurizio Lazzarato calls "immaterial labor" and what some thirty years ago has been called "controlled labor," control of automatons/machines. What kinds of aesthetical strategies do we need if we want to respond to or even interfere in such processes? What kinds of thoughts and feelings play a role in the change of the field of networking itself?

Arguing against the ranting and pretentious German classical stage, Brecht wrote, "With us, everything slips easily into the insubstantial and unapproachable, and we begin to talk of *Weltanschauung* (worldview) when the world in question has already dissolved. Even materialism is little more than an idea with us. With us, sexual pleasure turns into marital duty, the delights of art subserve education, and by learning, we understand not an enjoyable process of finding out but the forcible shoving of one's nose into things. What we are doing has nothing of an enjoyable *Sich-Umtun*, and to prove ourselves we do not say how much pleasure we had with something, but how much sweat it cost us to do it" (item 75).

5. Montage—shortage and glut. Montage originated from a situation where there was an acute shortage of material (Lev Kuleshov, the early period of film production in revolutionary Russia). What kinds of strategies are feasible in dealing with a situation where there is an overabundance of material? Can one even still practice montage with a glut of material, or are not reduction, cuts, and omissions called for? And what could this reduction

look like in a networked permanent production of glut/overabundance?

6. The Alienation Effect itself. Alienation as a concept to be posed against the imitation that is everywhere. Among other things, we suggest investigation of filmic strategies like, for example, the jump-cut for experimenting with discontinuities in digital design work. Irritate the simulation of continuity through the intervention of the machine itself. "The new alienations should only take away the stamp of the familiar of the social influenceable processes, which today protects them from interventions" (item 43).

7. Guy Debord—Situationist's gestures like *dérive* or *détournement*. Quite complicated because computers are very unstable systems and so, too, is network communication. Therefore, as we are dealing with extremely unstable systems, might not the creation of stability become a subversive tactic—at least for a while?

Perhaps we have to take care for the strengthening of local identities?

> As we cannot invite the public to fling itself into the story as if it were a river and let itself be swept vaguely to and fro, the individual events have to be knotted together in such a way that the knots are easily seen. The events must not succeed one another as indistinguishable but must give us a chance to interpose our judgement. (*Organon*, item 67)

We conclude this part with a thought by Klossowski again: industry claims as the actual principle of all its initiatives that all human phenomena—and each and every natural phenomenon as well—are eminently suitable to be treated as exploitable material, and consequently, all variations in value and all uncertainties of experimental experience are subject to this. The case is the same with the both spiritual and animal characters of erotic emotions, which are valued essentially as a power of suggestion.[5]

II.

The relationship of media to interfaces is dialogic. The understanding of this dialogical relationship is a result of an interpretation of the Cartesian dream. It stands against a background of "contra-modernism" despair at the lack of purity of linguistic communication. Dialogical philosophy formulates an ethical claim which, according to Otto Rössler's reading of Descartes, is prefigured in the *Meditations*. The precondition of dialogical thinking is that the relationship of the ego to the person opposite is only held to be a community if each party accepts the other, without reservation.

día-

The proverbial Cartesian doubt and the prefix *día-* have a common linguistic root, traceable to the Greek morpheme *dís*, which means "twofold, in two, asunder."[6] The assumption of an initial unity, which supposedly preceded the twofold, is problematic. There is no etymological evidence to support this assumption; rather, its antithesis suggests itself: unity is the result and synthesis of the twofold and manifold. The possibility "to think the twofold in one" rests on the premise of the twofold and manifold. The linguistic researcher Wilhelm von Humboldt has pointed out that at earlier stages in the history of language, the special character of the twofold was emphasized through the fact that it had its own grammatical number, the dual.[7] The dual makes it possible to denote objects that appear in twofold form in such a way that allows their essential difference from things that are single or manifold to be signalized. The dual makes it possible for the "other," which faces the "one," to be described in its otherness and not as a mere doubling of one's own ego: "The 'thou' is not similar to the 'I,' but confronts it as an opposite, as not-I: here the 'second' is not a mere repetition of the first, but is qualitatively 'other.'"[8]

The history of philosophy permits a relativization of the notion of unity. Unity first began to engage European thought when the cities of ancient Greece found themselves confronted by a superior enemy power that was completely different to themselves, the empire of the Persian king Kyros.[9] However, it is problematic to analyze the relationship of the twofold to the onefold as a hostile conflict. This analysis can, in agreement with Marshall McLuhan, be construed as an effect of the media technique of writing, which has fundamentally changed the minds of the ancient Greeks. The violent changeover from an oral to a written culture was heightened, according to McLuhan, by the invention of the printing press with its movable type, and this led to the mechanical age of the division of labor and industrialization. In modern science, the written word and letterpress printing have resulted in extreme specialization and fragmentation of knowledge.[10] It is this fragmentation that determines the relationship of the onefold to the other, and it is a common prejudice that it was the concepts of Descartes's modern philosophy that systematically accelerated this fragmentation and splitting up of given entities into ever more divisible elements.[11] In the history of the reception of Descartes's philosophy, both monological and dialogical approaches can be found that attempt to piece the fragments together. Those that argue from the side of the Cartesian ego and postulate its claim to knowledge as absolute can be termed monological; dialogical are the approaches that take the other in order to reflect on a rapprochement between the one and the other. Therefore, at the heart of the Cartesian ego, an interface, and thus a dialogical principle, is revealed.

Descartes

Descartes's *Meditations* appear in the "fair" perspective, as does the "twofoldness" of his thought. According to Otto Rössler's reading, Descartes can be liberated from the veil of mechanical prejudices and the reputation of mere fragmentation.[12] Here Descartes's concept of the ego is seen as something that operates at the dividing line between God and the world. It operates in such a way that the other is accepted as free and independent of principles that limit it. Otto Rössler understands the Cartesian "self" as occupying an intermediate position between the empirical and transcendental realms. Descartes doubts whether it is possible to distinguish between dream and reality. While asleep, he sees a book in which his future life is set down. When he wakes up, the book is not there. He falls asleep a second time and sees the book again, but this time the page about his future is missing. Descartes concludes that it is not possible for both of the dreams to be true. Either there is a book in the world of dreams containing important information or this book does not exist in the world of dreams. The two dreams do not tally. They are contradictory. They cannot be understood in their contradiction in one and the same world. The Cartesian ego can thus certainly assume that the world seen in dreams is contradictory. This is a decisive difference to the world that the Cartesian ego perceives in the waking state. Statements can be made about the world experienced in the waking state that concur. One can assume that it is formed according to reliable laws; in short, this world is determined.

Determination implies the possibility of taking hostages in a double sense. Determinism is at the same time of this double kind.[13] For on one hand, the ego can assume that its own perceptions are determined. This raises the question of how these determinations are formed or, to use a modern idiom, programmed. On the other hand, the ego can assume that the surrounding world is also determined and consists of machines. Again the question is raised about how these machines are programmed. The answer to both questions depends on one and the same position of the Cartesian ego.

Descartes assumes that an omnipotent programmer has set up the determinations. However, the quality of the program can be determined by the ego. Descartes understands by "quality" an answer to the binary question, regardless of whether the program was set up with intent to deceive. Is the world a hostage of the program? The Cartesian ego receives an answer through an astonishing twist. It decides to regard the surrounding world—and in particular the humans in this environment—as free beings and not as programmed machines. On the basis of this decision, the question is thus also decided as to whether the transcendental programmer designed the world with intent to deceive.

Through its own decision, the Cartesian ego arrives at an answer about the structure of the world. This decision prefigures dialogical thinking, namely, that concedes freedom to the other and, proceeding from the assumption of this freedom, constructs a common stock of meaning of the world. The Cartesian ego is at the interface between the transcendental world of the Great Programmer and the empirical world.

The ego is double-sided. One side faces the surrounding world of machines and the other the supernatural world of God. Descartes discovered two things: first, that this ego can manipulate both of these sides, and second, that the manipulations of one side have consequences for the other. Both sides are connected via the interface. If the switch is set to "mistrust," then not only does God become the culprit who alone deceives the Cartesian ego but inevitably the worldly side becomes a questionable machine that is subjugated to the will of the programmer. However, if the switch is on "fairness," then God is the benevolent and great Other, and at the same time, the creatures in the world perceived by the senses are magnificent in their diversity. The Cartesian ego thus becomes a manipulable medium and the opposite of pure. The Cartesian medium is, in Rössler's interpretation, an impure medium of change that consciously slides in between the divine light and the diversity of the world that it illuminates. The Cartesian ego can regard itself as an interface and thus manipulate. It can switch back and forth between a fair view and a mistrustful view vis-à-vis God and the world. In both cases, the interface changes the incoming and outgoing information, like through smoked or crystal-clear glass. Thus the Cartesian interface is never a pure medium: it either dims or illuminates. For Rössler, the interface is a paradise machine. It is driven by fair decisions.

Media Studies

At this point, the use of the word *medium* demands further explanation in the context of current debates in media studies. The editors of *Kursbuch Medien*[14] decline to ascribe a historically solid and substantial meaning to the media. We are fully in agreement and would like to support this with a brief look back at the etymology under the presumption that etymology will clarify the relationship between *media* and *interfaces*.

The word *medium* has its roots in Greek and Latin. It described a spatio-temporal position, for example, the space between above and below, between the rising and the setting Sun, and also referred to a public meeting place. Since early modern times, it is customary to understand a medium also as an incorporeal context in which something may take place. The definition of this in the *Oxford English Dictionary* reads, "Any intervening substance through which a force acts on objects at a distance or through which impressions are conveyed to the senses: applied to the air, the ether, or any substance considered with regard to its properties as a vehicle of light and sound." This usage led to the definition of *medium* as it is used in the natural sciences. In the methodological discussion of media studies, Friedrich Knilli has drawn our attention to the definition of *medium* as it is used in physical science—"an intermediate agency continuously pervading space."[15]

At the beginning of the twentieth century, the relation to etymology became more lax. Two competing epistemological opinions on the relationship between media, consciousness, and the world evolved—one connective and one divisive. The philosopher Ernst Cassirer understood media as symbolic forms that build a bridge between consciousness and the world, whereas the mathematician Hermann Weyl stresses the distance between symbols and consciousness, on one hand, and the world, on the other. Cassirer's reflections are based on the premise that the world is only given in human consciousness in the medium of symbols, which he discusses in its historical differentiation in his *Philosophy of Symbolic Forms*. The symbols used by humans are a medium in which consciousness and the world reach a synthesis. Cassirer makes a distinction between the media of mythical figurative expression and the medium of linguistic articulation and between both of these and the abstract use of symbols in science. Each medium forms the perception of the world and demands examination of what is considered to be true or false in accordance with the media.

Interestingly, Cassirer only began to use the word *medium* in methodological discussions in 1923. Two years before that, Hermann Weyl had published an article on the renewed crisis in mathematics[16] where he used the expression "Medium des freien Werdens" (medium of freely coming into being) to describe the continuum of real numbers. Weyl's argument rests on the assumption of a fundamental split between consciousness and the world. In contrast to Cassirer, he does not regard symbols as mediating bridges. Symbols and combinations of symbols are separate from reality. A further contrast to Cassirer is that Weyl speculates about a reality behind the familiar world that is accessible through mathematics and physics. One might say that the familiar world lies between mathematical consciousness and the true world.

The different epistemological positions of Weyl and Cassirer result in two different notions of media. For Cassirer, media are forms in which syntheses between consciousness and the world can arise. Media are thus meta-forms. For Weyl, media are indicators of a fundamental cut between consciousness and the world. This cut is a possibility for determining the "more real" reality behind the visible world.

Cassirer and Weyl were familiar with each other's publications. Their differences—between a mathematician and a philosopher—are an indication of the function that media studies might possibly have. Cassirer's life's work is the "heroic" attempt of a philosophical observer to keep abreast with developments in the natural sciences. In 1921, he published an account of relativity theory, which was read through by Albert Einstein before it went to press. And on the day that he died, he wrote an essay on the debate on causality in the reception of quantum mechanics. Hermann Weyl acknowledged Cassirer's achievements and expressed his admiration during a speech given at Columbia University, New York, in 1954, while emphasizing that his own position was a different one. Weyl did not share the opinion that one can speak of a uniform development in the way that humans utilize symbols. In his *Philosophy of Mathematics and Science*,[17] Weyl stresses the differences and makes it abundantly clear that the "heroic" age of Leibniz, where philosophy and mathematics cross-fertilized, is now over.

Weyl's recognition and emphasis of the fundamental difference between philosophical concepts and mathematical symbols in his philosophy lays the foundation for a new dialogue. He distinguishes between two types of philosophy. First, there is the historical account of the foundations of modern science, and Cassirer's work is his first example. Second, there is the "independent" thinking of philosophers such as Fichte, Schelling, and Husserl, who arrived at questions that modern science also poses, for example, the position of the human subject between finite empiricism and speculations about infinity, between determination and freedom. Weyl stressed the difference between philosophical metaphysical thought and constructive scientific laws but never tired of calling for science to retain metaphysical belief.

For Peter Weibel and Otto Rössler, natural science is media studies. This is plausible if one shares the opinion that the place formerly occupied by philosophy as mediator and communicator of scientific results is vacant. Weyl argues that philosophy asks the same questions as science, but in a different way, and therefore cannot mediate and communicate dialogically as long as generally (following the preceding argument, monologically) the precondition is to gain an overview of mathematics. Weyl questions whether it is in fact possible for a mathematician or philosopher to take up a position above the other's discipline. Media studies cannot judge per se from a superordinate position, for it is bound to the "in between" of the positions. Research on media obliges one to think in terms of relationships. This ranges from spatial and temporal relationships to relationships between disciplines and relationships between humans, machines, and the world. The starting point of media theory is the point where a relationship between discrete objects is assumed. Proceeding from this assumption, it examines the construction of the relationship whereby the interface is of paramount importance. We call this relationship a medium in which the interface decides between the "one" and the "other." The medium only becomes visible through differences. (In our opinion, this definition is already implicit in Weyl's understanding of the continuum, although it is generally ascribed to Luhmann's systems theory.)

The differentiations are grounded in the dialogic principle. As long as the other or the diverse is not assumed to be the basis of the established relationship, the discourse of media studies remains empty. In Rössler's interpretation, the Cartesian ego only succeeds in producing a consistent worldview when it recognizes that it can only think of the world of humans and the world of God as consistent and interrelated and if it takes an ethical decision. On the cut between a "fair" and a "deceptive" world, which the Cartesian subject can make, depends the form that the relationship between world, transcendence, and subject will take. Before, the relationship existed only as an abstraction, as a simple medium of possible decisions. Through the decision, the medium is either illuminated or obscured, that is, it is perceptible to a greater or lesser degree.

Paradise Machine Lampsakus

Computers and telecommunications allow us to communicate with other people. However, the myriad of possible connections remain "empty" if contact is not made. Vilém Flusser, who, like Rössler, acknowledges his debt to the dialogical philosophy of Martin Buber, illustrates this with the example of a chess player. If you have played chess with someone who lives thousands of miles away, then the possibility is always given that the other can enter into your life again and ask you to play. One must reckon with the entrance of the other without being able to count on it. The unexpected entrance of the other becomes a possibility that is always given. In Flusser's view, when it does enter, God also enters. The entrance of God—here the last piece falls into place between Descartes and the modern era—is always the entrance into the world and thus the chance to escape from self-referentiality.

Here there is an important distinction between media theory and media praxis. For Rössler, the interface becomes a paradise machine if it is set to fairness, that is, when the interface colors the center between the worlds so decisively that the world appears as free and not as contradictory at the same time. It follows that the interface can be programmed by ethical choices. The interface set to fairness can, for example, be controlled in such a way that the ancient city of Lampsakus is rebuilt on the internet. What we know of the original Lampsakus can be extended and expanded. Up to now, all that is known about this digital Lampsakus is that there, information and knowledge can be obtained for free in a sunny place. To date, Lampsakus is a possibility for gaining knowledge in a tranquil and orderly fashion. The possibility exists to South Americanize Lampsakus. However, if we can entertain the thought that knowledge can be obtained not only in libraries but also through dance and rhythmic forms, then there is a great chance that image makers, dancers, and others who have overcome their obligations to the alphabet will help to build Lampsakus with their arts and skills. It is with this hope that we dedicate this chapter to José Carlos Mariátegui, mediator to the Peruvian supporters of Lampsakus.

Translation by Gloria Custance

Notes

1 *Webster's Collegiate Dictionary*, s.v. "interface."
2 Bruno Schulz, *Die Republik der Träume* [The republic of dreams] (Munich: Hanser, 1967).
3 Pierre Klossowski, *La monnaie vivante* (1994), German version, *Die lebende Münze* (Berlin: Kadmos, 1999), 10–11.
4 The items quoted here are all taken from Brecht's *Short Organon for the Theater*. Through the different numbers, it should be easily possible to identify the fragments in the English translation.
5 Klossowski, *Die lebende Münze*, 20–21.
6 *Oxford English Dictionary*, s.v. "dia": "pref. before a vowel di-repr. Gr. dia-, di. The prep día through, during, across, [etymologically related to dís... twice."
7 Wilhelm von Humboldt, *Über den Dualis*; discussed in Ernst Cassirer, *The Philosophy of Symbolic Forms*, vol. 1, *Language* (New Haven, Conn.: Yale University Press), 156.
8 Cassirer, *Philosophy of Symbolic Forms*, 1:24.
9 Massimo Cacciari, *Gewalt und Harmonie* (Munich, Germany: Hanser, 1995).
10 Marshall McLuhan, *Understanding Media: The Extensions of Man* (Cambridge, Mass.: MIT Press, 1994).
11 Carl Schmitt, *Politische Romantik*; discussed in Dierk Spreen, *Tausch, Technik, Krieg—Die Geburt der Gesellschaft im technisch-medialen Apriori* (Berlin: Argument, 1998).
12 Otto E. Rössler, *Das Flammenschwert* (Bern: Bentelli, 1996).
13 Rössler, 15.
14 Lorenz Engell and Joseph Vogl, preface to *Kursbuch Medienkultur: Die maßgeblichen Theorien von Brecht bis Baudrillard* (Stuttgart, Germany: Deutsche Verlags-Anstalt, 1999), 10.
15 Friedrich Knilli, "Medium," in *Kritische Stichwörter—Medienwissenschaft*, ed. Werner Faulstich (Munich, Germany: Wilhelm Fink, 1979), 230.
16 Herman Weyl, "Über die neue Grundlagenkrise der Mathematik," *Mathematische Zeitschrift* 10 (1921).
17 Herman Weyl, *Philosophie der Mathematik und Naturwissenschaft* (Munich, Germany: Oldenbourg, 1966). First German version published 1926; first English edition 1949 by Princeton University Press.

7 The Art of Design

(Manifesto) On the State of Affairs and Their Agility

> *The manifesto is a genre of text that allows for a certain expediency of expression, that enables, on given occasions, at given times, the pointed articulation of complex relationships in the form of directives for thought and action. Whenever I have had occasion to write a manifesto, I have always considered myself its first addressee. This variation of a manifesto was written on the occasion of an art and design exhibition that was organized in 2014 in Berlin by the New York–based agency Shutterstock.*
>
> *This essay was originally published in 2014.*

⁂

Electrician. What would night be without you!
—Hannes Jähn, *Bogendruck 4*

1.

Every commission is finite. It has a beginning and an end; otherwise, it would be undefinable. Free artistic will emerges from insight and the sensation that the world as we experience it is limited and full of disruption, imperfection, dissonance. One of the privileges of art is the ability to transform the suffering that that causes into a productive design process. Formative energy means the capacity to transgress the finiteness of our existence into a more open pluriverse. "Organizzar il trasumanar" (organizing the transgression of boundaries) is a beautiful paradox to describe a core dimension of Pasolini's work as a poet and director.[1]

2.

The contrast between a defined framework, which in the mental realm can also be called consciousness, and the artistic liberty of an individual player is not an irresolvable contradiction. In fact, it is the tension-filled complementarity of opposites. One is unthinkable without the other. The freedom of individual will is not only compatible with the idea of a world of conformity; it is an inherent element.

3.

Freedom is primarily an experience. Free will—in particular, will that is transformed through imagination into aesthetic action—is only realized when I act, think, judge, create, design, fight, destroy, love. In its most concise, pointed form, it can be expressed in one sentence that expresses the essence of the art of design. It is the ability to choose what my will really wants. In other words, the free will to alter the world aesthetically is the medium through which the subordinated (the subjects) become creators. Designers are projectors. In the best cases, what they create are worlds of projection, ones that are different from the world we live in.

4.

The art of design is an act that requires the utmost agility. This agility is not identical to the type of mobility that the economy we call global demands of us day in, day out. Our mobility is not an object of exploitation. It strives to make do with a minimum of possessions and aesthetic ballast and cultivates an existence of peregrination and luxurious ambling, indulging in wonder and surprise.

5.

Concepts traded on the global market of strategies, like globalization, have their origins in semantics that have nothing to do with the various arts of design. We need different terms, a different orientation. The quality of the worldwide relationships we need to develop and cultivate is primarily poetic in nature. Categorical instrumental reason is foreign to this type of elegant worldliness. Under the conditions of the global networking of our work, the art of design could become a mondial theory and practice.

6.

Knowing which way the wind is blowing is once again becoming essential to survival. Learning from the pirates and traders of the seas and skies requires not only deep understanding and navigational skills as fundamental cultural techniques. The winds and the clouds that herald turbulence or storm must be read as notations. Berlin lies on a plain and will thus forever remain provincial. It is the contact to oceanic mind-sets and approaches that opens design to the world.

7.

If there is doubt and choice, a decision for possibility *(potentia)* is better than a pragmatic preference for reality. Worlds of possibility are worlds we can imagine, create, and achieve—ones we like, wish for, even yearn for and enter into venture with.

8.

For centuries, modern science, technology, and art outspent themselves trying to make the invisible visible, the imperceptible perceptible. This process has advanced far through the dating of nature and the cybernetics of social intercourse. The more the technical world is programmed to make the impossible possible, that is, functional and effective, the more worthwhile the attempt to confront the possible with its own impossibilities becomes.

9.

We live in a permanent Test Department, to reference a 1980s London rock band. Ideas and concepts are developed and tested for their market viability. But in artistic and design processes, the experiment always takes precedence over the test. An experiment is free and brings with it the possibility of (honorable) failure. A test is bound to previously determined objectives and targets and serves the creation of products.

10.

The appeal of alchemy laboratories in early modern times was not primarily to fabricate gold from common minerals. It was in the gripping experience of perfecting the imperfect. The metamorphosis of the transformer was just as important as that of the material.

11.

The art of design realized with advanced techniques should squander itself less on renovating or decorating what is rotten and more on the infinite and indispensable experiment of creating a world other than the existing one. Even if the shifts we are able to produce are measured only in micromillimeters or nanoseconds, the noblest task of design artists is to give back some of the time life has stolen from those who experience and enjoy our works. That requires us to allow ourselves the time the art of design requires.

12.

Outside the middle: the enormous effort necessary to occupy the centers of technical and cultural power is not worth it. Focusing on the gaps and the peripheries promises greater liberty, pleasure, and wonder. It does not preclude occasional excursions through the center toward other edges. On the contrary, a peripheral existence benefits only those who understand the quality of the center.

13.

Dual identities are in many respects the basic equipment of art activists and the tools we need for its design. In an economic sense, that means both mastering guerrilla tactics and understanding how the trader thinks and acts. For people involved with complex systems, it's not enough to be a philosopher and poet. I fear they will not achieve a sustainable existence without experience in arranging and steering.

14.

Machines and imagination need not be irreconcilable contradictions. Being the *homo artefactus* that we are, we can employ them as two different, complementary ways to understand, disassemble, and reconstruct the world. One penetrates the highest spheres of programmed worlds only through the powers of fantasy and imagination. But fantasy and imagination are well advised not to cast aside calculation and computation whenever they are essential. A mind-set of comfort has no place in the advanced art of design.

15.

One doesn't need to be an engineer or programmer to use advanced technologies to produce objects and processes that stimulate and motivate. But it is a tremendous advantage to understand how engineers and programmers think and work. Collaboration is impossible without respect for the other's work. The best artistic design projects are the result of collaboration. Unlike partnership, collaboration needs no rules for joint action; it is temporary teamwork.

16.

In the future, it will no longer be enough for design artists to be only operators or magicians. Accessing the world through design requires both an act of intervention and the laying on of hands: in the best case, by magical operators or operative magicians.

17.

To avoid an existence that is too little within time and is thus paranoid, or that is too much caught up in time and thus indulges in bitter melancholy, it can be helpful to practice conscious fragmentation. We organize, inform, publish, debate, and amuse ourselves in telematic networks. We rhapsodize, reflect, relish, believe, and trust in other, distinct situations, alone or at times with others. That leads to a balancing act: in the span of a single life, we need to learn to be able *to exist online* and *to live offline*. Otherwise, we will simply become interchangeable functionaries of the world we ourselves have created. We must not hand that victory over to cybernetics, the science of optimized control and predictability of complex events.

18.

The noblest task of all design arts remains to sensitize others who are different from ourselves and to do so using the particular means available to us: aesthetics. That will not change, regardless of the techniques and media through which we express ourselves.

19.

Not everyone can be a genius, but there are endless opportunities to avoid making a fool of yourself.

Translated by William Chaney

Notes

My sincere thanks to Georges Devereux and to Walter Benjamin, Josef Čapek, Hinderk M. Emrich, Vilém Flusser, Édouard Glissant, Ludwig Harig, Arnold Metzger, Pier Paolo Pasolini, Fernando Pessoa, and *Test Dept.*—Berlin, July 2014.

1 This wordplay is referring to Pasolini's famous book with poems, *Trasumanar e Organizzar*, published in 1971. Giuseppe Zigaina uses it as a title for his contribution to the catalog *p.p. Pasolini oder die Grenzüberschreitung—organizzar il trasumanar*, ed. Giuseppe Zigaina and Christa Steinle (Venice: Marsilio, 1995), 47–88.

FIGURE 8.1. Jean-Luc Godard, *Histoire(s) du cinéma* (1988–98), part IA, opening sequence.

8 "Too Many Images!— We Have to React"

Theses toward an Apparatical Prosthesis for Seeing— in the Context of Godard's Histoire(s) du cinéma

When Jean-Luc Godard released the first two parts of his Histoire(s) du cinéma *in 1989, it was clear to me that the master of European avant-garde cinema was working on compiling a kind of final testament to the whole of his work in the context of an archaeology of world cinema—methodically and by means of his very own medium of film. But there was here too a very palpable medial paradigm shift taking place, within which the video project began to develop: from optical-mechanical to electronic, from analog to digital, from local to globally interconnected discourse. Contrary to the full-throated proclamation of an "iconic turn" in art historical discourse, here the* Histoire(s) du cinéma *are interpreted in the context of a turn toward the literarization of audiovisual communications.*

This essay was originally published in 1991.

⁂

Why do highly intelligent artists and art critics—who these days need to be media critics as well, and all three functions are, I think, no longer even separable—why then do audiovisual innovators at the close of the second millennium of our time, at the fin de siècle of cinematography and television, continue to produce, with gestures as varied as those of Jean-Marie Straub and Danièle Huillet, Alain Resnais, Jonas Mekas, Patrick Bokanowski, or Jean-Luc Godard and Anne-Marie Miéville, so many "sound-images"— *sonimages*? Why go on projecting these into public spaces or broadcasting them into intimate, private ones? Out of sheer vanity? To secure a comfortable retirement? Because they can't help it? Have they, who were socialized in the context of, among other things, the debates between Claude Lévi-Strauss and Jean-Paul Sartre, the disputes over the structuralist *It* and the historical subject of Marxism, who witnessed the penetration of structuralism by psychoanalysis and the occupation of psychoanalysis by the cybernetic idea world of Norbert Wiener (Godard's first film, *Alphaville* (1965), was about computer fiction and computer reality, about poetry's struggle against a universal machine)—have they, who came to artistic prominence alongside the *nouveau roman* with its uncertain boundaries between the fictional and the real, under the influence of Roland Barthes's *Mythologies*, the development of a semiotics of film, the import of Marshall McLuhan from Canada to France, the apparatus theory of Jean-Louis Baudry or Jean-Louis Comolli's "machines of the visible"[1]—have they, their own idea world thus impressed, still not heard that we have abandoned all finalities, such as that of the representative in representation? Have they not heard that images on the analogy of iconography are utterly anachronistic? Do they not sense their abyssal mustiness, their fundamentally reactionary character? This is what we would like to convince them of, those of us who are thinking, acting, and, above all, *writing* at the level of image technologies. Panicked—or so it often seems to me—following the codes found in panicked motion.

Why do I concern myself, in the context of a question, or rather many questions, all orbiting a new center of gravity, questions having to do with cybernetically based/constructed architectures, spaces and objects, or "puppets," questions that revolve around binary code, everywhere proclaimed as the universal code, questions that outline the contours of new paradigms in media design, like that of interactivity[2]—why do I concern myself, in light of all this, with such an apparently hopelessly antiquated artifact like the video recorder, a technical object that, from the perspective of cyberyuppies or cyberpunks, belongs more to the lost world of Lascaux cave painting than to the audiovisual now-time?

Both lines of inquiry invite answers that point in the same direction, answers of a provisional nature but that are intended to occupy, fill out, or rather mark with a bit of archaic material the space for thinking the floating, the difficult to grasp:

1. We are clearly already exposed to the danger of losing ourselves prematurely in the flood of traditional technical images, of no longer wishing to differentiate, no longer being able to interpret (which in fact is nothing but the artful form of differentiation), no longer being able to select. Anything that, in its microstructure, appears to us in the mode of visual representation, no matter whether completely analog or digitally processed, we deem a suspicious, unseemly

FIGURE 8.2. "Let every eye negotiate for itself." From Godard, *Histoire(s) du cinéma*, part IA, opening sequence.

"image." These images *seem* to us all the same. We are not prepared at first to perceive the macrostructure—a mistake that we, as those who speak and write on the level of literally organized discourse and its equally dispositive structure, don't want to make, because if we did, then we would have to stop (at least publicly) speaking and writing.

2. If we acknowledge the shift in priorities—which we are to observe, therefore also co-create—within the reference zone of signs, in favor of self-reference and to the detriment of the remainders of the real (which I have been unable for a long time to imagine as anything but medially tangential), and if with this we bid farewell to the real, we are running the risk of drowning in an empty tub. At the moment, at least as a thought experiment, I'm trying to make some headway with the idea that, with these new technologically co-constituted audiovisual worlds, we are dealing with a multiplication of references and thus with their considerable expansion. (*Expanded Cinema* is

the name of Gene Youngblood's brilliant and still unrivaled 1970 book that first anticipated and analyzed developments like these.³) The audiovisual-telematic has evolved into an inseparable element of reality (which, to be sure, is not identical with the real, if the trio of the symbolic, the imaginary, and the real is to remain a useful mode of organization for our psyche). And that reality has since been quasi-extended into the monitors and projection screens that are always and everywhere present. These in turn emit into the real, to great effect. The boundaries between the different zones of reference have thus become more fluid, more difficult to recognize at first glance—which certainly presents a challenge to the interpretation of these visual (and acoustic) worlds, though it may not necessarily entail an epistemological quantum leap.

3. To be sure, the image technically reproduced by means of traditional instruments never really stood a chance of being interpreted, even of being duly regarded, particularly not when participating in the illusion of motion—that is, not in its form as a sounding time-image. This becomes especially apparent to me when I attempt, with my students, to interpret the audiovisual material of an evolved technodiscourse, for instance, advanced 3D surface animations. It scarcely occurs to us that the floating, algorithmically organized signs, the numerical images—however antirepresentational they may be in their isolated, static frames, which, to be sure, not all of them are—follow the same syntactical pattern, feature the same narrations as those that industrial cinematography and television overdeveloped decades ago. But all of them, each and every time, can be made sense of narratively by using Christian Metz's *Grande Syntagmatique,* which in fact he developed for the purpose of interpreting conventional cinematic syntagms. (And often it's even simpler than this: already commodified communications like, for instance, the internationally acclaimed and award-winning animation *Tin Toy* [by Pixar/Lasseter], turn out to be, upon closer examination of their construction, moving cartoons of the most trivial Disney variety, their mechanical shot/reverse-shot dramaturgy basically cinematographic junk.) Even in the technical vision created by blind machines, turbulent structures are seldom found. This has a lot to do with:

4. We cling, so far as images are concerned, to the individual image

above all, isolate out for our consideration the frozen fraction of a second, which is then clocked electronically at 50 or 60 Hertz in order to create the illusion of motion. (A classic case of the confirmation of the Heisenberg uncertainty principle in cultural–scientific discourse. For with this isolating gaze, we are creating *one* reality of the cinematic, which is manifestly distinct from other possible realities, such as the one we perceive.) We have, in my view, barely even begun to cope with this leap into the visualization of time, into the construction of space-time that is not only characteristic of the telematic but already present in the cinematic as well. Conversely, the great innovators of cinematography have never been particularly interested in the individual image as a quasi-molecular component of the cinematic organism. Their cinematic work, as creative thought experiment, did not and does not take place primarily on the set where images are recorded but in the editing room, where the syntax of the cinematic is organized, where separations are performed and connections made, where the real *relationship work* is done and where meaning is invested by virtue of montage. "The feeling man shoots, the thinking man edits"—or so Nam June Paik once trivialized the interventional practice of electronic image-making. Analytical vision via media machines: "Only an intellectual cinema will be able to decide the conflict between the 'language of logic' and 'language of images.' ... Such a cinema alone will have the right to exist among the wonders of radio, television and the theory of relativity." Eisenstein wrote this in 1931. And at the end of this text, he infers the radical consequences that we encounter time and again (and again and again) in the sphere of telematically based artistic praxes today: "This will be the contribution of our entire epoch to art. To art, which has ceased to be art, on its way to the goal of becoming life."[4]

5. We have learned to live with our technical "cultural prostheses." Some of these have grown entwined with us, may no longer be severed from our bodies. (It is worth remarking that we now carry many of the new gadgets and artifacts of enhanced media communications directly on our bodies, from the Walkman to the Body-Sound-Producer, the Watchman, the personal video, the pager, the Private Eye, even face and headsets, data gloves and data suits for the exploration of cyberspace.) To counter the strategies of semblance that are reflected in

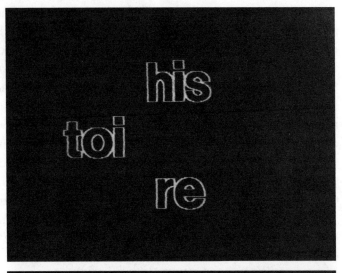

FIGURE 8.3. The fragmentation of the word *histoire* into its components *his*, *toi*, and *re*, each with its own meaning, is only one layer of the game. Also, the text becomes image. The semiotic status of the letters as symbolic signs is revoked, and the boundaries between the characters become fluid. The second and third *toi* appear, the fragments start to flicker, and with the rhythm, the symbols also become a musical material. From Godard, *Histoire(s) du cinéma*, part IA.

illusions of motion, for those who are on the receiving end of these emissions, requires technical means of objectifying and making the semblance available. When the illusion of movement has increased in velocity as breathtakingly and time annihilatingly as it has in our present audiovisual culture, we need a crutch to help us reenlarge time again, to artificially re-create its spatial dimension. The dual nature of the commodity form is still in effect, in all its ambivalence, from this perspective—in the context of the art market, too. Take, for instance, Peter Weibel's *Inszenierte Kunstgeschichte*[5] in Vienna: I was not there when and where it took place, though I was still able to engage with it via various electronic reproductions.

I.

I do not want simply to try to verify these claims, conjectures, and theories. But, by disassembling Godard's cinematic text (understood in the broadest sense) from a few particular angles, then interspersing my text with his, and by linking them in this way, I hope to illustrate these claims, conjectures, and theories and open them up for discussion. Godard's videotape lends itself especially well to this, because his reflections on making sound-images—which are themselves organized in symbols, indices, and icons, in letters, sounds, and images—already presuppose a cultural prosthesis like the VCR. Without one, my interpretation of his tape could not exist. The mode of production and the mode of reception form an indissoluble context. And this is because, thematically, there is something in their middle that seems to me to play a crucial role in our own brains as well, namely, the charged relationship between the *fleeting* (meaning also the fugitive) and that which can be *grasped*, via a process of reproduction that makes something accessible to us, so that we can reflect on it in a time that is organized differently.

For the *Histoire(s) du cinéma*[6] have at least two opposing phenomenologies, each offering something correspondingly and entirely different for perception and interpretation. As a singular event within the constant flow of broadcasting (or on the cinema's rented time, as contracted in the ticket), they are an outright attack. They take the ordinary panicked organization of the output of sound-images to extremes. At the very least, we stop seeing when confronted with this tremendously accelerated history of film, with story superimposed on story after story. In the first two parts of the collection, which will be discussed here, the retina is subjected in just one hundred minutes to roughly 150,000 images. Presented and perceived in this unparalleled fleeting mode, the *Histoire(s)* appear like an illustration of a phrase that functions as a sort of leitmotif throughout Godard's 1968 *Le gai savoir* (*voir*, "to see"; *savoir*, "science/to know"), his first great film-essayistic discourse on the history of images: "If you want to see the world, close your eyes!"

Naturally there is an immanent resistance to this, on the part of those whose vocation and calling it is to produce cinematic illusions of motion. To engage in the destruction of images by constructing them is a paradox. This is the point at which deconstruction begins. The filmmaker's revolt, the attempt to create a form of *effective countersemblance,* expresses itself above all when, from out of the oscillating event of broadcasting or cinematography, a *literary* process appears; when, from out of the act of rushed perception, an act of reading can emerge, an act that is no longer bound to the trajectory of time in flight but seeks its own tempo and trajectory that, in certain circumstances, allows the film to rest. It is this mode that the following interpretation inhabits.

II.

In the beginning, there was seeing. We've heard this sentence, or something like it, often from Godard. At the beginning of the *Histoire(s),* we can only watch it, superficially iconized: an eye, covered, protected by a photo camera poised at the ready, a close-up shot of the stony monocular vision of the Enlightened and all-monitoring central-perspective gaze, apparatively sublated, as it is in every camera with normal optics, while at the same time referencing a cinemythos, for it is the eye of the voyeur and the hunter from Hitchcock's *Rear Window* that is armed with the optical apparatus. (One-eyed figures recur constantly in the *Histoire(s)*: Nick Ray, Fritz Lang, for instance; Godard himself worked for a while wearing a stylized pair of glasses, one lens of which he had darkened. And, for him, the photographic tradition was essential to the cinematographic. Especially in part II of the narrative, he alludes to this again and again.) The camera's eye is swiftly replaced by a technically unprotected eye, and suddenly there flashes a pair of eyes: the machinological and the physiological image receptors. But if we look closely, we see that the film image is not showing us the eyes of one human being; these are two different eyes (one of which could be an enlarged image of the pair belonging to Giulietta Masina that later resurfaces several times in a second layer of image and that will also reappear in my text). As spectators in front of the screen, we become implicated in a complicated interplay of seeing and being seen, of action and *passion* (in the dual sense of the word—suffering and fervor—which Godard often deploys in his cinematic work). Like a graphic fanfare, in bold white lettering, the writing as image, the titled emblazoned before our eyes, signaling the method of *this* historiography: "Let every eye negotiate for itself." Construction plan and instructions for use. References and self-references, right from the start. The boundary between inside and out, for distanced looking, for seeing as that which one can "hold at a distance" (Merleau-Ponty), the possibility of which the eye is supposed to guarantee—this blurs with the very first images.[7] We are no longer certain (of anything). We've set ourselves up for a risky adventure. History as illusion, in Dietmar Kamper's sense as well: *it's a gamble,* a risky adventure.

Immediately following the natural organ, another part of the audiovisual technical system appears, one that a certain cinematic philosophy considers to be cinematography's most important artifact: close-ups of a 35mm editing table on which a strip of color film is positioned, gliding back and forth. An image of the mechanical—which follows the take as the filmmaker's mode of perception—replaces the image of eyes. Following reception by the eye's receptors is perception of the sights of offer: processes of storage and manipulation, machinal allegories of mental processes too, without which the (hi)stories of the cinema could neither be present nor aurally or visually perceptible.

"Communication," writes Godard in his *Introduction to a True History of Cinema,* "becomes possible when something re-emerges that had already entered in."[8] An impression turned back to the outside: it is in exactly this sense that the *(Hi)stories* are realizations of a communicative process—communications about film and about history, insofar as these had been captured on film, had been stored in successive images; communication about cinema above all, about the images that can be attached with it, the fragments of language, the musical sequences, sequences of song, banal noises, gunshots, cries of despair and of love that had impressed Godard and us as well from out of their own history; traces left in our imaginations and fantasies, pushed aside or obscured by our everyday thinking and doing into

FIGURE 8.4A. "The cinema substitutes for our gaze..."

FIGURE 8.4B. "a world more in harmony with our desires." From Godard, *Histoire(s) du cinéma*, part IA.

this underlying storage layer, seldom or never retrieved, for want of occasion, now (re)activated by reproducing machines; and, above all, the work of the reconstructor. (Hard work, in the Godardian sense. "One can work to love, or one can love the work," as he put it in his film *Passion* [1982].)

"Cinema truth... Factory of dreams factory": Godard as analyst of the dream machine—that is a role that suits him and that he quite openly enjoys, not only in the *Histoire(s) du cinéma*. Together with Anne-Marie Miéville, he plays it conspicuously in the recent history of European film/cinema. In his audiovisual reconstruction of this history, the filmmaker wanted not least to make himself the object of analysis: "I imagined... that, starting from this past, I could see my own once more, like a psychoanalysis of myself and my space within the cinema."[9] And, with the very first colliding combinations of image and sound in the *Histoire(s)*, a phrase from *La gai savoir* thrusts itself into their midst like a reminder: "Chance is structured like the unconscious."

"Cinema substitutes," reads a heavily emphasized title from the beginning of the film.

"Cinema substitutes"

"For our gaze"

"A world that corresponds to our desires."

And we will link this with a sentence that we'll hear a little later on: "The cinema projected, and men saw that the world was there."[10]

No prior history of cinema or film has taken us so clearly and resolutely right to the point of the sublation of the long maintained and closely guarded dualism between illusion and reality.

III.

Godard's means of reconstruction are apparatuses in a comprehensive sense. Insofar as these means are machines, they are visible, insistently shown to us right from the start. In the very first scene, and then over and over again: the editing table across which the film strips are pulled at varying speeds. The sounds that are modified in this process of expanding and compressing time are not background noise; the film flapping against the feed rollers, the whirring sound of the projector running and the acoustic fade when it stops, the familiar and beloved sounds of the cinema technician express in equal measure the pleasure in this work process and the violence released in the

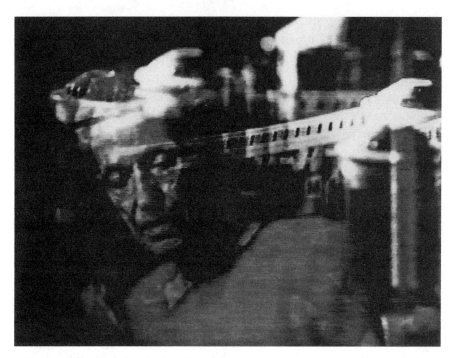

FIGURE 8.5. Nicholas Ray from Wim Wenders's film inserted into the image of the mechanical heart of the cinema: the editing table. From Godard, *Histoire(s) du cinéma*, intro, part IA.

work of reconstruction. An apropos beginning: images of Nicholas Ray in Wenders's *Lightning over the Water*, distorted by time lapse, as he suffers of advanced stage cancer: one of the most remarkable violences to have been added to the history of film in recent decades. On a gorgeous photo of Nick Ray Godard then writes in white letters: "Père, ne vois-tu pas que je brule?" (Father, do you not see that I am burning?)—the microphone, a nice big studio mic that, hung on the gallow arm, gradually visibly obtrudes into the foreground of the image, in front of the narrator, who stands beside the bookshelf puffing contentedly on his cigar, staring absently at a projection screen the viewer does not see but must imagine, as he begins to comment on the images (offscreen). The typewriter, this artifact of discourse par excellence; Godard chose an electric one, half word processor, half conventional typewriter, on which the typed text is first displayed on a small screen and can then be printed by pressing the key command and waiting just a short moment. In time with the hammering of the keys appears the myth of cinema personified: Chaplin, with Paulette Goddard, in a still from *Modern Times*, which is then interrupted by a cinematographic recording apparatus and, again, Godard at the typewriter. Icon and text-become-image are in turn linked together, correlated with one another in the interest of the object and the method of its treatment. This all has to do with the rules of the game.

Godard begins to speak the text, murmuring more to himself than communicating with anyone else. Some of his words are transformed into titles; others remain completely unintelligible. Entire sentences, phrases, incomplete sentences, words like signals. In rhythm with the hammering of the daisywheel head, human heads flash into view, filmmakers in the broadest sense, heads in which images spring up, from which images well forth, heads belonging to those who wrote the history of cinema. Irving Thalberg, for instance, the man who had fifty-two movie ideas a day—Godard juxtaposes him against the shabby creativity of television directors; and while he speaks over Thalberg's photo about television, the photo decomposes into horizontal lines, is fractalized televisually. Meanwhile, cinematographic recording apparatuses appear again and again, particularly cameras, the artifact with "two rolls, one of which fills as the other is emptied"; in the second episode, as he is trying to close in on the origin of the cinematographic, Godard complements this image with the following observation: "As if by chance, in the video industry the left reel is called the *slave* and the right the *master*," servant and lord, both dependent on each other, the one inconceivable without another.

Odd: The only machine that this writer and director does not show us is the one that he used increasingly, with a bewildering virtuosity, even in his cinematographic films: the tape recorder. Does he mean to keep this his secret? Without a multichannel high-performance acoustic recording device, it would be impossible to produce that dense meshwork of indexical, symbolic, and musical sounds that powerfully constitutes the aesthetic idiosyncrasy of the *Histoire(s)*. He was also conceptually experimenting with this particular style of layering of audio material in the short video he was making at the same time, *Puissance de la parole* (1988), a work commissioned for French telecom, which Godard used as a pretext for the *Histoire(s)*.

All of these media machines, whether shown to us or not, are quite overtly *time machines*—just like the video recorder on which he saves his images and from which he retrieves them, as if from an audiovisual synthesizer.

The tape recorder and video recorder belong in the same phylum of artifacts. Their common technological germ cell is the Poulsen telegraphone.

And they urge a common, symbiotic aesthetic. In them, the acoustic and the visual are no longer separate. In form, both are possible on the basis of essentially identical magnetic strips, the only difference being that a substantially higher relative velocity between the read/write head and the recording material is needed to read and write images than is required for sound. The technological identity of the process in the same medium (mis)leads to their being treated identically. In more advanced audiovisual production, images increasingly become sounds and sounds image, as on Godard's tape their sublation is exemplary. Both are the raw materials for a cinematic claviature. *Acoustization of the visual and visualization of the acoustic.*[11] Perhaps this is a vanishing point in the process of acceleration of the visual and its coupling with musical materials: the production and perception of *sonimages* as rhythm-works. The technocultural avant-garde of the 1920s was already at work on a project of this kind. But with a common (digital) code for image and sound, and with an identical medium, it undergoes an escalation and is transformed from an avant-garde practice to an everyday means, a pop commodity.

IV.

Even when interpreted in the mode of reading, in the *Histoire(s) du cinéma*, we are dealing with a multiply coded communication. On the surface, we perceive a vibrant kaleidoscope, a family album too (including legends, in the dual sense of the word: captions below the images as well as narrative accounts), there to be leafed through by the well versed, by those who know their film history: there's Chaplin, for instance, and Nicholas Ray, Jean Renoir, Lydou and Jean Vigo in *L'Atalante*—"Jean Jean Jean"—the Jeans are piling up, culminating with Jean Cocteau: "Le Testament d'Orphée"... these *rendezvous* with men and above all women of the silver screen: Adele Jane Faith Joan Ginger Rita... The show on offer like a gigantic picture puzzle. You can quiz yourself on how much you know of the history of the dream factory, who you recognize. It is a labyrinth as well, because no matter how great your effort, the search for identity leads nowhere. A parlor game for the bourgeois *cinéastes,* designed for private viewing among friends, the like-minded, and in this sense the medium is video specific in that the show on offer addresses itself to an intimate form of the public and leads to hours-long interruptible and repeatable play, for the picture puzzle only fully

FIGURE 8.6. Godard's first sketch for integrated *Histoire(s)* of Cinema and Television. The latter (TV) disappeared during the development of the project. From *Filmkritik,* no. 242 (February 1977).

Histoire(s)

du Cinéma et de la Télévision

unfolds once the fleeting cinematographic and televisual mode of reception has been left behind.

But this phenomenology of the tape is also video specific: sequences of images, scenes, and shots are fractalized into fractions of seconds. Shots from early silent film. Fritz Lang. Murnau. Griffith. Renoir. For a few swift moments, Hitchcock's Cary Grant flashes on screen in a steel-gray suit, menaced by planes in *North by Northwest*. (Planes appear over and over again, ominous, like Hitchcock's birds and Van Gogh's crows, most impressively in a sequence in which some bombers glide over the beautiful face of Rita Hayworth: "Only Angels Have Wings," which could also be read, in the particular arrangement of this shot, as "Angels Only Have Wings.") A cocked machine gun. Gangster wars and the war of American cinema against that of Europe (and vice versa). Vertov's Lenin. Eisenstein's Olga. A screaming poster, signifiers of market culture. A windmill vane from which a revolutionary has been hung. A dance scene from the 1950s and back again swiftly to the early days of cinematography. A pin-up girl or a Fra Angelico painting... The visual quotes follow so quickly one after another that one begins to blink, one has to pause, to interfere, in order not to go mad or just blank out: rewind—slow motion—mute—quick search, forward and back—pause. Godard really challenges us to use the audiovisual time machine by keeping the visual signs floating. He has organized sound-image experiences, separated the old and recomposed the new. This communicative gesture is also addressed to the audience. It comes across the screen like a carefully written invitation to interactivity, to dialogue with the reconstructor. At the interface of media-machine and media-human, the observer loses his customary and comfortable status. He becomes a *player*. He is summoned to subvert the montages on offer, to circumvent, to offset them.

And in this mode, too, the red threads are spun out.

The *Histoire(s)* do not represent the history of film. They objectify, as product, only *one* of the infinitely many possible ways to construe the reality of the history of cinema and of film. In this sense, they are models. A virtual reality, too, of a kind all its own—"potentia" the way Heisenberg meant it, the history of film in its "wave function." To become *one* particular history of cinema (Godard consistently uses the numerical designation), it must pass through the act of recording. "No record, no measurement," writes Nick Herbert in his proposal for a "new physics." "Only those interactions in nature that leave behind permanent traces (records) count as measurements...."

Only record-making devices have the power to turn multi-valued possibilities into single-valued actualities."[12]

Godard's mode of production exploits the artifactual visual prosthesis that lifted Japanese industry into its own mode of mass production; but he also assumes in those he addresses the presence of a mental video recorder. He plays with the displacement and lag between the processing of acoustic and visual stimuli. He seems to be calculating on everyone further editing his auditory and visual offerings. He seems to address himself directly to those mental videos with which everyone is in a position to make any film into his own film. Therefore the *Histoire(s)* are also a film of the imaginary viewers' films: "Films that were never made" (Godard), or better, films as they are constantly being made in endless number, only invisibly, because rather than having been transferred to a material medium, they've remained in the imagination.

The *Histoire(s) du cinéma* are thus written in several ways as apparatus-communication: as communication material for the mental apparatuses of the persons involved in the film process and those associated with it but also as apparative offerings for communication by means of apparatuses, by means of technical *prostheses* in the Freudian sense.

It is part of the hegemonic nature of cinema, as *dispositif*, that the machinery that generates the visible is generally not shown to those who are watching; the machinery of illusion is hidden, remains invisible, as the theorist of apparatus and filmmaker Jean-Louis Comolli put it. This is one of the most thrilling tensions in Godard's *Histoire(s)*: that they drag into the light of the electron tube what is generally kept in the dark. This rendering visible appears equally impossible without apparatuses, without crutches. The video recorder or player is in the age of infinite reproducibility one of the most prominent prostheses for seeing.

Godard seems aware of this in every respect. He uses it to calculate his cinematic offerings. The sociocultural mode of address (the intimate, the private, the radically subjective, the literalized . . .) corresponds wonderfully with the aesthetic structure of the tape. Godard and the audiovisual time machine: the real-time of the cinematographic illusion of motion practically no longer exists in its electronic processing. Godard brings its constant flow to a stutter. He allows the swift motion to collide with a countermotion, arrests the running film figures, decelerates the rapid gesture, freezes it, makes the objects moving at high kinetic velocities (for instance, Hitler's

FIGURE 8.7. Jean-Luc Godard, reconstructing and analyzing at the device.

hand while he speaks) into artifacts that he can reanimate, their movements modified, accessible to visual analysis, as if observed through a stroboscope. This extreme alteration to the linearity of film-time, which he keeps up throughout the entire tape, lends it a new, singular status, gives it (back) its spatial character. The images come across as an attempt in the permanent extension of (film) history against the increasingly sprawling, ever-fatter present. In this context, we read the force in the title: "It is time for life to give back to cinema what it stole from it."

V.

The individual images that we see on the monitor are conventional, familiar, yes, sometimes even *representative*. No wonder, since they are in fact borrowed from many kilometers of objectified and commodified film history as well as from art history's catalogs and picture books. (Godard appears to be well stocked by modern antiquariats with gorgeous volumes full of some of the greatest works of European painting. He quotes shamelessly from Renoir, Matisse, Goya, Seurat, Van Gogh, Gauguin, and many others, segmenting and cropping the images just as he needs them to be; it appears effectively like an act of revenge on what a cinematic projection does to the filmmaker's artfully crafted images.) The surprise, the astonishment, the provocation (in both seeing and thinking), come about as result of the connections, the constant superimposition of layers, the confrontation between the film frame and the surface of the image that seem not to belong together, though they are linked by titles and a dense, breathtaking network of sound and music that nevertheless clashes with the visual expression and plunges us over and over again into fantastic and intellectual adventures—a syntactical complexity and turbulence for which the optical and acoustic punctuations and conjunctions are the decisive bearers of meaning, the "and" more important than the elements that it binds together.

This acoustic and visual practice of coupling can hardly be described in words. Gilles Deleuze tried once, fourteen years ago, in a commentary on Godard's video *Six fois deux*:

> Neither a component nor a collection, what is this AND? I think Godard's force lies in living and thinking and presenting this AND in a very novel way, and in making it work actively. AND is neither one thing nor the other, it's always in between, between two things; it's the borderline, there's always a border, a line of flight or flow, only we don't see it, because it's the least perceptible of things. And yet it's along this line of flight that things come to pass, becomings evolve, revolutions take shape.[13]

In the *Histoire(s)*, the instruments of a developed video and sound technology are at Godard's disposal for the aesthetic structuring of this practice of conjunction. (In addition to a respectable budget from the cable channel CanalPlus and his collaborating partners.) Soft, protracted dissolves in which the individual layers of image are gradually superimposed or superseded, collages, color mutations, images in staccato rhythm, play with masks and die-cutting. Even when played back by a simple home video machine, the multiply superimposed traces of sound do not become background noise. Russian and Italian revolutionary songs, swing and concerto music, opera and chorus line, Leonard Cohen and Ton Steine Scherben—songs recognizable from recent music history are interwoven with quotes from French or German newsreels, laid over or under the insistent, monotonous voice of the director, interspersed with scraps of film dialogue, excerpts of poetry, works from literary history, business proceedings. Differentiable, audible with and against the image. Film/cinema and its history can finally be experienced as what they would like to be: points of intersection in which divergent discursive praxes and expressive forms encounter one another, and away from which they then drift, to meet again in other places, at other times.

Digitally based image technologies will have even more to do with the art of connection than those grounded in analog, more to do with the creative, for instance, mathematical-logical production of relations in which we are still so poorly versed, so far as producers as well as viewers and critics are concerned. The immobile and touchable element still dominates in our aesthetic understanding, as opposed to relations in perpetual motion.

VI.

The *dispositive* function of the beautiful semblance is based on the separation of aesthetic production and reception, which, in the context of art, means the production of visual art that has no equivalent in the production of a new art of viewing, which also implies that production and reception operate with separate materials, with separate techniques.

In this sense, Godard's videotape is a simulation: it presents a model for how the arts of cinematic production and reception can correspond with one another, linked in a network in which, at both nodal points of the aesthetic process, the same basic machinic and mental technologies are equally available.

Therefore also a model for a future, enhanced, interactive computer aesthetic, something along the lines of Gene Youngblood's concepts of the "renaissance amateur" and the "metadesigner"?[14] Even if, as a consequence, we approach ever nearer the utopia of a boundary-crossing between art and life? Perhaps. But I think that, even with conventional audiovisual technologies, we are still very much at the beginning of realizing a new aesthetic conception of this kind, the archaic depths of which we need to sound in order that our panicked algorithms don't leave us grasping at nothing.

"Cogito ergo video": The first part of the *Histoire(s) du cinéma* ends with this inscription, white on a black ground, like an image. Shortly before, the film returned to the point at which it began, with the eyes of Fellini's Giulietta Masina superimposed with the face of Edmund from Rossellini's *Germany,*

FIGURE 8.8. "I think, therefore I see." From Godard, *Histoire(s) du cinéma.*

Year Zero, along with Godard's commentary: "Why trust my eyes when I look... why not first examine my eyes?"

At the beginning of parts 2A and 2B of Godard's archaeology of cinema masterpiece, he writes the title with a fat, noxiously squeaky felt marker on white cardboard bearing the name of the production company. The first sentence of the film provides a caption: "To give a precise description of that which has never taken place is the work of the historian."

Translated by Lauren K. Wolfe

Notes

The title is a quote from Jean-Luc Godard's 1968 film *Le Gai Savoir (The Gay Science)*. The parts of this text that are drawn directly from Godard's videotape are the result, among other things, of in-depth discussions with Nils Röller around the year 1990. His ideas are present here.

1. Cf. Klaus Bartel's article regarding several of these "factors of socialization" in the context of recent French thought: "Kybernetik als Metapher: Der Beitrag des Französischen Strukturalismus zu einer Philosophie der Information und der Massenmedien," in *Kultur: Bestimmungen im 20. Jahrhundert*, ed. Helmut Brackert and Fritz Wefelmeyer, 441–74 (Frankfurt am Main, Germany: Suhrkamp, 1990).
2. This too is a prominent albeit deceptive signpost in the present labyrinth of concepts and media. To introduce it alongside the expanded artifacts and technical objects of computation means at the same time to deny the basically interdependent character of conventional media processes. Technically mediated communication has always been interactive. With a view to computer-based audiovisual technologies and new qualities of exchange, I therefore prefer to speak of "expanded interactivity" at the interface of media machines and media people.
3. Gene Youngblood, *Expanded Cinema* (London: Studio Vista, 1970).
4. Sergej M. Eisenstein, "Der Kinematograph der Begriffe," in *Der Querschnitt: Aus den Heften 1930 bis 1933* (repr., Frankfurt, 1981), 369.
5. In 1988, Peter Weibel made an exhibition for the Austrian Museum for Applied Arts in Vienna, in which he had produced all artworks, wrote all texts for the catalog, and curated the exhibition. See Peter Weibel, *Inszenierte Kunstgeschichte—Mise-en-Scène of Art History: With an essay by Jean Baudrillard* (Vienna: MAK, 1991).
6. The *Histoire(s) du cinéma* were commissioned by the French subscription television channel CanalPlus. They were planned as a series of at least eight parts, each of different lengths but all under an hour. I am referring only to the first two parts here (those completed by 1990).
7. Cf. Maurice Merleau-Ponty, "Das Auge und der Geist," in *Philosophische Essays*, 13–43 (Reinbek bei Hamburg, Germany: Rowohlt, 1984), 19.
8. Jean-Luc Godard, *Einführung in eine wahre Geschichte des Kinos* (Munich, Germany: Hanser 1981), 43.
9. Godard, *Einführung in eine wahre Geschichte des Kinos*, 16.
10. Citations where sources are not otherwise provided are from Godard's recorded commentary.
11. This line of thinking took solid shape in large part because of conversations with the philosopher Dietmar Kamper (1936–2001), during a conference at Städelschule, Frankfurt in 1990, to which this text dates back.

12 Nick Herbert, "Nur Werner allein hat die nackte Realität gesehen: Vorschlag für eine wirklich 'Neue Physik,'" in *Ars Electronica 1990*, vol. 2, *Virtuelle Welten*, ed. Hattinger et al. (Linz, Austria: Ars Electronica, 1990), 42.
13 *Cahiers du Cinéma*, no. 271 (November 1976). English version in Gilles Deleuze, *Negotiations 1972–1990* (New York: Columbia University Pres, 1995).
14 Cf. Gene Youngblood, "Metadesign: Die neue Allianz und die Avant-Garde," in *Ästhetik des Immateriellen?*, vol. 2, *Das Verhältnis von Kunst und neuen Technologien*, ed. Florian Rötzer, 77–84 (Cologne, Germany: Kunstforum International, 1989).

II
Particular Archaeologies

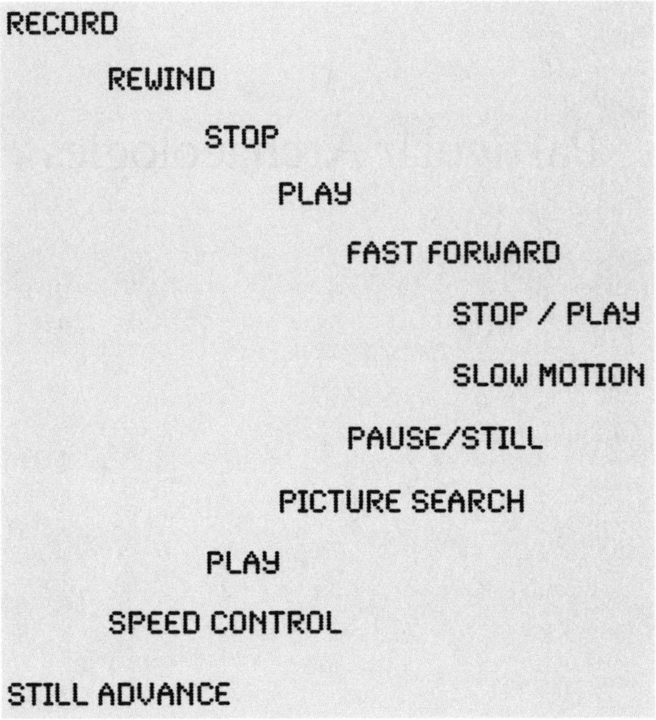

```
RECORD
        REWIND
                STOP
                    PLAY
                        FAST FORWARD
                            STOP / PLAY
                            SLOW MOTION
                        PAUSE/STILL
                    PICTURE SEARCH
            PLAY
        SPEED CONTROL
STILL ADVANCE
```

FIGURE 9.1. Schema: S. Zielinski.

9 The Audiovisual Time Machine

Concluding Theses on the Cultural Technique of the Video Recorder

While teaching in the early 1980s at the Institut für Sprache im technischen Zeitalter (Institute for Language in the Technological Age), founded in 1961, my media-archaeological curiosity was initially focused on a single artifact within a single, concrete, media-technological system—the electromagnetic recording of images (and sounds) on videotape. This was a challenge, both concretely and methodologically. The field of media archaeology did not yet explicitly exist. What I analyzed here for the first time as a particular cultural technique was an important instrument for the later development of an expanded hermeneutics in the concrete form of a precise philology of exact things—such as technical artifacts are. The following text is the concluding chapter of a comprehensive study On the History of the Videorecorder *(*Zur Geschichte des Videorecorders, *1985), which was republished in 2010 in unrevised form.*

⁂

> We experience ourselves as acting and thinking beings for whom the consciousness of time encompasses all that humans are and can become. Time, therefore, is not just the concept of a given content, it is the apprehension of reality in the tension between temporality and eternity. Time is the mode of consciousness that communicates and experiences itself as something different.
> —WILHELM DUPRÉ, *Time* (1974)

> There is no rewind button on the Betamax of Life.
> —NAM JUNE PAIK

1.

Television is in many essential respects a time medium. In both its natural and human dimension—that is, related to matters of individual perception—it is involved in the interplay of spatial and temporal coordinates, of which the latter are the more fundamental.[1] The production of electronic images, their transmission and recomposition in the receiving television apparatus, is a process that occurs at extremely high speed. At the receiving end, viewers experience it as a movement whose visual rhythms are identical to those they perceive in the outside world.

The principal technological function of the video recorder is to store television signals by converting their frequencies into electromagnetic impulses, which are then inscribed by means of one or several magnetic heads onto a magnetic tape, read off for purposes of reproduction, and transmitted to the receiving apparatus in the shape of frequencies. If the goal is to produce visual signs that are (virtually) identical to the original images captured by the television camera, these reading and writing operations have to take place at very high speeds, which is the reason for the broad frequency range of television signals.

The first practical proposals to solve this problem were based on analog reasoning: video recordings were to operate according to the same principle as tape recordings, which have a fixed magnetic head read the tape along its direction of movement. Problems arising from the broad frequency of the television signal were to be handled, first, by narrowing the gap between the magnetic heads and, second, by *drastically* increasing tape speed. But because the narrowing of the gap ran into mechanical limitations and, more importantly, because tape velocity has obvious spatioeconomic implications—the higher it is, the more material is used up—this particular solution remained impractical for a long time. Models from the early 1950s sported reels the size of wagon wheels but only offered a few minutes of recording capacity.

The development of cross-track recording and later of helical scan technologies fundamentally altered the solutions based on analog modeling. By having the heads first rotate at a right angle to the direction of the tape and then in the same direction, tape usage was reduced by meters, if not kilometers. What was decisive for the quality of the visual reproduction was no longer the speed of the tape but its relative speed (scanning speed minus tape speed) as well as the breadth and arrangement of the magnetic head gap and the surface structure of the tape material. Changing the temporal coordinates of the head–tape relationship was the germ cell for the success of the video recorder. Historically speaking, it had an additional economic effect, given that the degree of tape usage is not only a mechanical and practical problem. In the long run, it also determines the equipment's economic feasibility for the user.

Technologically, the video recorder is a time machine in a double sense. It stores the flow of television signals and makes it repeatable by allowing users to alter its temporal structure. The fact that this can be achieved, at least superficially, in a satisfactory way is based, in principle, on the ability to intervene in the space–time structure of the technical arrangement of the magnetic recording.

2.

The video recorder is a time machine in a double sense, with regard both to its original conception and to the use-values that were projected onto it. It was intentionally developed as a device capable of manipulating the temporal process known as television communication. There is, therefore, a specific overlap between the technological structure of the artifact and its primary cultural function. With this in mind, the start-up of Ampex's quadruplex installations as a time shift machine for the North American radio stations amounted to the origin of a modified cultural technique of the ways in which television was transmitted and perceived. It enabled organizing the whole of the rigidly structured daily flow of audiovisual messages in such a way as to synchronize it with the equally rigid and wholly organized daily life of the viewing audience. With the newfound ability to broadcast programs in three or four different time zones across the United States, it became possible to create an identical, television-based temporal experience for all the inhabitants of a giant subcontinent.

In due course, further time-manipulation functions were developed that served to expand and perfect the ongoing correlation between technological and cultural genesis. The transfer of complete homogenous program sequences—in view of today's computer-based editing capacities, it seems almost inconceivable that thirty years ago, the parceling and reconstitution of tapes was a practical impossibility—was followed by the adaptation of VCR-based film production techniques. Audiovisual reproductions of television broadcasts no longer needed to be undertaken in line with the chronology of the actual or staged event. It was possible to record and process ever finer set pieces in virtually any arbitrary order for any broadcast. This included the concentration of ads and video clips, whose autonomous segments, cinematographically speaking, were so foreshortened they were hardly noticeable as filmic moments but instead appeared as photo- or rather computer-graphical elements of a rhythmic patchwork. Given that recorded material had become so reusable, established film production practices were superseded not only in a purely economic sense. The video recorder facilitated synthesization features typical of the electronic direct broadcasts, such as constantly changing camera perspective and distances as well as the specific techno-aesthetic materiality of the television image, with their storage.

The improvement and modification of the tape material and the increasing erasure of imperfections—the so-called drop-outs—that arose in the process of reproduction, in combination with the mechanical and electronic refinement of the equipment, led to the creation of canned content that, at least superficially, can no longer be distinguished from a live broadcast. It enhanced the illusionary potential of television broadcasts as reproductions of reality. Events that had been reproduced twice (by the camera and then by the magnetic tape) were now immediately transmitted to the viewer, especially with regard to the difference between the time of the event and that of the broadcast or reception. Once it was no longer necessary to develop the canned material, the processing the raw materials into programs ready for broadcast was greatly accelerated; subsequently, the gap between these two points in time was minimized.

One of the results of this process was the introduction of a new perceptual dimension into television communication. It had already been realized, albeit in rudimentary fashion, by the Ampex Instant Replay machine HS-100,[2] which allowed sportscasters to make use of new techniques of deceleration and slow-motion dilation immediately following the recording process. In the case of spectacular occurrences—such as catastrophes deliberately or accidentally witnessed by television—this enabled the constant repetition of recorded events within only a couple of hours. But another consequence may have been of even greater significance for the way in which television presented and gave meaning to reality. The video recorder not only made it possible for the broadcast time to almost coincide with the event in question but it also allowed for the previewing of the recorded material to take place without any significant temporal delay. News reports turned into artifacts that were now subject, as it were, to the controlled immediacy of the medium. In addition, techniques of electromagnetic recording and immediate replay facilitated the addition of audience effects. Spontaneity itself—including applause—could now be regulated.[3]

3.

If we apply the potential for social impact as a criterion, today's experimental video avant-garde appears to be further removed from its programmatic claim of the 1960s, that "the new medium can only be used in a meaningful way if it activates its users rather than burying them underneath a flood of images,"[4] than it was in its early phase. Nonetheless, video art without a doubt witnessed the most intensive engagements with the distinct features of magnetic image recording. This connects art to the special practices of observation and control to be found in scientific investigations, the military, and police surveillance. The decisive difference is that in video art, reproduction is not limited to the electronically facilitated doubling of visual events, but both levels—visible reality and its storage—are juxtaposed and confronted in mutually interpretatory ways.

"Time structuring" is more than a necessary "point of departure"[5] for the arrangement of motion in film. It is, as it were, the externalized essence of video-artistic work. Following the tradition of the Fluxus movement, Nam June Paik likes to present himself as an agent provocateur whose dedicated signature engagement with the time–space relationship can be traced back to the distinction between painting/sculpture, on one hand, and poetry, on the other, established in Lessing's *Laocoon*.[6] In his own words, "video art imitates nature, not its appearance or substance but its inner temporal construction... that is, the process of aging (a certain type of irreversibility)."[7] Many video installations, sculptures, tapes, and actions coupled with performances can be seen as attempts to realize this program—works by Paik himself, or by Douglas Davies, Dan Graham, Friederike Pezold, Ulrike Rosenbach, Bill Viola, and others.

To focus on one particularly revealing example by the New York artist Viola: in *Reasons for Knocking at an Empty House* (1983),[8] the viewer is confronted with the reproduction, presented in austere black-and-white aesthetics, of a literally deranged situation. An anonymous male person is confined to an empty room furnished with no more than chair and a table, completely isolated from his surroundings. A fixed, unchanging camera perspective captures the man's increasingly desperate attempts to adjust to this extreme spatiotemporal arrangement. The passage of time is only indicated by the changing light outside the window and inside on the interior landscape of the enclosed room. Although the latter never changes for the viewer, it gains a dynamic, increasingly threatening character. What is so remarkable about this attempt at an electronic depiction of private isolation and its consequences for the individual concerned is the fact although the material of three days and nights has been cut down to nineteen minutes, there is an extreme discrepancy between the viewing time of the audience and its subjective temporal experience. The edited time manages to convey something of the real duration of the recorded event.

The reason why such experiments have had such a limited cultural impact in Germany has less to do with their idiosyncratic aesthetic construction than with the fact that German television, following a short—and comparatively euphoric—acknowledgment of video art products and attempts to make them part of everyday broadcasts,[9] soon closed itself off against such projects. As a result, the latter's public exposure is limited to the relatively enclosed domains of galleries, museums, and festivals, where artists and communicators stay among themselves.

4.

U.S. American television exists on the basis of selling airtime to the consumer goods industry or to corporations offering commercial and ideological services. These clients, in turn, assume that their messages, which after all are designed to stimulate circulation, will be received by the viewers and at least in part give rise to acts of purchasing. For many years, there was no reason for either broadcasters or advertisers to doubt the effectiveness of this mutual connection. In regular intervals, ads were compressed into commercial breaks lasting a few minutes—with market leaders focusing on the more attractive programs—and a well-honed, incessantly tested and modified arrangement of visual and acoustic stimuli guaranteed the audience's attention. Detailed studies of viewing habits, an integral part of commodity-based television, tended to confirm the audience's ongoing loyalty to electronic messages of consumption.

The first cracks in this harmonious relationship appeared when pay TV programmers started to offer ad-free shows, especially in the shape of feature films. In the course of late 1970s and early 1980s, a growing number of viewers paid for the privilege of enjoying uninterrupted programs, though initially it was no more than a supplement to the regular programming interrupted by ads.[10] The industry, however, was further troubled when already in the early years of mass-market VCR usage, additional changes became apparent. "Video Recorders Change Daily Habits"[11] was the headline of a late 1970s *New York Times* article reporting on sample surveys indicating that for viewers owning VCR units, "prime time" was declining in importance. Hugh Beville, a director of the Broadcast Ratings Council, expressed in greater clarity the imminent threat to networks: "Will home cassette players affect ratings?"[12] he asked, and set out to describe the numerous possibilities and opportunities video recorders offered to the viewers eager to emancipate themselves from the "temporal tyranny" imposed by broadcasters. His contribution was above all a plea to the latter to take into account changed viewing habits when conducting audience surveys.

In May 1980, A. C. Nielsen, the market leader in audience research, published its first systematic survey. The result left the trade press in no doubt, as is already obvious in the *Backstage* headline "VCR Affects TV."[13] Apart from the fact that there was a considerable overlap between VCR owners and pay TV subscribers, and that the replay of industrially prefabricated video was of no more than marginal importance, the Nielsen study highlighted the ubiquitous practice of "timeshifting." Three-quarters of the households surveyed had recorded programs using the automatic timer; 67 percent had watched the recordings in the early evening, that is, during prime time, normally two or three hours after the original broadcast.[14] Further survey results released by Nielsen two years later seemed to turn the television world on its head. The *Washington Press* summarized the results: "Millions of Americans are eagerly taping shows and either wiping out or skipping over increasingly high-priced ads—a practice that is raising fundamental questions in the minds of advertising executives."[15] The study reported that more than a third of the viewers surveyed stated they deleted the ads in 75 percent of taped programs[16] to enjoy uninterrupted shows. The hardware industry had delivered the necessary easy-to-handle user features. For instance, VCR remotes were equipped with a "wipe" or pause button that could be activated at the beginning of the commercial breaks and released at the end.[17]

Leaving aside the ad-infested programs of future private broadcasters, German television has no reason to fear this behavioral change. No ads are (yet) featured during the most popular broadcast times, and in pre-prime-time programs, ads appear in blocks in between different programs. Nonetheless, the American example hints at something that affects public television's communication process at a much deeper level. We are dealing with basic interventions, enabled on the side of the viewer by the video recorder, into the specific features of the medium itself.

5.

"In all developed broadcasting systems the characteristic organization, and therefore the characteristic experience, is one of sequence or flow. This phenomenon, of planned flow, is then perhaps the defining characteristic of broadcasting, simultaneously as a technology and as a cultural force."[18] Here Raymond Williams is emphasizing that the distinguishing feature of radio (and therefore also of television), as opposed to other medial presentations, is the fact that the communicative process does not emerge in autonomous program segments but in the constant sequence of diverse offers, which viewers experience as a flow or a particular part thereof. Stephen Heath and Gillian Skirrow added a further feature by stressing that the flow of television is structured by specific regularities. The emphasis, however, is less on the serial character of many programs than on the constant recurrence of carriers of meaning between and during the individual programs. "Flow and regularity"[19]—if we accept this double determination to be the semiotic basis of what turns television into a "signifying practice"[20]—entails a number of important modifications concerning the real or potential use of video recorders, under the condition, of course, that its technological use-value is fully developed."[21]

1. To use the metaphor of "flow" when describing the communicative practices of television not only indicates the constant movement and fixed unilinear orientation of the communicative offers involved but also stands for their extreme ephemerality. Traditionally, the viewer is exposed to all three dimensions of the program flow with no possibility of intervention—excluding, of course, the possibility of interruption by turning off the TV set. Television reception, then, is the perceptual integration of the viewer into a sequence arranged and channeled by broadcasters. The basic function of the video recorder in the hands of the viewer is to interrupt this constant flow by means of recording one or several sequences, structuring the ways in which they can be *re*-viewed and replacing their ephemerality with temporary persistence or enduring availability. Without recourse to any other modifying technologies, the simple recording of program segments and their delayed playback serves to erode important elements of the aura of the traditional process of television. Video recorder–based reception turns

Table 9.1. Schematic comparison

Traditional television reception	Television reception by video recorder
(i) *On the level of the program as a whole, the program is experienced as...*	
an externally controlled flow with a fixed determination of the temporal sequence with continuous regularity.	an accumulation of set pieces whose temporal sequence can be manipulated/determined by the user or organized in discontinuous fashion.
Reception is a regionally, nationally, or globally collective experience.	Reception occurs on a solitary or individual basis.
It has a centralizing character.	It has a decentralizing character.
Live broadcasts synchronize event, time, and lifetime.	Reception-time, event-time, and broadcast-time are in all cases asynchronous.
Television and radio possess an aura of immediacy.	Video recorders foreground the character of technological mediation.
The program is ephemeral.	The program is always available.
(ii) *On the level of the individual program sequence,...*	
the sequence of segments is fixed (linear). Any interruption of the reception necessarily entails communicative gaps.	the sequence of segments can be arranged arbitrarily. Sequences may be stopped at any point without resulting in perceptual gaps in the reception.
The filmic rhythm is predetermined.	The filmic rhythm can be altered.
Image messages cannot be controlled post facto.	Image messages can be controlled at any point.
The temporal extent (broadcast duration) of individual segments as well of the entire sequence is fixed.	The temporal extent (broadcast duration) of individual segments and of entire sequences can be extended or stretched.

the flow of offers into deliberately targeted set pieces that offer completely different options for use. What in the absence of any recording facilities can at best only be recalled in memory may now be repeatedly enjoyed and scrutinized. Equipped with audiovisual magnetic tape machines, television approximates the act of reading, though without forfeiting its analog characteristics on a semiotic level. In the eyes of many cultural critics, one of the signified disadvantages of television and radio (when compared to reading) is that one was not able to leaf through the broadcast text. The technology precluded possibilities of repeating meaningful contexts; instead, it subjected viewers to the constant—and constantly changing—signifying practice of communication. These objections are now, to say the least, relativized.

On the other hand, new qualities of reception can impact the sequential positioning of electronic images. The repeatability of sequences now offered to users provides manufacturers with the opportunity to design more complex or complicated contexts of meaning that no longer need to be decoded at first viewing. Likewise, it encourages greater care and rigor in the design of audiovisual realities.

The very notion of "time-delayed television" alludes to the traditional temporal structure that underlies traditional forms of television-based communication, which was elaborated by the broadcaster in feedback with the quotidian social organization of the viewing audience. In the age of the video recorder, however, this is no longer a binding notion. Viewers are now able to watch taped programs whenever they want—specifically, at one point in time never intended by the broadcaster. This serves to demote the importance of prime-time programming. In addition, viewers may alter sequences and proportions according to their own wishes. Regular flows can be reorganized; intros recorded on a magnetic tape can be turned into end credits; whatever happens to precede the evening feature film could instead be used as a late-night soporific; two or three episodes of *Dallas* or *Dynasty* may be viewed directly following each other as one block, which may render viewers more sensitive to the series' monotonous dramaturgy. By the same token, the compressed reception of several parts of a historical miniseries, such as *The Germans in WWII*,[22] may help viewers better understand how its structure tends to resemble that of a popular nonfiction book more than that of a film event. Of course, these new patterns of reception do not put an end to the experience of a heteronomously determined temporal program flow, on which they in final analysis continue to depend. They may, however, contribute to a more disintegrating form of television communication.

One of the medium's greatest impacts—and one very typical for radio and television—is its power "to invest an occasion with the character of corporate participation."[23] Indeed, television history could be narrated in point-by-point fashion as a sequence of program highlights. The Nazi Olympics of 1936 would be an early part of it, as would be all subsequent Olympic Games and soccer World Cups; the coronation of Elizabeth II on June 2, 1953, the broadcast of which marks the beginning of the Western European TV union;[24] the premiere of the TV satellite Telestar on July 10, 1962; the first Moon landing on July 16, 1969, which was broadcast across several continents,[25] followed by all the other important events in the history of manned space missions; decisive high-profile heavyweight boxing matches; national and international political events that turned into media sensations; and so on. Even—and especially—in the age of cinematographically or electromagnetically canned content, such live broadcasts significantly contribute to the aura of the medium as a key player of collective organization. The delayed reception of a program that originated as a live broadcast suspends the temporal synchronicity of event and televised mediation, thus fundamentally modifying the medium-specific form of the mass audience. In the act of reception, the assembled synchronized community turns into an accumulation of individualized viewers.

Following John Howkins, we may well ask whether this qualitative shift does not in fact make a mockery of our traditional notion of audience, which derives from Aristotelian notions of the unity of time, place, and action, and can thus only partly be applied to television audiences. Matters change even more once we introduce the solitary reception of industrially prefabricated videos:

> The popularity of home video is undercutting the very idea of a TV audience. If a person tapes a program off-air, is he part of an audience? If he rents the same program on cassette, is he part of an audience? If he watched the tape with friends or by himself, is he more or less of an audience? ... If he watches a rented cassette and refers at the same time to the book of the cassette, is he an audience? Where do you draw the line?[26]

Suspending the temporal synchronicity between broadcast and reception is not only of significance for "live" TV events, such as talk shows, studio debates, concerts, quiz shows, or any program featuring direct audience participation. The many connections established between daily routines are not least established by short segments that are directly targeted at the time of day or reception situation and contain important elements of communicative integration. These elements—ranging from addresses like "good evening" to wishing viewers a pleasant Saturday afternoon or an enjoyable weekend to nightly recommendations—are of no relevance to video users, whose attention is focused exclusively on the canned feature film. They seem slightly absurd on other weekdays or at a different time of day. Of course, in the future, the exact timing of the taping of a program will offer viewers the possibility to eliminate unnecessary segways and introductions.[27]

Together with the new possibilities of segmenting, individualizing, and relativizing the aura of television as an event medium, its disposability and restructuring are the most important forms of intervention the audiovisual time machine offers to viewers (especially in relation to the entire program flow). The related "demythologizing"[28] of traditional radio and television, however, appears to be especially pertinent on the second level, that of individually recorded programs.

2. From a culture-technical point of view, it is a matter of grasping what hardware manufacturers call "features," that is, equipping gadgets with special user functions whose use-value extends beyond recording and replay. It is not only that magnetophones have been perfected over the last couple of years; their partial integration into remote control units tends to provoke, as it were, their increased use and enables viewers to make use of playback features while comfortably resting on their couch.[29]

What was already of significance on the more general level of programming applies to an even greater extent to individual segments: the producer's original rhythm can be completely altered, the flow of sequences brought to a stop in a single image. Whether this is done playfully or for more specific and substantial reasons is, initially, of little concern. With regard to questions of reception, the very act of pressing the pause tab most new recorders are now equipped with represents the exact opposite of the ongoing image flow. It temporarily suspends the basic signifying practice of television. Owners of fully equipped devices can arrest individual images in virtually photographic fashion or rapidly accelerate their flow to search of another. But there are more possibilities of changing the rhythm. The frame-by-frame mode allows users to virtually dissect all sequences. Slow motion enables us to extend and study scenes, gestures, or object movements, but even the simple acts of using pausing or stopping fundamentally interfere with the specific properties of filmic communication. Once again, it is of little importance whether these interruptions came about because users needed time to analyze and discuss what they saw or because they had to attend to some other business. Accelerate fast-forward and rewind enable more selective viewing, for example, in the case of feature films, in which case the actual reception of broadcast and taped material is restricted to a few chosen sequences (e.g., action highlights). Dramaturgically, the resulting concentrates resemble the substandard film montages produced for the home cinema audience of the 1960s and 1970s.

As detailed and targeted as these possibilities of intervention may be,[30] they entail a modification of the way in which traditional television communicates meaning: "The notion that one single viewer can control moving pictures... is now stimulating a new approach to the grammar of television and film."[31] For journalists, artists, and media analysts, this privileged access to filmic material has long been part of their daily routines. They use time-manipulation techniques to optimize the (re-)production of image worlds or as an indispensable tool for analysis. But the fact that these feature are now also available to many who hitherto were subjected to the heteronomously organized flow of regularly scheduled programs has thus far only been noticed, understood, and exploited by a very small number of producers of commercial educational and hobby cassettes (the so-called how-to programs). The same applies to the managers and designers of ambitious audiovisual advanced learning programs in large corporations or to market communication specialists. The latter increasingly prefer the media technique of the videodisc, not only because it is both more efficient in terms of sound and image reproduction and more cost-efficient when it comes to replication but also because the growing storage possibilities for coded data enable a more targeted access to individual picture segments, that is, for time-controlled use of filmic material.[32]

6.

Part of the fascination the video recorder holds for consumers may reside in the new ownership structure over cinematic commodities. It is not only the fact that consumers are now in a position to integrate audiovisual segments of time in the shape of industrial software into a private sphere characterized by the consumption of goods and services. Nor is it the ability to modify fleeting experiences offered by cinema and television in such a way as to turn them into events that can be restaged on a daily basis. The video recorder addresses the long-standing—and in the course of cinematic history continuously exploited—desire to gain possession of freely accessible reproductions of adored starts or popular scenes by collecting the entire breadth of film production.[33] What hitherto could only be experienced in public, in the social space of the cinema, is now up for purchase; it assumes the shape of a tangible product, the video cassette, that can be integrated into any domestic setting.

But in terms of the video recorder's use-value, what is at the center of the cultural technique it enables is the possibility of temporal processing—

more precisely, the processing of already reified time. It is significant for the three essential usage forms: delayed television reception, electronic image production, and its ability to act as a continuous flow-heater, as it were, for industrially fabricated video cassettes.

"There is no individual 'Now' anymore that unambiguously refers to 'Before' and 'After.' The subject no longer situates itself at a certain point in time; it only experiences duration. Temporal consciousness of time dissolves in the river of time like an oil spill on water."[34] This is how Heinrich Popitz et al. describe in *Technology and Industrial Work* the way in which the increasingly monotonous labor process determines how those caught up in it experience time. Keeping this causal connection between organization of work time and leisure time, and further taking into account that the flow and regularity of broadcast programming tend to extend the experience of labor-regulated time into the remainder of the day, it becomes possible to explain the fascination exerted by the audiovisual machine in the following way: with the aid of the video recorder, users are able to bring about a temporal experience that they have lost at work—or rather, that they never possessed in the first place. Users reproduce temporal experiences—especially those related to leisure and vacation time—which they can then manipulate, for instance, by extension or simple repetition. They organize cinematographically prestructured time (in the shape of TV programs or feature films) according to their own desires. They enjoy the experience of retrieving, saving, extending, condensing, or doubling television time, for instance, by witnessing a live performance as they are also taping a program on another channel and watching them parallel—while the recorder's digital clock, in turn, merely marks the passing of linear time.

To be sure, these engagements with formed time are qualitatively not identical with the full unfolding of our temporal dimension in all its rich subjective expressive qualities, including "condensed time and loss of time; waiting time and event time; collective experiential time and boredom; time directed toward the future and time that processes the past,"[35] which Oskar Negt describes as constitutive for the production of a "temporal sovereignty as a conscious human activity."[36] After all, in this particular case, the possibilities of engagement are tied to an object, a film, or a television program that significantly contributed to shaping our experiential potential. Nonetheless, "emancipation and violence," the "emancipatory possibilities and dangers that come with the appropriation of such easily available technologies lie so closely together, that the criteria for evaluating them can only arise from the social context in which they happen to be embedded."[37] Within the sociocultural framework of traditional television communication, and against the background of the accustomed social distribution of apparatuses, we tend to evaluate the audiovisual time machine also as a useful artifact that may contribute to emancipatory processes. In the close interplay between work time and its remainder, the video recorder can be interpreted as a cultural technique that helps compensate deficits produced by the industrialization and technologization of daily life.

The loss of liveliness, flexibility, and self-determination in the (still) primary sphere of professional life as well as in the duty-filled secondary sphere fuels a growing desire to organize the remaining blocks of time according to one's own subjective ideas, and be it only within the narrow yet nonetheless socially significant context of television. Provided by the industry and ready for purchase, the audiovisual time machine represents a technological aid. For those who can afford its purchase, it extends the overall scope when playing with electronic reproductions of the image world. It is no contradiction but the other side of the ambivalent nature of these media technologies that the apparatus can also be used for nothing more than the extension of television-based experiences.

It is, no doubt, difficult to interpret the decentralized and discontinuous reception of television by means of the video recorder.[38] By means of turning what is designed with a collective reception in mind into an accumulation of singular and deranged acts, the artifact may disrupt or at least irritate the comprehensive daily integration imposed by television. However, with its focus on the domestic sphere, the intimate surroundings of one's own four walls, the usage of the artifact also works toward the increasing desocialization of lifeworlds, which Raymond Williams described in terms of a "mobile privatization."[39] The video recorder is an element in a system of technologically mediated communication, in which collective and social components are becoming increasingly anachronistic and for which we still lack an adequate description.

The images produced by Marey's photographic gun, Anschütz's quick viewer, Edison's kinetoscope, and, last but not least, the "living" late nineteenth-century images were the adequate medial expressions of the new relationship between humans and things.[40] Initially, the rhythm of the projection of perforated filmstrips was a source of communicative pleasure for those who produced social wealth working on assembly lines. Maybe the audiovisual time machine will one day be interpreted as the appropriate medial-apparative expression of a society whose integrated individuals have begun to treat mobility and privacy as the supreme virtues.

"Naturally, we all know what time is; it is the most familiar thing of all."[41]

"Si nemo ex me quaerat, scio; si quaerenti explicare velim, nescio."[42]

Translated by Geoffrey Winthrop-Young

Notes

1. From among the wealth of attempts to describe and define time as well as the relationship between space and time, which in the last couple of years have been complemented by numerous detailed analyses—e.g., see Christoph Asendorf's *Batteries of Life: On the History of Things and Their Perception in Modernity* (Berkeley: University of California Press, 1993) and Wolfgang Schivelbusch's *The Railway Journey: The Industrialization of Time and Space in the 19th Century* (New York: Berg, 1986)—we will only refer to two basic contributions, each of which presents very different concepts: Wilhelm Dupré, "Zeit," in *Handbuch philosophischer Grundbegriffe*, ed. Hermann Krings et al. (Munich, Germany, 1974), and Carl Friedrich von Weizsäcker, *Die Einheit der Natur*, 2nd ed. (Munich, Germany: DTV, 1982), esp. 143–46. On the definition of television as a time medium, see Friedrich Knilli, "Medium," in *Kritische Stichwörter zur Medienwissenschaft*, ed. Werner Faulstich (Munich: Fink, 1979), 237–38.
2. The device was an early variant of an analogue picture disc developed by Ampex in collaboration with ABC's research division and marketed in 1967. At the time, the disc was able to record thirty seconds of program. Segments of less than four seconds could be accessed, and reproduction was possible at any speed. Cf. the chapter on "Ampex History" in *Ampex: The Other Word for Innovation* (Redwood City, Calif.: Ampex, 1979).
3. Gerhard Eckert et al., *Knaurs Fernsehbuch* (Munich, Germany: Droemer-Knaur, 1961), provide a couple of contemporary examples illustrating how Ampex machines changed television's artistic production process; see esp. 319–28 on "Artist and Cans." Schmolke offers an impressive account rich in imagery: "With the help of technology we were able to outwit time. But

as we know, time had its revenge. Every second of a recorded program, be it composed of sounds or images, is at our fingertips. Technology enables precision programming, and the producers submit themselves to this perfection with professional pride. I recall an instructional film about the production and broadcasting of a German news program, in which this masochism dictated by seconds flipped over into a planned military victory over time. It was reminiscent of the mechanical heroics to be found in many films that associated the revolutions of rotary printing machines with the tank engines of the Africa Corps." M. Schmolke, "Rundfunktechnik in ihren Wechselbeziehungen zu Politik und Wirtschaft, Programm-Macher und Hörerschaft," in *Mitteilungen des Studienkreises Rundfunk und Geschichte*, no. 1 (1983): 29.

4 Bazon Brock, "Eine Zukunft dem Video? Fragt die alten Männer!," in *Videokunst in Deutschland 1963–1982*, ed. Wulf Herzogenrath (Stuttgart, Germany: Hatje, 1982), 126.

5 Wulf Herzogenrath, "Videokunst: Ein neues Medium—aber kein neuer Stil," in Herzogenrath, *Videokunst*, 15.

6 See Gotthold Ephraim Lessing, *Laocoon: An Essay on the Limits of Painting and Poetry* (Indianapolis, Ind.: Bobbs-Merrill, 1962).

7 Quoted in Herzogenrath, *Videokunst*, 23.

8 Viola was able to produce the videotape in the television laboratory of PBS station WNET/13 in New York. It was presented, among other venues, at the fourteenth Internationales Forum des Jungen Films in the Arsenal Kino on the occasion of the thirty-fourth Berlin Film Festival in 1984.

9 A high point was reached in June 1977 when two regional stations (WDR/Westdeutscher Rundfunk and HR/Hessischer Rundfunk) devoted nine evenings to the broadcast of videos presented at the Kassel Documenta, which in that particular year had put special emphasis on video art. Furthermore, see K. Wesker, "'Bitte schalten Sie um!' Aspekte zum Verhältnis Video und Kunst," in *Tele-Visionen, Medienzeiten: Beiträge zur Diskussion um die Zukunft der Kommunikation*, ed. S. Zielinski, 57–64 (Berlin: Express, 1983).

10 E.g., see I. Horn, "Fernsehverbreitung und Fernsehnutzung in den USA: Neuere Ergebnisse der US-Zuschauerforschung," *Media Perspektiven* 6 (June 1982): 400.

11 *New York Times*, March 29, 1979, C1 and C3.

12 Archive of the Television Information Office (TIO), New York, P.D. Cue, No. 3, 1979, 102.

13 *Backstage* 2 (February 1980): 79.

14 *Backstage* 2 (February 1980): 79.

15 *Washington Post*, May 27, 1984, F1.

16 *Washington Post*, May 27, 1984, F1.

17 Akai, for instance, offered this feature with its VHS Recorder VS-9800. Ads emphasized the use-value of the pause feature, but, grotesquely, they avoided mentioning that it was primarily targeted at commercials. "The remote control pause is a feature you'll use time and again. With it you can edit out any unwanted material while recording a program, or interrupt the action whenever you like during playback." Quoted in *What Video?* (February 1981): 6.

18 Raymond Williams, *Television: Technology and Cultural Form* (London: Fontana/Collins, 1974).

19 Stephen Heath and Gillian Skirrow, "Television: A World in Action?," *Screen* 18, no. 2 (1977): 15.

20 Heath and Skirrow, 8.

21 It must be emphasized that some of the following reflections, which should be read as a thesis, reach beyond empirically verifiable data. To my knowledge, thus far, only the Burda study of 1982 has collected quantitative data about the usage of VCR's operating functions. It reported an astonishingly high use of the time features (timers, pause function, speed-up, slow motion, etc.). See Burda-Marktforschung, *Daten zum Videomarkt* (Offenburg, Germany: Burda, 1982), 21.

22 A six-part TV series by Joachim Hess and Henric L. Wuermeling, broadcast by the ARD (First National German Television Program) in May 1985; used here as an example only.

23 Marshall McLuhan, *Understanding Media: The Extensions of Man* (New York: McGraw-Hill, 1964), 328.

24 The program was simultaneously broadcast in Great Britain, France, the Netherlands, and West Germany. Further on the development of "Eurovision," see Gerhard Eckert, *Das Fernsehen in den Ländern Westeuropas: Entwicklung und gegenwärtiger Stand* (Gütersloh, 1965), 66–74.

25 E.g., see Walter Bruch, *Kleine Geschichte des deutschen Fernsehens* (Berlin: Hauder & Spenersche, 1967), 89.

26 John Howkins, "Intermediary: Television Never Had an Audience," *InterMedia* 11, no. 4–5 (1983): 96.

27 This refers to the planned introduction of program-specific numerical codes that will be added to TV guides and thus enable the exact programming of the VCR even if the show in question should be postponed.

28 A. Kooyman uses the term in a generic sense to assess program changes triggered by the new technologies of TV production, distribution, and reception. See Kooyman, "The New Television Culture," *EBU-Review* 35 (January 1981): 34–35.

29 Quantitative media research uncovered the problem that the VCR's remote control unit also impacts normal program reception, which can now be consumed with the help of a playback channel. Brian Wenham, the BBC's

program director, describes the new situation VCR-equipped viewers find themselves in: "It was this last quite unforeseen piece of technology by-play that finally brought the Broadcasters' Audience Research Council (BARB) to its knees and forced it to admit that it could only guess what was going on." *InterMedia* (1983): 28.

30 For a more detailed discussion featuring several examples of the video recorder as an "intervention machine"—though without reference to the sociotechnical development of the device—see S. Zielinski, "Der Videorecorder als Eingreif-Maschine: Vorschläge zur besseren Verwendung des Apparats," in *Kabelhafte Perspektiven: Wer hat Angst vor neuen Medien?*, ed. Klaus Modick, 98–105 (Hamburg, Germany: Nautilus/Nemo, 1984).

31 J. Chittock, "Das britische Kino unter der Video-Bedrohung," *Media Perspektiven*, September 1983, 72.

32 See Rainer Bücken, "Bildplatte: Schulungsmedium der nahen Zukunft," *Congress und Seminar* 8 (1982): 20–23. The account features impressive examples for the degree of acceptance of the videodisc in U.S. American education and marketing, for instance, with General Motors and Ford. Further, see Franz Netta, "Die Bildplatte in Verkaufsförderung, Information und Schulung: Telesect 1000, ein neues System zur aktiven und selektiven Information," *AV-Branche* (1984): 9. Netta is managing director of Telemedia Ltd., with which Bertelsmann was able to secure its position as market leader in the videodisc business (together with another affiliate, the software producer Sonopress). Bertelsmann appears to be building on the very successful strategy of diversification it pursued in the 1970s. However, video hardware manufacturers are reacting to the competing videodisc technology. For instance, the Sony Corporation offers for variants of its U-Matic systems (Vo-5630) a set of add-ons that allow users to turn media technology into "interactive video." Programming and search functions facilitate the targeted search for program segments. A video responder is able to answer preprogrammed questions, and a printer allows for the sequential and temporal control of the learning process. See Sony Deutschland Ltd., Communications Products Group: Sony Communications Systems—Interactive video, Cologne.

33 See also Thomas David Boehm, "Pleasures and Treasures: Das Film(kunst)werk im Zeitalter seiner privaten Besitzbarkeit," in *Tele-Visionen—Medienzeiten*, ed. S. Zielinski, 29–34 (Berlin: Edition Express, 1983).

34 Quoted in Ute Volmerg, *Identität und Arbeitserfahrung: Eine theoretische Konzeption zu einer Sozialpsychologie der Arbeit* (Frankfurt am Main, Germany: Suhrkamp, 1978), 104. In a subchapter of her comprehensive dissertation, Volmerg addresses the destruction of the "consciousness of time and space" (102–5). I am here pursuing the analogy that the observed phenomena are not only of relevance for the understanding of classic industrial

labor processes but to a large degree also apply to other professional activities whose rhythms are determined by technological devices and infrastructures.
35 Oskar Negt, *Lebendige Arbeit, enteignete Zeit: Politische und kulturelle Dimensionen des Kampfes um die Arbeitszeit* (Frankfurt am Main, Germany: Campus, 1984), 207.
36 Negt.
37 Negt, 227 and 251.
38 This is a conscious use of the terms Kinder employed to describe the practices of MTV, which primarily broadcasts commercial blocks in the shape of video clips. See M. Kinder, "Music Video and the Spectator: Television, Ideology and Dream," *Film Quarterly* 38, no. 1 (1984): 2–15. This type of program offering can be seen as a planned response to changed reception practices and desires, quite independent of their specific musical content. The offering consists of a collection of short rhythmical sequences that can be arranged in any arbitrary order and for which the synchronicity of event time, broadcast time, and reception time is irrelevant. In addition, it can be easily used in connection with mobile reception devices like the Walkman.
39 See Raymond Williams, *Television: Technology and Cultural Form* (London: Fontana, 1974), 260ff.
40 For the silent movie, see the brilliant analysis in Asendorf, *Batteries of Life*.
41 *On the Phenomenology of the Consciousness of Internal Time,* trans. John B. Brough (Dordrecht, Netherlands: Kluwer, 1991), 3.
42 Augustine, *Confessions* 11.14.17, is talking about time: "If nobody asks I know; if I wish to explain to someone who asks, I do not." With thanks to Nam June Paik.

10 War and Media

Marginalia of a Genealogy, in Legends and Images

In a rare collaboration with the Erich Maria Remarque Peace Center, the European Media Art Festival in Osnabrück (Germany) organized a turn-of-the-century exhibition as provocative as it was impressive: Bilderschlachten—2000 Jahre Nachrichten aus dem Krieg *(Battles of images: 2000 years of reports from the war). I wanted to support this project with a form of text with which I had already been experimenting in the context of other exhibitions: a dense montage of discursive texts and images could be used to excavate and expose meaningful fragments of a larger historical context. Thematically, I placed a great deal of value in the idea that critique does not exhaust itself in the realm of the visible but that acoustic media have an important role to play in the context of war.*

This essay was originally published in 2009.

※※※

The Basics: Body and Message

Couriers and messengers are the oldest known bearers of communications. In ancient Rome, as in every ancient empire, there existed well developed systems that put these cumbersome bodies to use. At the height of Roman power, there was the *cursus publicus,* a system of roads provided with a series of relay points that extended from African Carthage in the south to London in the west, Byzantium (Istanbul) and Antiochia in the east, and Vienna in the north. Along the *līmes,* the Roman army commanded a densely packed chain of outposts that they used for telegraphy by torch relay (Figure 10.1).[1] Persia and China also had similarly expansive and well-elaborated territorial networks. Between Jerusalem and Fallujah (in present-day Iraq), not only were new Moon signals for the Jewish calendar sent by torch-telegraph but so were military communiqués in times of armed conflict.

A functioning communications network was in effect a time machine. It enabled early warnings for impending military attacks and informed local populations of catastrophes and unrest in distant lands, though it also allowed for the transmission of glad tidings, like news of births and weddings.

FIGURE 10.1. Watchtowers for optical telegraphy by means of torch along the open streets of the Roman Empire. From Volker Aschoff, *Geschichte der Nachrichtentechnik* (Berlin: Springer, 1984), 39.

It linked remote military and administrative outposts with the hegemonic centers of power and, in so doing, secured the survival of their regents.

The benefit of the courier system, however, was also a drawback: with a courier system, messages need not be encrypted. The sufficient conditions for a successful act of communication were simply that the message had a portable form and that its chosen code could be understood at both the sending and the receiving ends. The main drawback, depending on the intended purpose, was that the speed of transmission was contingent upon fatigue in the bodies of slaves, messengers, horses, or pigeons. Were the bearer of the message able to read, understand, and speak, then this bearer could potentially also become a conspirator.[2]

Descriptions of systems of transmission that decoupled the message from the body of the messenger and transferred it to a technical means (paper, textiles, metal, optical, or acoustic waves) appeared rather early on. With a variety of forms of signals to use as proxies for the written or recorded state of affairs, attempts were made to send messages more swiftly than the embodied messenger could move, at the speed of light, say, or the speed of sound. When Stentor speaks with an "iron voice" in Homer's *Iliad*, this is often read as indicating that by circa 700 B.C.E., metal voice amplifiers like the megaphone were already in use. In Archimedes' history of technology, there are references to so-called heliographs, which send flashes of light across great distances by using mirrors to reflect it; these are also mentioned in Xenophon's *Hellenica*, which dates as far back as the fourth century B.C.E.[3] Following the Warring States Period (476–221 B.C.E.), China's expansive transportation network evolved a proper territorial telecommunications system. During the Han Dynasty (from 206 B.C.E. through 220 C.E.), stations were posted every fifteen kilometers throughout the entire territory. This particular interval of distance was chosen in order that a diversity of physical signals—whether by carrier, flag, drum, or smoke and fire—could be communicated quickly, broadly, and unequivocally. The entire Chinese empire, under the Tang Dynasty, was thoroughly saturated with stations like these, which operated like motels: a person could dine there and stay overnight or switch out his carrier animals, and at each, there was an official responsible for the traffic in communications.[4]

With each of these methods, however, only very brief and succinct messages could be relayed. These were not yet the systems of writing into the distance (telegraphy) that later would enable the transmission of limitlessly complex information. Before such a system could be constructed, two problems needed to be addressed: messages would have to be encoded and the sender and receiver would have to be synchronized. In the second century B.C.E., Polybius came up with a sophisticated and practical semiotic proposal that messages be transmitted successively, letter by letter, by means of torch signals, whereby each discrete position of the torch corresponded to one of the twenty-four letters of the Greek alphabet (Figure 10.2). In order that the signals could be clearly identified, Polybius suggested that televiewing tubes in the form of long diopters be used, though obviously these were not fitted with lenses. This was telematics in a pure form. In his proposal, Polybius refers to Aeneas Tacticus, to whom he attributes the fourth-century B.C.E. design for a hydraulic semaphore system that consisted, on both the sender and receiver sides, of identical vertically erected water containers that were each horizontally inscribed with the exact same lines of information; the water fill level would be adjusted to indicate the line of information that was to be communicated. The two sides would be synchronized by means of an optical or acoustic signal that had been agreed upon by both parties in advance.[5]

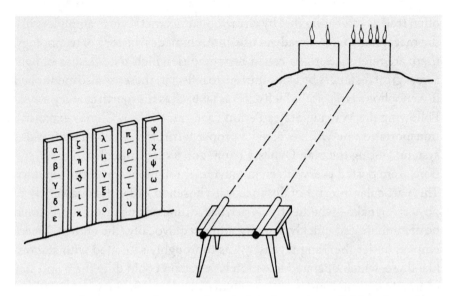

FIGURE 10.2. "Transmission of messages, letter by letter," according to Polybius (second century B.C.E.) by means of torch, writing tablet, and diopter as televisual devices. From Aschoff, *Geschichte der Nachrichtentechnik*, 49.

Muristos: Hair-Raising Sounds

The invention of an instrument called a "broadly sounding trombone"—or, literally translated from its Arabic description *(al argin al buqi)*, a "trombone organ"—was ascribed to an enigmatic author by the name of Muristos, who is purported to have hailed from Greece and who, in the truest sense of the word, haunts ninth-century translations of Arabic literature. The instrument is said to have had a powerful sound that, according to the unanimous testimony of a number of different writers, could be heard from as far as sixty miles away and was used to frighten off approaching enemies from afar. The German ancient historian and early expert on the history of Arabic-Islamic sciences, Eilhard Wiedemann, who about one hundred years ago translated many of the relevant Arabic language texts into German, summarizes Muristos's elaborate description of the instrument as follows: the apparatus consists of a vessel that is partially filled with water. (According to Muristos, the container holds up to 550 liters, such that the dimensions of the instrument must have been enormous.) A wide sound tube is inserted in its lid. Air is blown into the water through tubes leading in from the vessel's sides. A

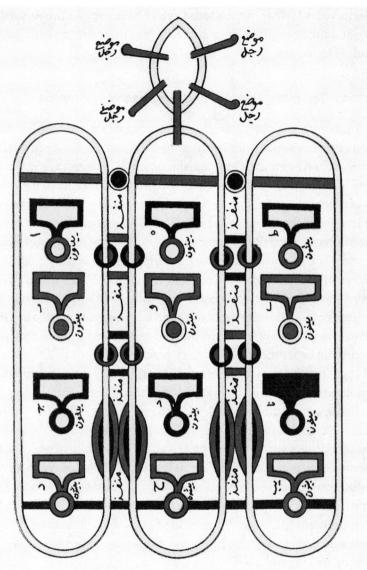

FIGURE 10.3. One of Muristos's pneumatic organs, as they appeared in translations of Arabic literature prior to the turn of the first millenium. From Henry George Farmer, *The Organ of the Ancients: From Eastern Sources* (London: William Reeves, 1931), frontispiece.

powerful sound is thereby produced, which then escapes the vessel through the sound tube on top. If pipes are placed on this tube, it yields (different) sounds (Figure 10.3).[6]

Fontana: Terrifying Images and Aggressive Automata

The fourteenth and early fifteenth centuries saw foundational works written on experimental research into explosive materials for use in warfare and for ceremonial bombast. One such document is an extravagant manuscript dating to circa 1420, which was composed at the University of Padua. It consists of seventy pages of dense, colorful sketches showing sundry explosive artifacts, both terrifying and decorative. The legends accompanying the illustrations are partially written in code, though the code is not difficult to decipher. The inventions and their descriptions apparently were intended as secret. The manuscript, which is now held by the State Library of Bavaria, acquired in the hands of a previous owner the title *Bellicorum instrumentorum liber* (Book of the instruments of war) and was originally written by Giovanni Fontana (circa 1395–1455), who could be considered a pioneering figure in early artistic experimental laboratory practice.

From the curiosity cabinet of this inspired Paduan pyromaniac and cryptologist emerged an illustrated design for an apparatus that could project a figure so dreadful and diabolical that the mere sight of it would cause advancing enemies to flee in fright. The apparatus consists of a small, easy-to-handle lantern with a curved glass wall on the surface of which the figure to be projected is painted. Behind this is located a light source that produces an enormous shadow on an outside wall. The designer's imagination was a little bizarre. The devil figure meant to terrify is explicitly coded female, with prominent breasts and exaggerated pubic hair; it features dragon-like talons, a monstrous horned head, and a threatening posture, and it wields a giant spear (Figure 10.4).[7]

In the genealogy of self-moving apparatuses that we call automata, there is another technological masterwork that also unambiguously reads as female: *la strega infuocata*, or "the fire-spitting witch" (Figure 10.5). On page 63 of the manuscript, Fontana not only placed her *en scène* but provided visual instructions on how to build her, which are detailed on page 64 in much the same manner as they might appear in a manual. The mechanics of this explosive, kinetic monster may be described as follows: the wooden witch

FIGURE 10.4. Fontana's female devil projection for frightening away aggressors, from the beginning of the fifteenth century. From Eugenio Battisti and Giuseppa Saccaro Battisti, *Le macchine cifrate di Giovanni Fontana* (Milan: Arcadia, 1984).

FIGURE 10.5. Fire-spitting witch automaton from Fontana's *Book of the Instruments of War* (c. 1420). From Battisti/Battisti, *Le macchine cifrate di Giovanni Fontana*.

shoots out from an opening in a wall of one of the fortification's defense towers and glides downward along a track. According to the first-century prototype in Hero of Alexandria's theater of automata, the witch's movements are controlled by a rope that is coiled to rotate in a determined direction around a piece of wood that acts as a programmed winch. The figure's head and limbs are moveable, as are the attached enormous bat wings and an oversized tail. The witch is also equipped on an unspecified part of its body with an arrow-shooting apparatus. The figure is brightly lit from within, fireworks explode from its ears, and fire also spits from its mouth. The internal mechanism is housed inside a sturdy wooden frame so that the automaton does not fall apart as a result of the vibrations and explosions to which it is exposed.[8]

From Bread Pans to Torpedoes

In the early modern era, Padua served as one of the bridgeheads between the Occidental and Oriental worlds of knowledge. Its scholars and engineers profited from the richness of Arabic-Islamic culture, even as they shamelessly exploited it. In his pictographic representation of the instruments of war, Fontana at one point makes explicit his connection to forms of eastern knowledge: on the second page of his manuscript, he refers the design of a drivable "ram" to an instrument belonging in the Arabic tradition. His borrowing or plagiarism is even more apparent when describing flying, rolling, and water-traversing rockets—the latter called "Chinese arrows" by virtue of their origin.[9] Here Fontana largely fails to provide his references. The construction of "auto-mobile torpedoes," which were built of concave iron sheets the Arabs had originally used for baking bread, was the special area of expertise of Hasan al-Rammah, whose relevant texts appeared between 1275 and 1295 (Figures 10.6 and 10.7).[10] The scholar Fuat Sezgin—himself from Istanbul and a key contributor to the creation of a new museum in the former Constantinople, whose aim since 2008 has been to support the revival of Arabic-Islamic cultural consciousness—proceeds on the assumption that Fontana was also familiar with still other, even earlier sources from the deep time of Arabic-Islamic explosive war technologies.[11]

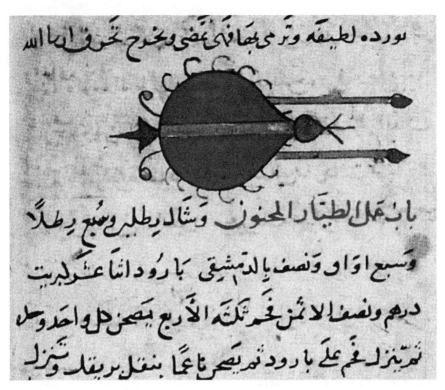

FIGURE 10.6. One of Hasan al-Rammah's bread pans repurposed as torpedoes, from the late thirteenth century. From Fuat Sezgin (in collaboration with Eckhard Neubauer), *Science and Technology in Islam*, vol. 5, section 12, *War Technology* (Frankfurt am Main, Germany: Institute for the History of Arabic-Islamic Science, Goethe University Frankfurt, 2003/2011).

Rotating Artillery: A Medial Paradigm

There is a conjecture that has become paradigmatic of postmodern media theory: fast-rotating mechanisms for the successive recording of moving objects—like Jules Janssen's astronomical *revolver photographique* and, above all, Étienne-Jules Marey's photographic shotgun and the stop-and-go of cinematographs—were developed in the nineteenth century roughly simultaneously with Samuel Colt's invention of the drum revolver. Artifacts designed to shoot images and apparatuses designed to kill are in direct contact and serve the construction of a genealogy of media that receive their original impulse from destructive military technologies.

Around 1500, Leonardo da Vinci conceived of a revolver with sixty-four barrels, but in this respect, he was only an early projectionist—or, in other words, a designer. In Konrad Kyeser's text *Bellifortis*, supposed to have been

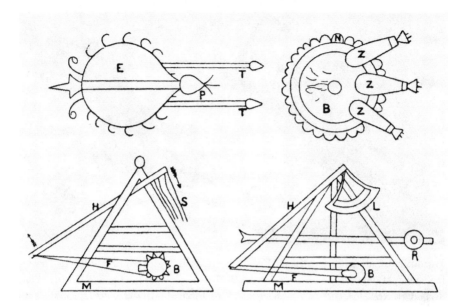

FIGURE 10.7. In Romocki's illustration (p. 71), he explains that the streaming lines around the pans indicate that the torpedoes were capable of moving through water. From S. J. Romocki, *Geschichte der Explosivstoffe: Sprengstoffchemie, Sprengtechnik und Torpedowesen bis zum Beginn der neuesten Zeit* (Hildesheim: Gerstenberg, 1983). Facsimile reprint of the 1895 Berlin edition.

begun in the 1390s and completed in 1405, prominence of place is given to projectile artifacts that work in rapid succession, firing off one projectile after another. Among these, one reads of a cannon organ equipped with packets of three or more flue tubes, which are mounted one above another at differing levels (Figure 10.8a). Kyeser uses the term *revolvere* to designate any instrument designed to destroy that features six cannon barrels arranged around a rotating cylinder (Figure 10.8b).[12] But the origin of the rotating stop-and-go mechanism is not essentially located in military technology. Wheel clocks with similar mechanical features were developed in Europe around the turn of the thirteenth to the fourteenth century, and we know that the mechanism that laid the foundation for this dates back to the early ninth century in the deep time of Arabic-Islamic culture. The course of development of the gear wheel, as the primary element in the production of rapid, steady, uninterrupted movements, points all the way back to the high cultures of antiquity, where it served primarily to aid in pulling water up from the depths to make the parched land flourish—quite the opposite of destruction.[13]

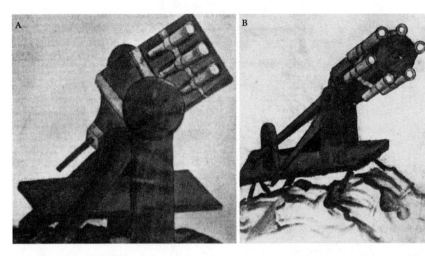

FIGURE 10.8. The weapon that Konrad Kyeser refers to as the "death organ" as well as the "revolving artillery" from 1405. From F. M. Feldhaus, *Die Technik der Vorzeit, der geschichtlichen Zeit und der Naturvölker: Handbuch für Archäologen und Historiker, Museen und Sammler, Kunsthändler und Antiquare* (Leipzig, Germany: Engelmann, 1914), 406.

Secret Messages: Della Porta

In early modern Europe, not only were the stage-setting principles of modern optical and acoustic technologies created and developed. In a context of fierce, warlike conflicts spanning decades and of sophisticated practices of secret intelligence, a number of strategies were also developed for the encryption of messages. Giambattista della Porta, founder of the experimental Academia Secretorum Naturae, was one of the early champions of these emerging cultural techniques.

A decisive motivating factor for his preoccupation with the art of *criptologia* (the title he meant to give one of his final, ultimately unpublished works) was the threat posed by censors and inquisitors. Already in 1612, three years prior to his death, della Porta's patron Federico Cesi wrote in a letter that it would be advisable for him to address his letters to a straw man, "for, were one to write to Porta, the letters may not be all that secure."[14]

Among many other strategies, della Porta suggests a way to encrypt messages by resolving a word's spelling into two discrete values: two intersecting horizontal and vertical lines could be drawn so as to produce nine right-angled fields; then the reduced number of twenty-one letters of the alphabet would be drawn within these fields in a stipulated sequence. Each

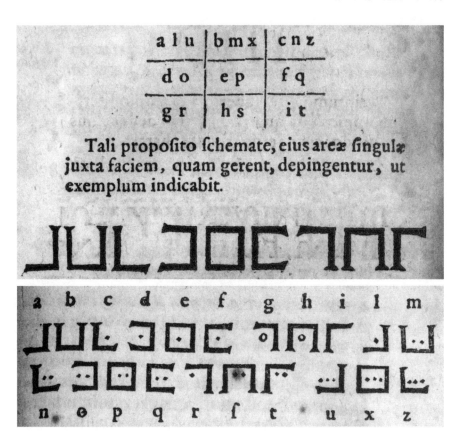

FIGURE 10.9. Della Porta's bivalent code from 1561 for the transmission of secret messages, with examples of spellings. From Giambattista della Porta, *De occultis literarum notis* (Strasbourg, France: Lazarus Zetzner, 1603).

of the upper three fields would contain three letters, while the remaining six fields would each contain two. The cryptogram is not written as text but instead takes the form of graphic notation. Two, three, or four lines are drawn at right angles to one another, and these define the respective field of letters. A second geometrical form—the dot—then defines the letter in each field that is to be selected. Since the fields consist of up to three letters, *one* dot indicates the first letter in a given field, *two* dots indicates the second, *three* dots the third, and so on (Figure 10.9).

The cryptogram consists of a simple dot-and-dash code, as would later be used in telegraphy. The only difference is in the notation. Whereas with Morse code, two discrete characters follow one directly behind the other, in

FIGURE 10.10. Acoustic early-warning system in the seventeenth century. From Robert Fludd, *Ultriusque Cosmi Maioris scilicet et Minoris, Metap(h)ysica, Physica atque Technica Historia*, vol. 1, *Tomus Primus: de Macrocosmi Historia in duos tractatus divisa* (Oppenheim, Germany: Theodor De Bry, 1617).

della Porta's system, the two are overlaid or interconnected. In this way, the act of reading becomes a practice of swift and precise pattern recognition.

Eavesdropping on the Enemy: Fludd

In the premodern era, Robert Fludd—a trained physician, Rosicrucian, and free mason—published one of the last great attempts to cram between the two covers of a single book all that one could possibly describe about the world. His *Utriusque Cosmi, Maioris scilicet et Minoris, Metaphysica, Physica, Atque Technica Historia* appeared in two volumes between 1617 and 1618. In it, he discusses the great questions of theology, music, the art of combination, and natural philosophy; but the books also contain fantastical technical marginalia—like, for instance, equipment with which one is able to cross a river underwater and thus undetected by the enemy, and an acoustic early-warning system made up of delicate bells that ring at the approach of enemy troops (Figure 10.10). The book's first volume, dated 1617, also contained a lengthy chapter on "military algorithms."

FIGURE 10.11. Board game and box lid to the game Radio-Sende-Spiel from the period of the Second World War. Zielinski Archive.

The Micropolitics of Power

In September 1939, the Nazis began their mad war against all that was not identical with themselves or that failed to promote their singular identity. The radio was their most important mass medium—the universally affordable "people's wireless" (Volksempfänger VE 301), also known as "Goebbel's snout," the artifact that would become the expression of boorish propaganda and escapist entertainment. It was over the radio that the population learned of the invasion of Poland on September 1, 1939, and was simultaneously confronted with a decree by the "Reich Ministry of Public Enlightenment and Propaganda": listening to foreign broadcasts was now an offense punishable by imprisonment and forced labor. In the event of high treason—such as listening in secret to the communist broadcast Radio Moscow—the punishment was death, and this was indeed enforced in several instances in the course of the war. Radio retailers provided few signs for the dial that could be rotated to receive transmissions from outside broadcasters; the text printed on the stiff white placard read, "Think first: Listening to foreign broadcasts is a crime against the national security of our people. On order of the Führer, it is punishable by penal servitude."

To get the population to agree to such nonsense, a short while later, a board game called the Radio-Sende-Spiel (Radio Transmission Game) was put on the market (Figure 10.11). The board comprised two differently colored spaces—one red, the other blue—both of which functioned much like the "Chance" spaces in the Monopoly board game. When a player landed on a blue space belonging to a Nazi broadcast station, the player was invited to sing a "pretty little song" or a "soldier's song" or to give a weather report, or was otherwise awarded with some triviality. If, however, a player landed on a red space marked as a "foreign broadcaster," then the rules of the game were unequivocal:

> ATTENTION! ... Listening to foreign broadcasts is a punishable offense. Whichever player occupies this space must state aloud: I have committed a criminal offense! I must pay a penny fine and begin again at Start.

Convalescence Television and Deadly Cameras

Wir senden Frohsinn—this was the title of the first monograph on the theme of "watching television under fascism" that was published in the post–World War II Federal Republic of Germany.[15] Its author was Swiss media theorist Erwin Reiss, who, with Friedrich Knilli, also published the first introduction to film and television analysis in 1971. The title is taken from one of the most bizarre television programs in the history of the medium. The full title is *Wir senden Frohsinn—Wir spenden Freude!* (We broadcast good cheer—we bestow pleasure!). On March 12, 1943, the program aired for the hundredth time. When the show was on air, the studio welcomed a steady stream of the biggest names in the entertainment industry at the time, many of whom continued to entertain audiences after the war. In the "total war" of German fascism against the rest of the (nonfascist) world, the broadcast had the effect of a psychological salve. The wooden crutches of those who had lost limbs and suffered other serious injuries in the various theaters of war were left casually leaning against the broadcast-receiving apparatus manufactured by Telefunken (Figure 10.12). In some large-scale projection television theaters in Berlin, it was predominantly the wounded and disabled who were supposed to find spiritual sustenance in the televisual displays of artificial paradise.

But the real money was not in convalescence television. The television industry of the day truly profited from integrating its technical know-how into the death-machinery of the Nazis. For instance, in 1940, the Henschel company, located in the Schöneberg neighborhood of Berlin, tasked the engineer Herbert Wagner with improving the targeting accuracy of existing technical systems for the radio control of bombs. His work resulted in the production of a guided missile, the Henschel Hs 293. A camera had been

FIGURE 10.12. Convalescence television with mirror receiver by Telefunken. Zielinski Archive.

FIGURE 10.13. Bomber pilot at the remote controls (Tele-Commander) of the television bomb.

integrated into the remote seer and remote destroyer at the head of the missile; at the tail were antennae for the television bomb's control signals. The pilot of the plane that carried the bomb guided the weapon of destruction via a "Tele-Commander," a control column that essentially functioned like the joy stick on a computer (Figure 10.13).[16]

Wars in Living Rooms

What it means to transmit images of war via the private sphere's premiere electronic medium is not unequivocal. Television integrates the horror of destruction and military mass murder into the normalcy of everyday civilian life. Between beer and pizza, political and commercial advertisements, games and relaxation, one becomes accustomed to catastrophe—above all, those taking place far from home, far from the supermarket or the bar around the corner. Yet, because journalists and their employers strive to get as near as possible to the events of war, indeed to embed themselves within

FIGURE 10.14. Wolf Vostell, *Miss America*, 1968. From *Wolf Vostell, dé-collagen 1954–1969* (Berlin: Galerie René Block, Edition 17, 1969), 131.

it, to translate the naked terror and the despair of those affected as directly as possible into sound and image, television tends to concretize what would otherwise remain imaginary for nonparticipants and tends also to provoke moral outrage or dissent in the more sensitive among them. In the second half of the 1960s, television images of outrageous brutality began to amass, displaying the cruelty the U.S. military inflicted upon its unevenly matched opponent in Vietnam, without regard for the civilian population, even or especially the most vulnerable among them. Images like that of a mother holding the corpse of a small child in her arms contributed to the increasing popularity of the worldwide antiwar movement. It would be a while yet before Thatcher's England would learn about the highly effective ambivalence in televised images of war: the Iron Lady had imposed a blackout on media coverage of England's offensive war in the Falklands. Depictions of the wars that followed oscillated, in propagandistically strategic ways, between the concrete visualization of horror and near-total abstraction. In 1968, Wolf Vostell—who was himself interrupting the smooth flow of the medium as early as 1959, with his electronic *TV Dé-coll/age,* and who, in 1963, in a spectacular Happening in New Brunswick, crowned a television in barbed wire before laying it to rest in a grave—collaged a photograph of Miss America together with another image that has since become iconic: the young North Vietnamese man Nguyễn Văn Lém being shot in public, in broad daylight, by the South Vietnamese chief of police, Nguyễn Ngọc Loan, on February 1, 1968 (Figure 10.14).

Almost without Image, Nearly Speechless

The feature film *Daratt* (Dry season) tells a story of grace under the most dire of conditions. It was filmed in 2006, in the midst of Chad's ongoing civil war, and screened at Peter Sellars's New Crowned Hope festival as part of the Vienna Mozart Year 2006–7. The director, Mahamat-Saleh Haroun, does not train the lens of his camera on the tanks and guns of the civil conflict. He turns away from the objects of military reportage and refuses the standardized television view. His camera opens up a perspective on everyday life as it is lived beyond the armed conflict—calm, with minimal shots, barely any words. He stages this view as an unbelievable ethical and political utopia. The film deals with a kind of revenge to which one is culturally obliged and the possibility of subtly refusing it, and with a kind of compassion and grace that

FIGURE 10.15. Video still from Mahamat-Saleh Haroun's *Daratt* (Dry season), 2006.

is unparalleled. These ideas do not at all conform with conventional images of war. The video still (Figure 10.15) shows a scene shortly after a general amnesty for war criminals had been announced on Chad's national radio. Suddenly, gunfire is heard. In the image, there are no dead, no mercenaries, no soldiers, no weapons or war machines—only the shoes we must imagine the dead once wore.

Translated by Lauren K. Wolfe

Notes

1 For more detail, see Oskar Blumtritt, *Nachrichtentechnik: Sender, Empfänger, Übertragung, Vermittlung* (Munich, Germany: Deutsches Museum, 1988), 22–23.
2 In antiquity, the most brutal technique for preventing the messenger's knowing the message was to have it tattooed on the crown of his clean-shaven head. See my chapter on della Porta in *Deep Time of the Media: Toward an Archaeology of Hearing and Seeing by Technical Means* (Cambridge, Mass.: MIT Press, 2006); see also S. Zielinski, "Medien/Krieg: Ein kybernetischer Kurzschluss," *Medien im Krieg: Die zugespitzte Normalität. Sonderheft des Medien-Journal zum ersten Irak-Krieg* 1 (1991).

3 See Volker Aschoff, *Geschichte der Nachrichtentechnik* (Berlin: Springer, 1984), esp. 19ff.
4 See two papers presented by Nianzu Dai of the Chinese Academy of Sciences at the fifth Variantology Conference in Cologne, 2006: "The Transmission of Messages by Post Stations in Ancient China" and "The Origins of Advertisement" (as yet unpublished in Europe).
5 See S. Zielinski, "Von Nachrichtenkörpern und Körpernachrichten: Ein eiliger Beutezug durch zwei Jahrtausende Mediengeschichte," in *Vom Verschwinden der Ferne, Telekommunikation und Kunst,* ed. Edith Decker and Peter Weibel, 235–38 (Cologne, Germany: DuMont, 1990).
6 Eilhard Wiedemann, "Byzantinische und arabische Instrumente," in *Archiv für die Geschichte der Naturwissenschaften und der Technik* 8 (1918): 155.
7 See the encyclopedic work of Laurent Mannoni, Donata Pesenti Campagnoni, and David Robinson, *Light and Movement, Incunabula of the Motion Picture 1420–1896* (Gemona, Italy: Cineteca, 1995), 44–45.
8 See Battisti and Battisti's discussion in *La macchine cifrate di Giovanni Fontana* (Milan: Arcadia, 1984), 96–97, whose text I have expanded here and made somewhat more precise.
9 S. J. von Romocki, *Geschichte der Explosivstoffe: Sprengstoffchemie, Sprengtechnik und Torpedowesen bis zum Beginn der neuesten Zeit* (1895; repr., Hildesheim, Germany: Gerstenberg, 1976), 70.
10 See von Romocki, 68–69.
11 Fuat Sezgin, *Science and Technology in Islam,* vol. 5, section 12, *War Technology* (Frankfurt am Main, Germany: Institut für Geschichte der Arabisch-Islamischen Wissenschaften, 2004), 126.
12 In addition to Kyeser's own book, see F. M. Feldhaus, *Die Technik der Vorzeit, der geschichtlichen Zeit und der Naturvölker* (Leipzig, Germany: Engelmann, 1914), 405–6.
13 See Conrad Matschoß, *Geschichte des Zahnrads, nebst Bemerkungen zur Entwicklung der Verzahnung* (Berlin: VDI, 1940).
14 For more detail, see my chapter on della Porta in *Deep Time of the Media.*
15 Erwin Reiss, *"Wir senden Frohsinn": Fernsehen unterm Faschismus. Das unbekannteste Kapitel deutscher Mediengeschichte* (Berlin: Elefanten Press, 1979); see also S. Zielinski, *Audiovisions: Cinema and Television as Entr'Actes in History* (Amsterdam: Amsterdam University Press, 1999).
16 Image and context from Joseph Hoppe, *Ferhensen als Waffe* [Television as a weapon], Berlin Museum for Technology and Traffic (no date).

11 *Theologi electrici*
A Few Passages

> *Moving from the Jesuit polymath Athanasius Kircher through the work of Protestant reformer Friedrich Oetinger to the thought and work of Romantic scientist and physicochemist Johann Wilhelm Ritter, one finds a coterie of thinkers who conceived of God as a vibrating megabody and electricity itself as the presence of the all-powerful on earth. The historical phenomenon of* theologi electrici *helped me to better understand the manifest signs of a new, present-day religiosity, which I read in the early venerations of the internet as an all-powerful medial apparatus—a new religiosity that viewed electronic telematics as the new salvation. Walter Benjamin makes an appearance in this context, both as a figure of warning and as an important figure of transmission, between Ritter's fire-writing and a utopian space like Ted Nelson's Xanadu.*
> *This essay was originally published in 2005.*

※ ※ ※

The Question

The title reads a bit antiquated and seems out of place in the context of a contemporary discussion meant to have been inspired by Walter Benjamin, about the relationship between theology and politics. My approach is indirect. It begins with a questionable assumption. Did Benjamin see a utopia in political universalisms such as Marxism and—from my perspective as a media archaeologist—such as those in the new technologies of production, distribution, and the perception of art? Did he consider these a substitute for and alternative to religious universalisms such as Zionism? In some rather unlikely places in the writings of this philosopher and critic of art and media, an overt fascination surfaces with the work of the physiochemist Johann Wilhelm Ritter, who, around 1800, developed a natural philosophy of galvanism and electricity into a comprehensive worldview and, escalated in Romantic style, lived it as experimental practice. Benjamin's passage through Ritter's work was brief. In it, one finds him wavering between a universal and a pluralistic view of the things that make up and hold the world together, including politics and religion. Remarkably, though, his media critique remained completely unaffected by this passage through the thought of one of the most thrilling thinkers of the electric, who Goethe in a letter to Schiller referred to as "an astonishing phenomenon," a "true paradise of knowledge."[1]

Ritter can be interpreted as the executor of a monistic theology of the electric, in scientific experiment and in its exuberant superelevation. In the eighteenth century, a loose coterie of religious scholars trained in the natural sciences were deemed the "electric theologians." A powerful idea underlay their association: in the vibrations and voltages of the electric, they believed the divine expressed itself. In some instances, they produced their own sensational installations of electric phenomena as *displays* of divine omnipotence. These are also origins of a medial praxis that since the last fin de siècle has been called *media art*. Considered emancipatory at the time, the theology of the electric was a universalistic, but in essence also a deterministic, worldview. A phenomenon of nature that could be transformed and intensified via technology was elected to be the universal mover, or *Allbewegerin*.[2] But to think of the world as a single vibration at variable levels also meant to conceive of it as dynamic and situated in constant flux. Thus the utopian potential of *theologi electrici*. Its predominantly Jesuit-educated advocates met with staunch opposition from the mainstream of the Catholic church. They were accused of unriddling a mystery upheld as divine and of having aimed at depersonalizing God.

"Our social activity" has lost "its metaphysical seriousness," laments the I-interlocutor in Benjamin's fictional *Dialogue on the Religiosity of the Present*. "For nearly all those who are active socially, this activity is merely one aspect of civilization, like electric light. Sorrow has been undeified, if you will pardon the poetic expression."[3]

Centrosophia: Origins and Fundamentals of a Theology of the Electric

The theologians of the electric evolved a particular impact in the middle of the eighteenth century, in parallel with the unabashed technical domestication of lightning by Benjamin Franklin (1706–90), the storage of weak static electricity in Leyden or Kleistian jars, and the spectacular demonstrations of conductive and nonconductive bodies carried out on young boys and girls, who were hung from the ceiling as living media and had weak current applied to them, thus attracting light particles of matter or igniting flammable materials. In de Sade's double novel, constructed as the two poles *Justine* (1791) and *Juliette* (1799), concepts like these found their sui generis culmination. "Oh Juliette," says the matron right at the start of the first book,

> of all human virtues you will make vices and all vices will become virtues to you, then a new universe will appear before you, a blissful and consuming fire will course through your veins and ignite that electric fluid on which life depends.[4]

But Justine had to die of lightning. Excursus II of Adorno and Horkheimer's blazing critique of instrumental reason as deployed in mass media opens with a brilliant analysis of the figure of Juliette.

The *theologi electrici* drew considerable inspiration from the seventeenth century. The first systematic account of the phenomenon of magnetism in the passage to modernity is attributed to William Gilbert. His treatise *De magnete*, written in 1600, was a cool undertaking, describing in painstaking detail the actions of amber and lodestone and a number of other minerals whose properties had been known since antiquity, classifying them, collecting them under the concept "electrica," and, in a bold geological move, interpreting the planet Earth and its gravitational properties as one enormous magnetism. Here there was no room for religion. Gilbert was a doctor who, like the mathematician John Dee, occasionally attended the body and court of Elizabeth I, who was persona non grata at the Vatican. The most urgent of worldly problems that he wished to help in solving were those of navigation. His concern was with political and economic supremacy over the world's oceans and with precision measurement in the magnetic compass.

Things were a bit different in the case of the Jesuit polymath Athanasius Kircher, in honor of whose four-hundredth birthday Eugenio Lo Sardo and colleagues threw an opulent celebration in the Roman Palazzo di Venezia in 2001, for which reason I need not provide biographical details here.[5] In all kinds of fields of knowledge, keeping a close link between theology and natural philosophy, Kircher exhaustively elaborated one central idea: the world of experience is anything but peaceful and harmonious. It consists of a dissonant plurality, with fierce tensions and contradictions. All natural experiments and all of the arts are tasked, like religion, with establishing peace and harmony, with fabricating from the multiplicity a "contrary unanimity" or a "unanimous contrariety," as he referred to it in his baroque music theory, the *Musurgia universalis* (1650). His *Ars magna lucis et umbrae* (1645–46) traces the same idea of a bipolarity in optics. In his major geological work *Mundus subterraneus* (1665), everything on Earth is determined and regulated subterraneously by a "central phenomenon"—the power of fire. On this Empedoclean element of attraction, Kircher constructs his "*centrosophia.*" Kircher was deeply impressed and inspired by his passage by boat from Messina to Naples in 1636, during which time all three southern Italian volcanoes—Etna, Stromboli, and Vesuvius—were active and causing the sea to bubble and churn.[6]

The *Magnes sive de arte magnetica* of 1641 was Kircher's first ever published work. By analogy with fire, Kircher conceived of magnetism as a hidden force holding the world together at its core, as an elementary force of nature operating everywhere, from the great cosmic movements of stars and planets to the tiniest living plant- and animal-like creatures, even in the love existing between the poles of gender. The Jesuit Kircher chose a chain to serve as the emblem for this all-concatenating energy, a "*nexus unionemque*" that linked dead with living phenomena as well as each of the separate scientific disciplines one with another. The meanings reverberating in the Latin word *nex* not only yield the positive phylum of connections and linkages but also imply convolution or engulfment and entanglement or bondage in debt. *Nex* in Latin is violent death, murder, execution. *Nexus*, "linkage" or "entanglement," is also an affair of power, just like the strange invisible force of magnetism that attracts and repels. (*Vis* and *virtus* even more so are unambiguously connoted by power. I have never quite understood how artists could find anything worthwhile in *virtual reality*.)

Kircher, raised in ultraconservative Geisa near the German provincial town Fulda, was a faithful servant to his lord in Rome. He had no interest in agitating the Catholic doctrinal edifice administered by the Vatican. His own concept of magnetism as the energy perfusing the world adhered strictly to Robert Fludd's idea of the great "pulsator mundi," the meta-magnet God who set things oscillating and whose omnipotence governed the relations of "consortium & dissidium," of "reciprocal conviviality and quarrel."[7] But Kircher suspected the explosive power hidden away inside his ideas about magnets. The following is a citation from the foreword to his 1641 text:

> We are inquiring into the coherence of the entire universe and all living things in a new and singular manner. Whoever holds the key to this method, may he know that he will find open to him the door to knowledge of all

things hidden, indeed to that true wisdom to which philosophers aspire and that they call magic, and to the secrets of that true philosophy.[8]

A century later, two Pietists from Württemberg, Friedrich Christoph Oetinger and his student Johann Ludwig Fricker, along with the Czech, or Bohemian, Prokop Diviš from Helvíkovice were far less restrained. As "physico-theologians," they were working on reformulating the concept of nature and that of their God—or, to be more exact, of God in nature. Their texts were not published in luxuriant folios but occasionally censored, and for that reason, finding access to them can be difficult.[9]

Socialized in a strictly Jesuit context, the electric theologians of the eighteenth century attempted once again to think theology and natural philosophy as a unity. They combined the discovery of electricity and that of magnetic and galvanic phenomena with the idea of a God who was immediately present in the world and, consequent to this new conception of God, with a new understanding of the relations between "soul and bodiliness, spirit and matter, life and substance" (6). Veering away from, though at the same time complementing, the medieval metaphysics of light, they projected a new image of the divine:

> Magnetism and electricity appear as the clearest manifestation of the concealed presence of divine force in the world and in things, as the concealed power that permeates the entire universe, creating life, motion, heat. (7)

These electric theologians carried through to the end what Kircher had only hinted at, with radicality of a different kind:

> From the "Magnet God" derives... the magnetic force of nature. The depersonalization of theology urges... a practical equalization of the divine spirit, as the *vis magnetica dei*, with the all-ensouling force of nature. (14)

The possibility of such a construct was premised on the, at least in principle, imperceptibility of magnetism and electricity. "We do not possess an electric sense," Graetz writes later in the nineteenth century.[10] Electricity cannot be smelled or touched, tasted, seen, or heard. Only by permeating and in confrontation with matter is it experienced as heat, motion, vibration, or light: the "ideal appearance of a distance" in the present; if the voltage ratios are correct, then the current will flow, attraction and repulsion can unfold as archaic forces.

A key text, also regarding the Benjamin–Ritter connection that is of interest here, is clearly the 1764 dissertation by the Tübingen theologian Gottlieb Friedrich Rösler. Systematically following through on Kircher's basic physicotheological ideas, Rösler construes electric fire as the light of the first day of creation and declares it the *spiritus mundi*. Nature's electric fire is conceived as something inherent in nature itself. It becomes a "vital principle that constantly urges reconfiguration, that wants to actualize itself in ever newer forms of life." It is a universal phenomenon and declared an anthropological principle. It seems to this electric theologian "a blasphemy to assume that God would have created a dead clod of dirt that he then only after the fact inspired with spirit. The electric fire already dwelled within the substance of the clod from which God created man, the clod of dirt already possessed a sensible soul. Inspiration with spirit is not identical with the act of first ensoulment, but represents a second, belated act: the outfitting of the human being with the capacity to think, with the intellect."[11] Oetinger writes in his *Biblical and Emblematic Dictionary*, "It is a double life man lives, the sensible and the intellectual; the former is electric, the latter way beyond electricity: but the boundaries cannot be determined."[12]

References to Presocratic natural philosophers surface not only in the bodily or organic dimensioning of spirit and in the definition of magnetic energy as attraction and repulsion that dates all the way back to Empedocles but also in obvious echoes of the original Atomist mind frame. "In each body," writes Oetinger, for instance, "are enclosed the tiniest fiery parts which partake of the general formation of the larger body."[13] The overwhelming longing that drove the theologians of electricity was the same longing practiced in post-Socratic philosophy, a longing for a condition beyond separation, the same that had in the prior hundred years of its development firmly established itself in modern natural science as well. These figures believed they had at last recovered in electricity a symbolic force for the unity of spirit and matter, body and soul: God is electric, my soul is electric, nature is electric. Thus the creed of *theologi electrici*.[14]

Ritter's Experimental Fulfillment of the Idea of a Theology of the Electric and Benjamin's Point of Entry

That was the obsessive stuff of the life of the nineteen-year-old Ritter, the physicist and chemist from the town of Samitz, present-day Zamienice, when

FIGURE 11.1. Portrait of the physicotheologian Friedrich Christoph Oetinger, whose theosophy may be summarized as the theory of the bodiliness of the divine: in the visible things of nature, the invisible can be read. From *Zum Himmelreich gelehrt— Friedrich Christoph Oetinger,* exhibition catalog (Stuttgart, Germany: Württembergische Landesbibliothek, 1982).

in 1796 he went to Jena and became for a number of years the cult figure of Romantic poets and philosophers. Not much time was allotted him. By 1810, he was dead, having recklessly sacrificed his body to the laboratory in an effort to prove the universal meaning of galvanism and electricity. In a radicalization of the experiments Volta had conducted on himself, Ritter attached the positive and negatives poles of a voltaic pile to his hands, lips, tongue, temples, eyeballs, and other sensitive parts of his body and painstakingly recorded the effects, the different voltage states on the oscillograph of his body. In his final years, as he was working on, among other things, a "theory of incandescence," he believed he could no longer endure these tortures without seeking relief in artificial paradises such as opium and alcohol, which only accelerated his physical disintegration.

Four innovations are associated with Ritter in the technological/natural-scientific context:

- the discovery of ultraviolet light just outside the visible spectrum
- the discovery of the accumulator principle using the so-called charge column
- the discovery of electrolysis, the chemical decomposition of water and other liquids by running a continuous current through them (contemporaneous with, though independent of, several English and French physicochemists)
- the experimental discovery of the connection between electricity and magnetism, which he spent the final years of his life working on but was no longer capable of formulating[15]

His experiments with, observations of, and conclusions regarding tone and sound in connection with electricity are nevertheless extraordinarily important for the arts and their interaction with technical media. By radically shifting the point of observation in his investigation of the so-called Chladni figures, he was able to prove to a certain extent that the graphic patterns formed on the vibrating hardware resulted from variable voltage states and not—as originally assumed—from the relation of motion to rest:

> The body is hard only ... through its rigidity. Where ... there are different rigidity values, immediately there is also a differential in the electrical value of the body—electrical voltage itself.[16]

FIGURE 11.2. Portrait of the physicochemist Johann W. Ritter (1776–1810). From the Zielinski Archive.

Based on observations like these and on a profound sense of the unity of physics, life, and art, in 1805 Ritter developed a rather astonishing anthropology of the arts, one of the last works he was to publish. With human activity as his plane of reference, he categorized the development of the arts up through the present into four phases:

- architecture/urban planning
- sculpture
- painting
- tone/sound/music

The first three are arts of memory for Ritter. In urban architecture, the human being monumentalizes its activity, objectivizes it in sculpture, and in painting (which the physicist called *half-space* or *shadow-body*), the activity is more demanding, since it also requires the activity of an observer to complement the sensory data of the two-dimensional image. But only in the temporal form of sound does art truly come into its own—that is, for Ritter, to life.

Walter Benjamin was familiar with these ideas, though not from Ritter's excessive and difficult-to-read technical treatises, but rather from the aphorisms he had long been writing under a pseudonym, which he collected shortly before his death and which appeared only posthumously as *Fragments from the Estate of a Young Physicist.*[17] Benjamin, to his great pleasure, had got them cheap at an auction in Berlin and referred to them in his essay "Unpacking My Library" as "the most important sample of personal prose of German Romanticism."[18]

The "brilliant" Ritter and his mental eruptions into an intuited electromagnetism and into the anthropology of the arts occupy a prominent place in Benjamin's *Trauerspiel* book. Tentative, indirect, clearly sensing that he is at best only able to scratch at the surface of this peculiar intellectual universe and that not to embark in its theological dimensions would be to do it a grave disservice, he commences, in a staccato style of citation that is unusual for Benjamin as a writer, with the remark that Ritter's account opens a perspective "the penetration of which we must ourselves forego, for it would be irresponsible improvisation."[19] Benjamin is particularly fascinated by Ritter's "rapturous delight in sheer sound," this being characteristic of the spirit of the times around 1800, something Benjamin sees as an important clue to the ultimate decline of the baroque *Trauerspiel*. In his interpretation of the tones coalescing into graphics on the vibrating plates of the electroacoustician Ernst Florenz Friedrich Chladni, Ritter remarks on

> what the sound figure is to us inwardly: light-figure, fire-writing.... And so each tone carries its letter with it immediately.... This extremely intimate connection of word and script, that we write when we speak... has long preoccupied me.... Word and script are one, right at their source, and neither is possible without the other.... Each sound figure is an electrical figure.... I wanted... to rediscover or at least to seek the primeval or nature's script by means of electricity.[20]

The text from which Benjamin cites is the later edited version of an 1805 essay taken from the *Magazin für den neuesten Stand der Naturkunde* (Journal of contemporary natural history). Ritter had written it in reply to a letter from his friend Hans Christian Ørsted, the Danish physicist who has been exclusively credited with having discovered electromagnetism around 1820. The two men were in close correspondence with one another over this fundamental issue of physics. Benjamin seems to break off at the most gripping moment of Ritter's performance: "Every image is only a form of writing." Architecture, sculpture, painting, are "writing-after" or "writing-again." "In the context of allegory, the image is only signature, only the monogram of essence, not the essence itself in its covering."[21] Thus Benjamin summarizes the natural scientist's "theory of allegory." In his rhapsodizing language, the natural scientist himself found another expression for this, the formulation of which is heavily reliant on the Lucretian concept of the *simulacrum*. "The purpose of art: making present what is absent. Monument. But the beloved herself is more than her image." Lucretius wrote in the first century before Christ, "Nam sie abest quod ames, praesto simulacra tamen sunt" (For what you love is distant, the images are present).[22]

The critique of what lies behind is only marginally important for Ritter. For him, the "beloved" is first and foremost living nature, the one that any future physics should take as its object. In sound, as a special phenomenon within electric and electromagnetic circumstances, he senses the vibrations of a future art that for him is one with life and with physics (which he believed was *the* life science) and that for him is driven by bipolarity, as the principle of life:

> A new field will open—*time*. It too is organized, and only from the fusion of *both* organisms, of time and space, does the true and highest meaning of all life and being emerge. Change is everywhere, nowhere is there stasis. All things have their own time, and even this does not consist in tranquil succession, which exists nowhere.[23]

Thus writes the twenty-five-year-old shortly before he is appointed, without either graduating or completing his habilitation, to the newly founded Königlich Bayerische Akademie der Wissenschaften in Munich.

For Ritter, physics—particularly experimental physics—was a practice of making the imperceptible inner states and processes of matter's motion visible, audible, and felt in one's own body, of making available to sense experience that which is not directly tangible but present nevertheless, of translating this into data that in turn could be calculated and applied in technical artifacts. In his *Heterodoxies*, as he referred to his own essays, sound and light are ultimately one. "Hearing is a seeing from within, the innermost inward consciousness."²⁴ They are simply different expressions of a singular central phenomenon (Kircher)—the electric and its various states of charge and oscillation. For the experimental naturalist, it is above all a question of the distribution of quantities on an infinite scale. "When bodies oscillate *extremely fast,* they *glow,*" he writes a few pages down from the place where Benjamin left off with Ritter's text in his *Trauerspiel* book. The dynamic "light-figures" or "fire-writing" about which he rhapsodizes time and again are for Ritter extremely fast, high-frequency oscillations where sound passes over into the visibility of phenomenal light, which for him represents the "highest degree of reality."²⁵ When describing the opposite pole—the passage into low-frequency tones only barely or no longer audible to human beings—he chooses at length a remarkable comparison:

> The Earth's rotation around its axis—that is, the oscillations in its interior that are occasioned by the rotation—may produce a meaningful tone, and its orbit around the Sun another, and the Moon's revolution around the Earth a third, and so on. Here one gets the idea of a colossal music of which our own small music is but a very meaningful allegory.... This music's harmony can only be heard in the Sun. The entire planetary system is for the Sun a *single* musical instrument. Its tones may appear as sheer *vitality* to the Sun's *inhabitants,* but to the *spirit* of the Sun *itself,* as the truest, most exalted sound.²⁶

We can associate this powerful image with the eccentric economy of the universe that Georges Bataille developed in the 1930s, but it also turns up again in modified form in one of Benjamin's radio dramas. Benjamin dedicates the drama to the man who inspired the acoustician Chladni's experiments in sound figures and who provided a point of departure for Ritter's thought and experiments with "fire-writing" and vibrating "light-figures": the physicist, philosopher and author Georg Christoph Lichtenberg, who in 1777 was able with his Eidophor to record and copy positive and negative electrical voltages as figures of bright powder on a dark ground—light-figures written by electric fire. Benjamin's radio drama appeared in 1933. In it, the inhabitants of the Moon observe Lichtenberg on Earth. The "Moon beings" have three apparatuses available to them:

> first, a spectrophone through which they can hear and see all that is happening on the Earth; a parlamonium, by means of which they are able to translate human speech, the sound of which irritates the Moon's inhabitants, whose hearing has been cosseted by the music of the spheres; and finally, an oneiroscope through which they can observe the earthlings' dreams.²⁷

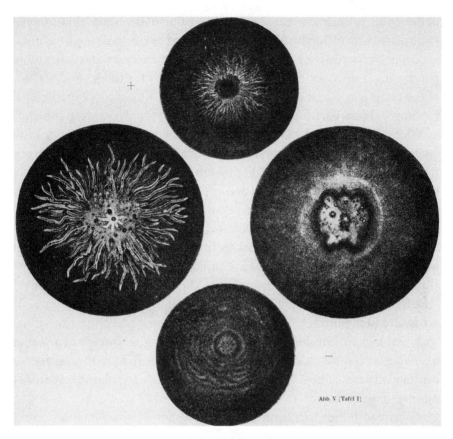

FIGURE 11.3. Sönnchen (+) und Möndchen (−) (little sun and little moon). Four of Lichtenberg's figures of electricity. From Georg C. Lichtenberg, *Über eine neue Methode, die Natur und die Bewegung der elektrischen Materie zu erforschen* (Leipzig, Germany: Geest & Portig, 1956).

"The Mass Is a Matrix"

From a media- and technological-historical perspective, Benjamin's artwork essay appeared right in the midst of the age of electronics, of telecommunicative and networked media. The year it was first published, broadcast radio was just twenty years old, and in Nazi Germany and elsewhere, test operations had begun with iconoscopes—electronic television cameras that no longer projected photograph for photograph but that wrote, point by point and line by line, planar light-figures in time. The fire- or light-writing that fascinated Benjamin in Ritter's text became an everyday technical reality in Hans Berger's electroencephalogram (EEG), which, by the end of the 1920s, the newspapers were already heralding as the direct "electrical transcription of thought," as the immediate "electrical writing of the brain."[28] But Benjamin's discussion of film is still very much shaped by the traditional technical image, or by successively transported photomechanical images, to be more precise. It is, moreover, astonishingly unaffected by the theory and practice of the avant-garde of the 1910s and 1920s: by Lev Kuleshov's experimental montage, Dziga Vertov's omnipresent *cine-eye*, the dynamic graphic art of Viking Eggeling and Man Ray's battles of materials on celluloid. With "shock" as the central category, another quality of Benjamin's interpretation is nudged into the subtext. For one thing, he explicitly refers shock to the Dadaist pictorial strategies of decomposition and reassembly; that said, an implicit presentiment may also reverberate in this, of the media already present and yet to come. Both etymologically and in everyday use, "shock" is associated with electricity and with the daily life-threatening dangers it posed at the time. One might take a similarly interpretive approach to the word phylum "emanation," "vapor," "air," the atmosphere, "soft shimmer," "trace," or "glimmer of light" that is contained in the concept of aura. Its proximity to fire imagery is as apparent as the semantic connections it has with that highest burning atmospheric layer, with the concept of *aether* as a vacuum with substance in an unspecific state, which, from the nineteenth century onward, is how it has been discussed as a popular theme in any number of publications.

"The mass is the matrix from which all traditional behavior toward works of art is today reborn in a new form." Thus begins section 18 of the first (1936) version of the artwork essay. It doesn't matter whether Benjamin was concerned here with the biological content in this remarkable concept (in Latin, *matrix* is another word for the mother animal, the uterus, the belly of the snake)—though, given the entire sentence, we may safely assume that he was—or whether he was referring to its mathematical and information-theoretical meaning, or both: his representation is borne by the hope that the artwork's absorption in the distracted masses and the dominance of the "tactile"[29] in the optical reception of film could produce a revolutionary change in the process of art. In recent decades, a utopian potential has consistently been identified in Benjamin's media critique, on the basis of expressions like these. Starting from the passage through his fascination with the fragments of Ritter the natural philosopher, whose understanding of the electric as a central phenomenon and whose outline for an anthropology of the arts Benjamin called a "virtual romantic theory," I would like to make a case for a rethinking of this media-theoretical potential, also and directly in regard to its involvement with politics and religion. Benjamin's hope is bound up in the strategies and tactics of classical modernism, firmly tied to the age of mechanics. He either ignores or recoils from the phenomenon of the electric. The few hints of his displeasure or impatience with the new electrically based and networked media, such as radio or telephone, that turn up in rather marginal texts allow one to surmise that following on this hope for change via the technical reproducibility of the artwork might possibly have been a fundamental critique of the electronic universe, like the one Adorno and Horkheimer outlined shortly thereafter, on the example of mass media, or like Günther Anders's fulminating analysis in the 1950s.[30] The author of the *Trauerspiel* book ardently stokes the aesthetic and semiological aspects of Ritter's theory of galvanism and electricity but recoils in irritation from their metaphysical excesses, their ideological and theological implications. He rejects the "unity of the manifold," whatever its central mediating phenomenon may be, as fiercely as he does the "thoughtlessness" of a "pantheism" grounded in technical knowledge, of a universalist understanding of technology *as* religion or a substitute for it.[31]

It's possible that his fascination with Ritter's *Heterodoxies* could bring a concept of technology into play that was proposed decades later by Vilém Flusser on the example of photography and by Félix Guattari as the subversive concept of machines: against the comprehensive hegemonic apparatus, to think incursion and the point of discontinuity that is linked with local, positive machines that do not allow themselves to be completely controlled.[32] These would be the relatively autonomous mechanical machines that Benjamin discusses in the context of photo and film apparatuses; and this would be the experimental practice, rich in tensions and contradictions, as well as the profligate life of Ritter, the Romantic physical chemist.

Electrical current is the controlled equalization of potential in the non-equilibrium of negatively charged electrons and positively charged protons. To conclude, allow me the liberty of a crude analogy. In Benjamin's *Dialogue on the Religiosity of the Present*, the I-interlocutor says, "But religions arise out of difficulty and need and not out of prosperity."[33] Free electrons move from lower to higher states of potential. Ritter's experimental and fantastical performances of the idea of an electric theology are full of such references and connections. But it is also characterized by dialogue with the other, which is a productive tension in difference, as a possible alternative to an affected and in the end convenient and fatal dualism that the theologians of the electric were attempting to sublimate with their concept of God. In a letter to Franz von Baader, Ritter formulates the ethical principle of his behavior, in a style reminiscent of the Jewish philosopher of dialogue Martin Buber: "What makes us good is not what we think up for ourselves, more so what we intend for others and what then falls back to us."[34]

Translated by Lauren K. Wolfe

Notes

1. Quoted in Erich Worbs, "Johann Wilhelm Ritter, der romantische Physiker: Seine Jugend in Schlesien," *Schlesien* 4 (1971): 227.
2. The concept is René Fülöp-Miller's, who wrote, four years prior to Benjamin's first version of the artwork essay, a wonderful book on the "fantasy machine of cinema" (*Die Phantasiemaschine*, 1931), which integrates cultural critique with psychoanalysis. I'm using it here also because of its nearness to Schelling's concept of *Allbeseelung* and the unity of natural forces.
3. Walter Benjamin, *Gesammelte Schriften*, vol. 2.1 (Frankfurt am Main, Germany: Suhrkamp 1977), 19. Published in English in Walter Benjamin, *Early Writings 1910–1917* (Cambridge, Mass.: Belknap Press, 2011).
4. Donatien-Alphonse-François de Sade, *Juliette oder Die Wonnen des Lasters: Erstes Buch* (Cologne, Germany: Könemann, 1995), 14.
5. See Eugenio Lo Sardo, ed., *Athanasius Kircher: Il Museo del Mundo* (Rome: De Luca, 2001).
6. All of Kircher's titles referred to here are discussed in detail and established bibliographically *in extenso* in my *Archäologie der Medien: Zur Tiefenzeit des technischen Hörens und Sehens* (Reinbek bei Hamburg, Germany: Rowohlt, 2002), translated into English as *Deep Time of the Media* (Boston: MIT Press, 2006). On sea voyages, see 32–33.
7. Rudolf Goeckel's formulation, very similar to Kircher's, dating from as early as 1609.
8. Athanasius Kircher, *Magnes sive de arte magnetica* (Rome, 1641), 7.
9. Apart from Oetinger's texts, there are scattered letters from him to Prokop Diviš and Fricker as well as dictionary and compendia entries in Catholic encyclopedias. In 1970, Ernst Benz wrote a dissertation on *Theologie der Elektrizität* (Mainz: Akademie der Wissenschaften und Literatur, 1971)—translated into English as *The Theology of Electricity* (Allison Park, U.K.: Pickwick, 1989)—and there is another more recent one in Italian by

Paola Bertucci, who is concerned above all with English writers (e.g., John Freke) of the late eighteenth century. It contains a chapter on "electric fire and the *theologia theutonica*"—so called because strongly premised on Jacob Boehme's theosophy, with its dualistic conception of life energies, which influenced Ritter as well. In the following passages, I am summarizing and honing Benz's work. The page numbers in parentheses refer to his dissertation.

10   Leo Graetz, *Die Elektricität und ihre Anwendungen zur Beleuchtung, Kraftübertragung, Metallurgie, Telephonie und Telegraphie* (Stuttgart, Germany: Engelhorn, 1883), xi.
11   Quoted in Benz, *Theologie*, 58.
12   Reedited with a preface by Dmitrij Tschiezewskij and Ernst Benz in *Emblematisches Cabinet* (Hildesheim, 1969), 400.
13   See Benz, *Theologie*.
14   Cf. Benz, 67.
15   Cf. Carl Graf von Klinckowstroem, "Johann Wilhelm Ritter und der Elektromagnetismus," *Archiv für die Geschichte der Naturwissenschaft und Technik* 9, no. 2 (1929): 68–85.
16   Ritter in a comment to Hans C. Ørsted. Johann Heinrich Voigt, ed., *Magazin für den neuesten Stand der Naturkunde* 9 (1805): 33–34.
17   Johann Wilhelm Ritter, *Fragmente aus dem Nachlasse eines jungen Physikers: Ein Taschenbuch für Freunde der Natur*, ed. Steffen and Birgit Dietzsch (1810; repr., Hanau: Müller & Kiepenheuer, 1984). Facsimilie print with afterword by Heinrich Schipperges (Heidelberg, Germany: Schneider, 1969). An Italian edition appeared in 1988 with the title *Frammenti dall'opera postuma di un giovane fisico* (Rome: Edizioni Theoria). An English translation appeared as *Key Texts of Johann Wilhelm Ritter (1776–1810) on the Science and Art of Nature*, ed. and trans. Jocelyn Holland (Boston: Brill, 2010).
18   Walter Benjamin, "Ich packe meine Bibliothek aus: Eine Rede über Sammeln," in *Medienästhetische Schriften*, ed. Detlev Schöttker, 175–82 (Frankfurt am Main, Germany: Suhrkamp, 2002), 181. Published in English as "Unpacking My Library: A Talk about Book Collecting," in *Illuminations: Essays and Reflections*, edited by Hannah Arendt (New York: Schocken Books, 1968).
19   Walter Benjamin, *Ursprung des deutschen Trauerspiels* (Frankfurt am Main, Germany: Suhrkamp, 1972), 240. Published in English as *The Origin of German Tragic Drama* (London: Verso, 2009).
20   Quoted in Benjamin, *German Tragic Drama*, 241; in Ritter's writings, this passage can be found at the beginning of the appendix to the *Fragmente* of 1810.
21   Benjamin, *German Tragic Drama*, 242.
22   Quoted in Ritter, *Fragmente*, item 619; Lucretius, *De rerum natura*, German ed. (Stuttgart, Germany: Reclam, 1973).

23 Ritter, *Magazin für den neuesten Stand der Naturkunde*, September 1803, 213–14.
24 Ritter, *Fragmente*, item 358.
25 Ritter in a letter to Christian Gottlob Voigt, March 26, 1804, quoted in Else Rehm, "Johann Wilhelm Ritter und die Universität Jena," *Jahrbuch des freien deutschen Hochstifts* (Tübingen, 1973), 206.
26 Ritter, *Fragmente*, item 360.
27 Walter Benjamin, "Lichtenberg: Ein Querschnitt," in *Drei Hörmodelle*, 51–86 (Frankfurt am Main, Germany: Suhrkamp, 1971). To my knowledge, Benjamin's text never was produced for radio during his lifetime.
28 Cf. Cornelius Borck, "Electricity as the Medium of Psychic Life: Psychotechnics, the Radio, and the Electroencephalogram in Weimar Germany," Max-Planck-Institute for the Advancement of Science, preprint 154, Berlin, 2000.
29 Walter Benjamin, "Das Kunstwerk im Zeitalter seiner technischen Reproduzierbarkeit," in *Gesammelte Schriften*, vol. 1.2 (Frankfurt am Main, Germany, 1974), 466. English translation as "The Work of Art in the Age of Mechanical Reproduction," in Benjamin, *Illuminations*.
30 Günther Anders, *Die Antiquiertheit des Menschen*, esp. vol. 1, *Über die Seele im Zeitalter der zweiten industriellen Revolution* (Munich, Germany: Beck, 1956).
31 Concerning the "unity of the manifold," see Walter Benjamin, "Dialog über die Religiosität der Gegenwart," in *Gesammelte Schriften*, vol. 2.1 (Frankfurt am Main, Germany: Suhrkamp 1977), 20. The passages on "thoughtlessness" of a "pantheism" are found on 20 and 22.
32 Cf. Henning Schmidgen, *Das Unbewusste der Maschinen: Konzeptionen des Psychischen bei Guattari, Deleuze und Lacan* (Munich, Germany: Wilhelm Fink, 1997).
33 Walter Benjamin, "Dialog über die Religiosität der Gegenwart," 25.
34 Cf. Wolfgang Hartwig, "Physik als Kunst: Über die naturphilosophischen Gedanken Johann Wilhelm Ritters," PhD diss., University of Freiburg im Breisgau, 1955, 89.

12   Historic Modes of the Audiovisual Apparatus

> *In the early 1990s, film and cinema historians around the world began to reflect more intensively on the massive shifts that were increasingly legible in the use of film and in the perception of filmic constructs. The classical cinema apparatus, as hegemonic structure, began gradually to give way. The Parisian journal* Iris *dedicated a special issue to the audience and to the public in the cinema. By the time this special issue came out in 1994, my study on* Cinema and Television as Entr'actes in History (Audiovisions) *had been on the market in its original German for five years. Some of this book's theses have been condensed and updated in the following essay.*

⁂

A topic such as this could precipitate jubilatory activity in reflections. We might be tempted to celebrate a self-present, human cinema subject as perpetual, that is, to celebrate an imaginary subject; we might be enticed into constructing a genealogy of progress, of a progressive transfiguration; we might be strongly inclined to construct a linear history, from the anarchic beginnings at the nineteenth fin de siècle, *How We Got into the Pictures* (Noel Burch), to the trends toward the disintegration of spectatorship at the twentieth fin de siècle in favor of the hegemony of seeing machines and cultural identities like electronic surfers or shipwrecked passengers in the naturally forceful metaphorical free flow of signifiers. We might succumb to the danger of ascribing mystic qualities to cinematographic spectatorship, or—what amounts to the same thing epistemologically but which would be almost worse—we might completely dissolve the mysteriousness of spectatorship as a theoretical concept into a generality, because of the arrogance of a penetrating transparency, and in so doing deprive the tension between dream and reality, wherein film has its location, of its foundation: the very existence.[1]

What I shall invite you to do with me for a short time is to inspect critically those shifts in the relationship between filmic subject and filmic apparatus, which in their basic ideas can offer resistance to the temptations and dangers I have just mentioned. This is done by setting certain important heuristic principles, all of which have no theoretical status as yet. The principles are as follows:

- The historical relationship between subject and apparatus cannot be described in terms of a continuity; for then we should merely duplicate what cinematic projection does all the time in principle, namely, it appears to transform that which in reality is discontinuous into that which is continuous. We presuppose this relationship to be discontinuous.
- Continuity, I maintain, is on a metapsychological level, situated above the various arrangements of film viewing and film experiences: the level of the need, the desire for visual illusions of movement and audiovisual illusionization as a whole, which goes back long before the hundred years of cinema history and extends beyond it.
- Although in many cases in discredit, I continue to try to work with the concept of the *dispositif*, as it was introduced into the cinema debate by Jean-Louis Baudry, not identical with Foucault's superstructural concept. This is because I can see no alternative yet that is in a position to grasp more adequately the interdiscursive event and experience time-space of film reception, which also includes the individual wish-machine.

> The Iris Colloquium aims at opening up the semiological field to contributions from disciplines such as history, economics and psychoanalysis in order to approach a more "real" fe/male spectator.[2]

If you add to this list of disciplines philosophy, ideology critique, and art history, then you arrive at the approximate construction that the Apparatus Debate of 1970–71 attempted to address: with the integration of ideas from Althusser, Husserl, Panofsky, Freud, and Lacan, into a critique of systems of representation and their organization of subject-effects. Cinema is itself an interdiscursive point of intersection. Thus the film-subject, too, which is itself highly *undisciplined,* cannot be found by any other way than by wandering between the disciplines, even at the risk of becoming discipline-less.

For our short stroll through the historical modes of filmic dispositions, combined with the attempt to keep the Apparatus Debate alive and to develop it further, we can certainly use Baudry's diagram (Figure 12.1) from his first essay on this context, "Effets idéologiques produits par l'appareil de base," as a starting point. So, to remind you of the debate:

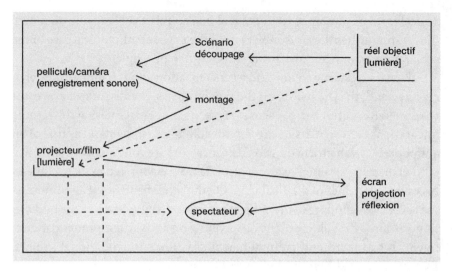

FIGURE 12.1. From Jean-Louis Baudry, "Effets idéologiques produits par l'appareil de base," *Cinéthique* 7–8 (1970): 1–8.

Actually, this point of departure is undisputed in all more recent ciné-subject heuristics that I am aware of: the spectator is in a physically immobile position and in a mode that is allied to hallucinations of the waking state and the dreams of sleep, situated between the projection apparatus in the strict sense of the term and the reflecting screen. Contact with real experience of "the outside" is cut off. The real is, as a rule, subjected to a fourfold process of transformation:

1. transformation into language in the scenario/découpage
2. transformation into images (in the sense of Cartesian geometry by means of the camera-eye) and into sound recording
3. the editing of this material
4. the illusionistic revivification through the readdition of light by the projector and the production of apparent continuity from the discontinuous material

I shall not dwell here on the weaknesses of this schematic model, which I see principally in the fact that although the "spectateur" is not conceived of as wholly passive in that Baudry understands the recourse to this situation in the sense of an activity of the wish-machine, he accords him or her no place of interdependence in relation to the "editing/montage" and models such a radical form of severance from the real that could never really be obtained. This distinguishes the ciné-subject from the psychopathological case of the schizophrenic or paranoiac in *degree* but not in principle.

What interests me so much in a generalization about Baudry's model is, on one hand, that he insists on the powerfulness, the discursive power, of the apparatus in the narrow sense per se, including its quasi-architectonic integration, as a force. Its relation to the film text I understand as that of an antipode, with which the transformation factors are continually at odds, in a kind of ongoing strained relationship (Comolli elaborated on this a little in his essay "Machines of the Visible").[3] On the other hand, what also interests me is the close binding of the subject's psychic activity/state to the body, to its position as the vehicle of the observer's gaze and the temporal existence, in which the social and cultural machinery hinds the individual subject. Existence is used here in the sense of Binswanger's *Le rêve et l'existence*, as the basis connecting "vital function" and "life history," connecting the individual/the isolated individual and also the universal/society.[4]

I would like to present six exemplary modes of such positioning and temporal fixing. I regard these as the load-hearing *Bausteine* for the whole edifice of film-subject identities in the past, in the present, or that can be anticipated for the nearest future. The concepts that I use in their classification must also be understood as assigning them to certain epistemological problems, which in the interaction of the formulation of advanced media theory touch upon questions of interest to the natural sciences, technology, and culture.

The Observer and the Participant in the Conception of Athanasius Kircher's Allegory Machine

In Kircher's *Ars magna lucis et umbrae* (The great art of light and shadow), first published in 1645/6 in Rome, second edition 1671, Amsterdam, the two viewer/spectator identities that will be fundamental to later developments are still found in a single symbiotic arrangement (Figure 12.2). There is no split subject; instead, Kircher constructs two subjects with opposing viewing points and corporalities. Subject A, which is to be illusionized, imaginized, is situated in the space of the illusion or projection. It goes there to set the wish-machine going. It looks into the mirror and at first sees itself. The *je* and the *moi* in mirror-inverted correspondence. But the imaginary *I* becomes

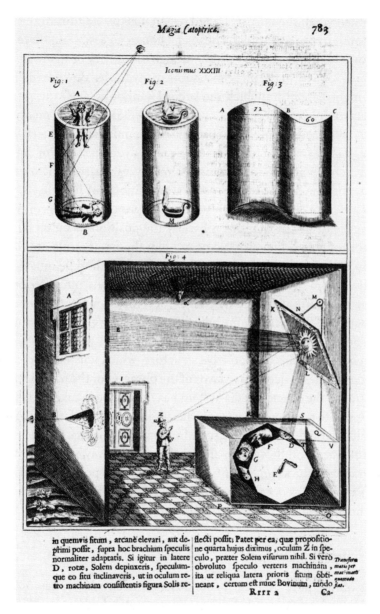

FIGURE 12.2. Kircher's metamorphosis apparatus for the allegorical transformation of an observer. On the upper side of the case containing the drum with images, one can see that a rectangle has been cut out so that the mirror arrangement can project the images. On the left, outside the chamber, is the disembodied eye of the voyeuristic observer B. The figure at the top of the page shows glass cylinders for Kircher's planned projection of figures that appear to float in the air. From Athanasius Kircher, *Ars magna lucis et umbrae* (Amsterdam: Johannes Janssonius, 1671), 783.

transmuted when the mirror is tilted, when it is turned toward the hidden projector: it becomes a sun, the head of a lion or an ass, when these images are thrown up by the allegory drum through the agency of an invisible third party. The "Other" in the *moi* can assume many forms. Bewilderment, possibly a nightmare, from which the subject cannot escape—despite being relatively mobile—inside this strange chamber. Subject A is incorporated into the architecture and the apparatus. By comparison, Subject B is at a remove. It is the outside observer, the secret voyeur, perhaps of the primal scene. Kircher places in it an unframed reality, as a part of the objectively real. It savors the deception process in which Subject A finds itself, yes—it may even organize/command this process as projectionist, as controller of the allegory drum. In fact, this is exactly the position taken up by almost the whole of classic film semiology and film theory as a rule. It is also the position of the analyst, who fancies himself or herself to be outside of the deception, who believes he or she can transmute the inner process into truth by transforming it into language. (See, for example, the drawing on the program of the colloquium: a distanced sidelong glance over the heads of the spectators.)

Kircher's model is a dispositive structure, which epistemologically fits premodern times; a decision in favor of the observer or the participant position vis-à-vis the special media system has not yet been taken. This model allows for both positions, that is, it indicates the division of roles that will be generally adopted on a massive scale in the years to come. There are those who see clearly and those who are dazzled. By the way, with regard to the iconography of the arrangement, I would just like to point out that the eye of the external observer has no body; it is detached, free of gravity, so to speak, hovering at the eyehole of peep-show architecture.

The Controlling Subject

Models of the "pure" Subject B, the observer of a reality (often of nature) located outside of the psychic and media apparatus, had been outlined long before Kircher. But due to the interdependency of experimental practice and formulation of theories, their establishment as a hegemonic mode came only at the beginning of the seventeenth century with Christoph Scheiner, Kepler, Galileo, and Descartes as the theoretical highlights. The camera obscura is really the ideal form wherein the world is created by the eye and for the eye. Projected onto the media apparatus standing opposite the subject, at his disposal, it becomes operable two-dimensionally for the purpose of its reversibility as image, as architectonic plan.

This controlling, disciplining, ideal observer-eye is—like Kircher's Subject B—principally bodiless, severed from its physical and psychic vital functions. It is completely the life-history eye. It is that artifact-like entity that we can identify in popular representations by the camera and that which through recent experimental brain research has actually now become replaceable in bioinformatics: by light-sensitive microsensors connected to the appropriate areas of the visual cortex.

I shall not elaborate further on this specific construction of the subject in modern (natural) science. This has been done many times before, one of the most recent outstanding examples being Jonathan Crary's *Techniques of the Observer* (the source of Figure 12.3):

> If at the core of Descartes' method was the need to escape the uncertainties of mere human vision and the confusions of the senses, the camera obscura is congruent with his quest to found human knowledge on a purely objective view of the world. The aperture of the camera obscura corresponds to a single, mathematically definable point, from which the world can be logically deduced by a progressive accumulation and combination of signs. It is a device embodying man's position between God and the world.[5]

However, it is important to emphasize one thing—and Kircher's mode of seeing does just this: that although this conception of the subject was hegemonic for a long time in the media approach to the world, it did not exist in a pure form. This constitutes the essential discontinuity of history. The concepts of a controlling, divine sight were continually counteracted/countered by apparatus and textual constructions that were connected with the imagination, the dark area of dreams and drives, the linkage of seeing and secrets: from the accessible and portable peep shows to the phantasmagorical performances at the turn of the eighteenth century. That which Martin Jay[6] maintains for art history is also valid without qualification for the history of the filmic: hegemonic constructions vied with subcultures of apparatus visualization. This friction is also something that determines our current engagement with computer-centered systems of the filmic, or rather, *should* determine it theoretically.

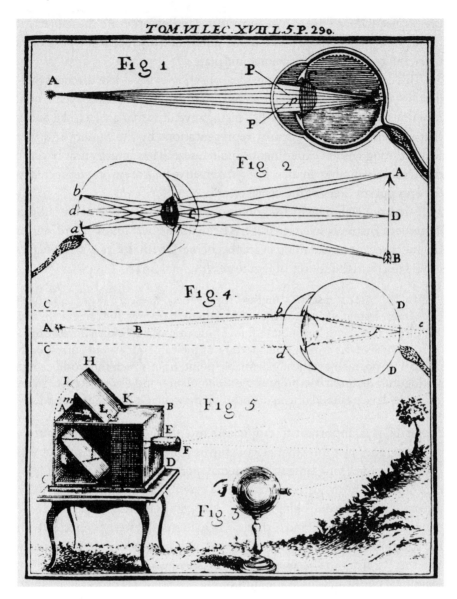

FIGURE 12.3. Comparison of eye and camera obscura. Early eighteenth century. From Jonathan Crary, *Techniques of the Observer: On Vision and Modernity in the Nineteenth Century* (London: MIT Press, 1990), 49.

The Subject in the Illusion and Reality Test

Completely and utterly nonuniform, extremely heterogeneous, appears to me to be how the subject is assigned, how it is excavated from the historical time-space, which cinema mystics still refer to as Pre-cinema (thus writing off its fascinating originality and independence). From the magical art of projection to the complex productions of the dioramas and pleoramas, the diversity of performances found within the panoramas, from the individual machines/devices for scope illusions (Wheel of Life, Praxinoscope, etc.) to the first audience-oriented special systems of moving technical illumination and projection (Elektrischer Schnellseher, Zoopraxiscope, Bioscope) and the first cinematographic arrangements, we are dealing here with plural changes: changes from positions of maximal involvement and extreme distance (for instance, in the case of those artifacts where the person to be illusionized and the person manually creating the illusion are identical: Praxinoscope, Mutoscope, Kinetophonograph, and related devices for individual use; Figures 12.4 and 12.5).

Although these represent such different organizations of the observer's gaze, there nevertheless appears to me to be a kind of center of congruence: the plurality of the subject—analogue to the dispositive heterogeneity—is located in suspension between the *reality test*, on one hand (here Walter Benjamin's definition is most apt, as, in my opinion, his heuristics of the technical reproducibility are an excellent conception for the phenomena of the nineteenth century, but not for the twentieth), and the *illusion test*, on the other. The latter-day double and/or split identity of the ciné-subject is prepared here in alternating hot and cold, so to speak, and thrashed out. Both the apparatus for producing the *impression of reality* and the extravagant production of *Schein* are equally the object of curious observation, as are the textures that they have to offer the subjects, whether the curious gaze travels through the apparatuses themselves, that is, the latter are used directly for illusionization, or whether—as in the case of the early cinematographic performances—the machines and their operators, background noise, projection surface, projection room, and spectator are not disposed in a firmly established order of proximity, forming a tense relationship to each other of mutual stimulation, of alternating actions and reactions. An open relationship, also with respect to the reality outside of the media, which invaded the illusion space not only in spectacular ways, such as the screen bursting into

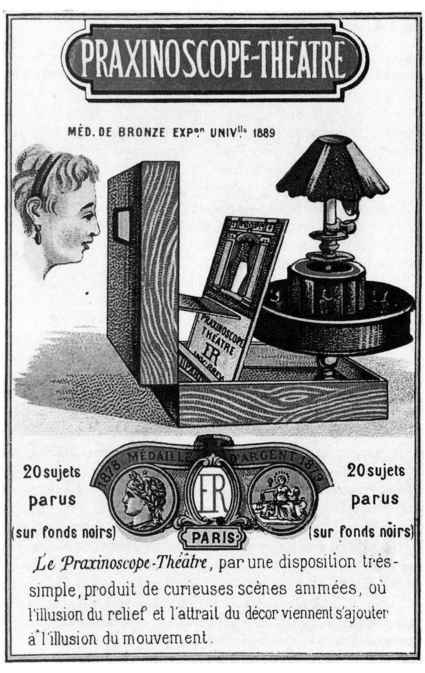

FIGURE 12.4. Charles-Émile Reynaud's individual Praxinoscope-Théâtre. From Magic Lantern Society of Great Britain, Zielinski Archive.

FIGURE 12.5. Optical phonograph by Edison/Dickson (1887). From the Zielinski Archive.

flames, but also through film breaks, extreme flickering before the invention of the cover wing of the shutter, the public house, café, or vaudeville atmosphere; it invaded this space, permeated it, and helped to structure it. For my part, I am unable to associate the human subject of transcendence with this. Rather, the subject of transport, of transit, of passage, where not even temporal submersion is excluded. Incidentally, the physical mobility of the very early ciné-subject corresponded to the mobility of the machines: before the time when it was primarily the film software that went on tour, it was the hardware that wandered (together with its closely related software—also a problem of insufficient compatibility).

The *Dispositif*

The space occupied by the event and the organization in classic cinema of the transcendental subject, that dark cube where the imaginary signifiers revel and celebrate their orgies in their encounter with the desires of the spectators, in which the ideal psycho-physiological architecture for Lacan's dramaturgy and dramatic art of the subject is seen in the dialectic of the experience of lack and compensation, that place of the immobile flesh of the spectator caught between the divine hand of God, pointing in the beam of the projector, and the winding sheet to which Godard compares the screen in his *Histoire(s)*—this is really a product of the twentieth century: its theoretical formulation began at a time when it had ceased to exist as the primary location of filmic activity; it was at best suitable for the transfiguration into a myth. It established itself parallel to the mass installation of telecommunication practices in the late 1920s, and underwent a polishing and perfecting as audiovisual *dispositif* at the time when fascism took hold in Germany and raged throughout Europe, at the time of the epitomy of barbarism and its aftermath of hot and cold war. The first important empirical study on the death of the cinema, John Spraos's *The Decline of the Cinema*, appeared as early as 1962.

What I would like to convey with this somewhat strange periodization, at least rudimentarily, is perhaps already clear in the very brutality of the formulation—the dramatization of the subject,

> in the cave (Baudry),
> in the dark cube (Barthes),
> in the machine of the visible (Comolli),
> and/or in the cabinet of mirrors (Lacan),

must not only be considered in terms of its healing, therapeutic function but also interpreted in connection with its extreme ambiguity in the unfolding of desires and the containment of desires. This, for example, occasioned the vehement engagement of feminist film theory with the first apparatus heuristics from Pleveny, Baudry, Comolli, and the relevant passages from Metz and led to new departures in the work of the filmic avant-garde, from expanded cinema, cinema for amateurs, and radical opposition to the basic apparatus to even cooperation with television. (Consider Godard's film work from 1968 to 1978, from *British Sounds* to *France tour détour deux enfants*.) We will focus on this in the fifth mode.

The Symbolic Turn toward the World in the Televisual Apparatus and the Decentralization of the Subject in the Video Space

The majority of dislocations that have been defined for the media subject in the shift from cinema to TV hegemony (insofar as such analyses have been conducted seriously),[7] I personally cannot make much use of. They are either superficial and do not stand up to closer techno-aesthetic examination (as in the posited shift from the discontinuity of cinematographic material to the alleged continuity of electronic material; in television, continuity is also production of *Schein*) or they are static, ahistorical, which goes back to McLuhan's differentiation between hot and cool media, between high and low definition, projection versus introspection, and so on. From the very beginning, television existed also in the version of a large-scale projection medium in the public sphere—its establishment as a medium of intimate privacy was above all an economic and sociocultural policy decision and not a techno-aesthetic or psychological one—the genesis of the televisual apparatus with its fine-grained structure of the images can also be described as a gradual but ever closer approximation to the cinematographic image—but also to the photographic image; and currently we are experiencing again a revival of televisual projection in the area of everyday culture, both in the public and the private space (not only with the variants of high-definition television).

For me, the most significant shift is connected to the specific corporality of televisual filmic visuality and is expressed markedly in the postures of the dispositive arrangement. The subjects' bodies in front of the monitor are disarranged and decentralized. They can be in motion in relation to the electronic picture frame or within its acoustic range, in the mode of passersby; immobilized, at rest, they describe the most different angles possible in relation to the axes in the televisual text (the ideal camera and action axes): from acute-angled displacement to turning aside, distancing of the body as the vehicle of visuality (Figure 12.6). Postures such as these can also be observed—in a less marked form—on occasions of monitor viewing in an anonymous collective. The eyes of the viewers are very wide open, attentive; the bodies are not with them, with the subjects, but somewhere or other in transit in the interspace between recipient and monitor.

The basis for these shifts in the direct sense are differences in ambience, in the architectonic conditions, and in the lighting in the illusion space,

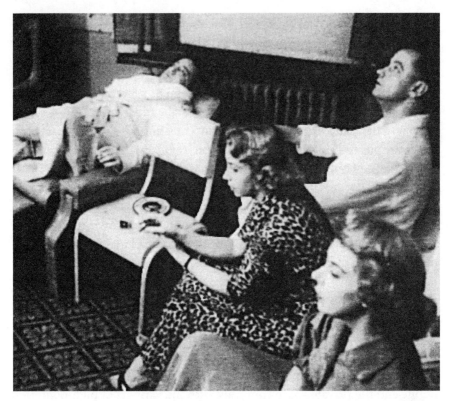

FIGURE 12.6. It could be excorporation. But it is incorporation, in the box: Jackie Gleason, Audrey Meadows, Art Carney, and Joyce Randolph. The set is turned off. From *Time Life*, March 1989, special issue: "TV Turns 50."

which stand in close reciprocity, on one hand, to the different kinds of text bodies of cinematography and, on the other, to televisuals. Arguing from Lacan's three intermeshed fields of the psyche as the arenas of the dramaturgy of the subject, the bodies of the TV-subjects sit or lie nearer to the symbolic, incline toward staying on the edge of the imaginary. (This is why the cinema myth loves Lacan so much: his imaginary also offers a kind of a refuge between the real and the symbolic, where one can lodge in a state of temporary well-being.) The dominance of the spoken word in television that this medium appears unable to do without, the tautological basic construction of its audiovisual signifying practice and the rhythmization, the text, as acoustics in Music Television, are, along with the spaces of TV use, open to the extra-filmic world, the essential preconditions for these shifts. The concept of television is now as ever a misleading promise of use-value.

Television evolved first and foremost as a culture of speaking, of listening or of not-listening; on the part of the aesthetic production, this has only slowly begun to change. And not least in the context of recent technologies of presentation.

It is a different case with the image- and music-centered medium of video. The filmic videotape (including the new multiple-choice offerings on videodisc like Peter Krieg's *Suspicious Minds—Die Ordnung des Chaos,* 1991), with its orientation toward the individual and small groups, or video art experiments in their media materiality move close to the boundaries of literature, on one hand, and of painting, on the other. With respect to the positionings of the subject, we are dealing here with basic characteristics as in the televisual: there is no binding but rather principally a mobilization of the seeing and hearing bodies and above all a distancing, which is not neutralized by the possibilities of participating in the text by machinic means but rather amplified by these.

Substitutional Presence, Incorporation, and Participating Observation in Present and Future Modes of the Filmic

With this level of that which is already here and that which can be anticipated regarding new practices of subject positioning, we leave the relatively firm terrain of historical reconstruction. But in a sense, we still remain on it, for in the present prehistoric age of computer-centered film reality, once again, we see the emergence of that heterogeneity and indistinguishableness we encountered in premodernity.

Again extreme contradictions meet and collide with each other: in the cinematographic mode, the use of applications, reproduction, and production modes like the IMAX (Figure 12.7) or OMNIMAX theaters, with their 70mm projection, sixteen sound channels, gigantic screens with breathtaking deep-focus images and perfect marginal sharpness, these carry the incubation techniques of traditional cinema to extremes while retaining and intensifying all the subject-effect operations, as formulated by Baudry in *L'appareil de base*. In the foreseeable future, the electronic image will not be able to compete with this, even though the qualitative jump from traditional TV to high definition over-1000-lines-TV and opulent quadrophonic sound is, in proportion, comparable to the distance dividing the 35mm standard from IMAX.

FIGURE 12.7. Cross section of a cinema. Courtesy IMAX Corp.

On the level of marketing strategy, the subject's use-value expectations of telepresence in the imaginary space are being utilized in connection with such applications. However, the actual meaning of this concept, shaped by NASA, of the *substitutional* presence of the media-being at a time/place, which is absolutely not accessible to him or her, only really unfolds in a cultural industry context with the appliances, programs, and protheses of so-called virtual reality. On a commercial basis, "W-Industries" offers in its "Location-Based Entertainment" centers of which only a few techno-freaks dreamed a short while ago: equipped with Data-Dessous, head-mounted three-dimensional image generators, and sensors for the human–machine interface, the filmic subject discards its status as an observer and voyeur and becomes a participating player in an audiovisual model based on algorithms in which potentially Cartesian geometrics are suspended, because—assuming that the primary identification works—the subject not only identifies with certain agents on the secondary level but, substituting itself for

FIGURE 12.8. From *Virtuality* (Autumn 1992).

the machine, assumes their active roles. This is an incorporation that is not possible in reality, and the boundaries between the symbolic and the imaginary are thus under further attack. Only "civilized people" who are intensely accustomed to symbolic interactions are capable of activating their imagination within such models. The connection, the contact to the material world, to the assurance of the presence of the real, remains even here a necessity. The agents who are to "drop out" of their own bodies are set down in, for example (Figure 12.8), a synthetic natural environment and above all limited in their mobility. Not as much as in traditional cinema, of course, but to fetter them appears to remain a condition sine qua non of cultural industry illusionization.

It is precisely at this point that, in the present, the artistic breakthrough aims in employing such techniques. To cite but one elaborate example, which also illustrates the indecisiveness of artistic subject positionings, as part of their project *Simulation Space: Mosaic of Mobile Data Sounds,* the interdisciplinary group Knowbotic Research constructed a walkable music databank (it received one of the 1993 Golden Nica awards at Ars Electronica in Linz). A real attempt at an endophysical artistic model. The acoustic fragments, which were collected from all over the world via E-net, are structured in the logic of self-organization by a complex program and then sent to the dark active space that simulates the data store. There they build groups, gangs (definitely in the sense of Guattari and Deleuze!), which continually break up and re-form into new units. The visitors to this installation enter this space and structure the acoustic space though the movements of their bodies, accomplished by means of an infrared sensor and a private-eye-viewfinder, which then provides the visual information on the acoustic material. In this mode, the artistic installation follows exactly the principles of incorporation and participation; indeed, the mobility of the subjects' bodies is both prerequisite to and constituent for the aesthetic process. Specularity and narcissistic identification no longer play any role in this time-space, the subject-effect of ego-transcendence, for the data and structuring machines remain out of sight.

However, the group of artists wanted to make the presentation available to an audience. So, on a screen located outside of the actual space of the event, the chaotic movements described by the acoustic fractals are optically represented, electronically projected. Here Subject B comes again into its own, as we have already seen in Kircher, as a collective observer and voyeur-subject.

This installation, with its double perspective of perception, brings me to my conclusion, which is brief but, I hope, not abrupt.

In the future, we will continue to have to deal with heterogeneity of filmic subjects and their positionings, where the ciné-subject will perhaps be accorded a more specific place; that is, it may have to strive for it. This differentiation is not a question of who is being addressed but what is being addressed and *how*. The modes of distancing, of observation and participating observation, of active intervention (so-called interactivity), together with the mobilization and decentralization of the subjects' bodies, will be the domains of all those fields that are *utterable*, that can be expressed in language, that possess either actually or potentially a symbolic character, are indebted to Foucault's *dispositifs* of knowledge and truth, and to sexuality, insofar as it is of a promiscuous and pornographic nature. The relationship of the world exterior to the filmic media is one of open borders or of identity, which amounts to the same thing.

By contrast, the ciné-subject submits of its own free will to the fetters of what is left of the real and of perpetual attempts in vain to get closer to the real through images and sound. Immobility of the body, traumatic suspension of existence in the darkness and anonymity of a freely chosen cube—these are the preconditions for the imaginative excursions into the realms of the impossible, that which stubbornly refuses to be able to be spoken: death and eroticism, the big and the small death. "A film is a gun and a girl." Godard lets us read this in his *Histoire(s) du cinema* (part IA, 1989), formulated from a masculine perspective. I am sure that he was not only being ironic.

Translated by Gloria Custance

Notes

1 The concept of "existence" is used here in the sense of Ludwig Binswanger, *Traum und Existenz* (Berlin: Guchnang & Springer, 1991). The short introduction is a conscious montage of various phrases that have exerted a strong influence on the discussion of film theory in recent years, from Baudrillard, Burch, Foucault, Lacan, and Virilio.
2 Taken from the invitation letter of the colloquium.
3 Jean-Louis Comolli, "Machines of the Visible," in *The Cinematic Apparatus*, ed. Teresa De Lauretis and Stephen Heath, 121–42 (London: Macmillan, 1980).
4 Binswanger, *Traum und Existenz*, 135.
5 Jonathan Crary, *Techniques of the Observer: On Vision and Modernity in the Nineteenth Century* (London: MIT Press, 1990), 48.
6 Martin Jay, "Scopic Regimes of Modernity," in *Vision and Visuality*, ed. Hal Foster, 3–28 (Seattle, Wash.: Bay Press, 1988).
7 See, e.g., Sandy Flitterman-Lewis's essay in Robert C. Allen's *Channels of Discourse, TV and Contemporary Criticism* (Chapel Hill: University of North Carolina Press, 1987) and—less elaborated—John Ellis, *Visible Fictions: Cinema, Television, Video* (London: Routledge and Kegan Paul, 1982).

13 "To All!"

The Struggle of the German Workers Radio Movement, 1918–1933

The critique of political economy and a critical theory of the consciousness industry were important, closely correlated paradigms of media thinking in Germany in the 1970s. But this single-minded working out of hegemonic relations began to bore me rather early on. My research into the Workers Radio Movement in the Weimar Republic, until its suppression by German fascists, began what would turn out to be a years-long search for traces of technopolitical and techno-aesthetic alternative movements in the tradition of interventionist media practices from below—guerrilla video, hacking, or critical engineering, as a contemporary variant on epically funded media practices is called.

This essay was originally published in 1977 in German in a much longer variant.

⁂

From Early Centralization to the Split of Arbeiter-Radio-Bund (Workers Radio League)

"From high above came words elated / Let Telefunken be created"—this is how in 1924 an old major remembered the 1903 merger, prompted by imperial and military interventions and mediations, of the two leading German electric corporations into the Telefunken Company. "Already in 1903 the young auxiliary weapon earned its first laurels. Six mobile stations participated in the South-West African campaigns.... What a beautiful sales area!"[1] In other words, the development of radio was tied to the suppression of colonial uprisings as well as to the first imperialist campaigns to reorder the world. German capital, after all, was "confronted by the fact that other countries had been on the spot much earlier; as a result Germany had to make do with partial successes at best." To quote Hans Bredow (1879–1959), erstwhile Telefunken chairman, later director of Reich Broadcasting Corporation, and, ultimately, recipient of the Federal Republic's Grand Cross of Merit, "only German radio was in the course of time able to satisfy German desires."[2]

FIGURE 13.1. Light wireless station in the beginning of the war, 1914. *Left,* sender and receiver. *Right,* engine and tower. From the Zielinski Archive.

Leichte Funkenstation bei Kriegsbeginn 1914
Links Sender und Empfänger, rechts Motor und Mast

FIGURE 13.2. Light wireless station, summer 1914. From the Zielinski Archive.

In 1918, however, many Germans had only one desire: peace and freedom.

When the brand-new Soviet government in November 1917 broadcast its widely disseminated message "To all!" announcing the German–Russian peace decree, it also revealed a hitherto undiscovered global playground to radio. In any case, when the former Telefunken chairman, who had participated in early tests using tube transmitters designed to cheer up entrenched soldiers, arrived at the Western Front in 1917, he discovered to his "horror, that the troops had been infected by the mood of the home front. There was incessant politicking and the momentum appeared to have been lost."[3] Following fraternizations on the Eastern Front and the High Seas Fleet mutiny, mass strikes, and antiwar demonstrations, Berlin had, on November 9, 1918, fallen to revolutionary workers. The fact that on the very same day the soldiers' councils of the radio operators constituted a Zentralfunkstelle (Central Broadcasting Headquarters; ZFL), seized Telefunken's international

FIGURE 13.3. Radio colonialism: two "native workers" in a village in northern Rhodesia producing the energy for a generator, which supplies high- and low-tension electric circuits to drive a radio transmitter and receiver of the Marconi company. From *Armchair Science,* July 1932, 264.

broadcast stations, established contact with the Soviet government, and broadcast a message "To All!" about the revolutionary events in Germany left the reigning bourgeoisie in no doubt that the *socially* "organized power"[4] of the medium posed a considerable threat.

During its brief rule of the airwaves, the German Workers Movement was able to put to use the technical and organizational wartime experiences in the service of a much broader, mass-media-based pursuit of its political aims. This posed such a threat that the Telefunken chairman himself took to arms and joined a regiment of volunteers[5] ordered by the Social Democratic secretary of defense to crush the revolutionary uprising. As the first postwar government was formed, he joined the civil service: "Resisting the radio operators' revolutionary aspirations was also of technical interest to me" (Figure 13.4).[6]

The war, then, not only produced a motley crew of advanced radio militarists but also left behind a reserve army of savvy radio operators who could have turned radio into a highly effective means of social communication capable of advancing a socialist republic. One precedent was set by the development of the medium by the Soviet government, whose principal

FIGURE 13.4. Hans Bredow's ID as a volunteer "against Spartacus." From Hans Bredow, *Im Banne der Ätherwellen*, vol. 2 (Stuttgart, Germany: Mundus, 1956).

radio station in 1917–18 transmitted two thousand to twenty-five hundred daily messages "To All!" Despite the fact that it did not yet broadcast the spoken voice, it had recognized the potential of this "newspaper without paper" and distance "as a powerful mass medium."

Although former Telefunken chairman Bredow, now in charge of the radio division of the Reich Postal Service, had already, on November 16, 1919, lectured reporters on the ability of the new medium to offer "the defeated German people affordable entertainment and information,"[7] the social balance of power did not make its installation as a mass medium appear "desirable."[8] It was only four years later, on the evening of October 29, 1923, when capital had created a more or less stable state and military apparatus, that an infantry regiment performed the national anthem with the infamous line "Deutschland Deutschland über alles" for the first time on air. Only after the Enabling Laws of October 1923 allowed the ruling classes to bypass parliament and create the necessary conditions for emergency measures, and only after the legitimate regional workers governments had been crushed by the military and the armed workers uprising had failed (not the least because of underdeveloped radio- and temporarily proscribed paper-based means of communications), did German radio come into being as an entertainment mass medium.

Capitalist commercial pressure—that of the broadcast industry in particular and of big business in general—presupposes that the masses have both objective communicative and subjective material needs. The former depend on the historical stage of the sociability of communication, while the latter appear in the shape of purchasing power. The inauguration of public radio catered to these needs; at the same time, however, social imbalances limited mass participation. As a result, there was a vibrant upswing of do-it-yourself initiatives: more and more workers took to assembling their own receiving devices. While this trend, which soon led to the formation of amateur associations, found its immediate outlet in widespread tinkering, the Telefunken leadership was still reeling from the impact of the November revolution. It not only had to deal with the politics of the seemingly apolitical amateurs but in addition—and against the background of the historical social development of social dynamics, which in the eyes of the government still made it appear necessary to outlaw the German Communist Party (KPD) for another four months, just four weeks after German radio had first been introduced—it somehow had to address and react to the interests of the workers themselves:

> These groups, intent on gaining possession of radio to use it for political purposes, relied on a league of operators allegedly comprising tens of thousands of former military radio operators. Many of them became radio amateurs after the war. In all likelihood, however, they knew little about the political intentions of those who manipulated them behind the scenes; they would have hardly participated in driving Germany even further into the abyss. In any case, the existence of such a league of radio operators made up of former soldiers was of great concern to military and governmental circles.[9]

To gain control of the amateur movement and counteract the influence of the workers movement in the radio domain, attempts were made to establish unified control over all middle-class amateur associations, the first of which, the Deutscher Radio-Club (German Radio Club), had already been founded on April 6, 1923. By July 28, 1925, the groups had been lumped together in the German Funkkartell (Radio Cartel), which, following suggestions from AEG engineers and with the support of the Reich Postal Office, was merged with the Funktechischer Verein (Radio Association) into the

FIGURE 13.5. Collective reception on a Russian train station after the October Revolution: "The stronger the Red Army, the sooner the war will be over!" From Pierre Miquel, *Histoire de la Radio et de la Télévision* (Paris: Librairie académique Perrin, 1972).

German Funktechnischer Verband (Radio Federation). This enabled centralized political coordination; at the same time, it set the stage for industry to appropriate and exploit amateur activities and experiences. The postal office granted middle-class associations the exclusive right to administer the exams necessary for the approval, operation, and evaluation of radio operating licenses *(Audion-Versuchserlaubnis)*. On the basis of the Enabling Act of March 8, 1924, the Reich president passed an emergency act "for the protection of radio" stipulating that the installation and use of a private receiving unit (with or without tubes) required a "trial permit" that could only be obtained by passing an exam. While the latter demanded a fair amount of comprehensive in-depth knowledge (spelled out in twenty-six focus points), it was probably well within the expertise of former military radio operators, including those who had been members of workers and soldiers councils. As a precautionary measure, it therefore was deemed necessary

FIGURE 13.6. Individual reception in a bourgeois salon with a luxury radio set. From Miquel, *Histoire de la Radio et de la Télévision*.

to add that "personal qualifications were of prime importance, that is to say... the personality of the applicant has to serve as a guarantee that there will be no adverse impact on the endeavors to promote radio."[10] Violations of this emergency act carried a prison sentence of up to five years (at the time, the maximum penalty for manslaughter was three years); indeed, the mere attempt was considered a punishable offense, and the police had the right to conduct searches at any time without having to wait for a warrant.[11]

One of the achievements of the Workers Radio Movement was to bring about the suspension of the law requiring a trial permit, which occurred on September 1, 1925. The Arbeiter Radio-Klub (Workers Radio Club) had been engaged in fighting the law and its associated exclusive bourgeois rights since its foundation on April 10, 1924.

In the early stage of German radio, the Workers Radio Movement was primarily engaged in spreading the technical skills necessary for the self-manufacture of the necessary equipment and in ensuring that workers enjoyed their share of reception. This, however, cannot obscure the fact that right from the outset, the activities geared toward unifying working-class amateurs and integrating them into the organized Workers Movement represented an essentially political struggle that in its various historical stages was carried out with different means. "Radio will be free of politics or it will not be at all.... The fate of radio is tied to its nonpartisanship."[12]

In the wake of the objections to the broadcast of the German national anthem, the Berlin-based station Funk-Stunde AG (Radio Hour) wrote an open letter to the press: "The Radio Hour is a thoroughly nonpartisan enterprise with no political agenda whatsoever. Its exclusive function is to provide radio entertainment."[13] Yet an important part of the first German Christmas broadcast by the Berlin station on December 25, 1925, had been a "Political Christmas Greeting" by Zentrums-Partei (Center Party) Reich chancellor Wilhelm Marx.[14] Comparing the station's political self-representation with its actual programming during the first quarter of 1924 reveals not only obvious political content but also further-reaching interests. Alongside daily stock market news, weather reports, trading prices, and entertainment music—that is, more or less indirectly political broadcasts—there was the eight o'clock news followed by evening concerts and dance shows. The news was accompanied by household tips (antiburglary protection, baby care, fashion, etc.) or foreign language lessons, but there was also a lecture by the Women's Committee on the War Guilt Lie. On March 20, 1924, former secretary of the interior and deputy chancellor Karl Jarres, a member of the Deutsche Volkspartei (German People's Party) that was backed by the banks and big industry and who, in October–November 1923, together with Konrad Adenauer and industrial magnate Hugo Stinnes, had been one of the leaders of a separatist movement in the occupied Ruhr region, opened a lecture series on "The Occupied Territory."[15] At the time, there were no more than 8,000 officially registered listeners, though "it [was] an open secret that the number of unauthorized amateur receivers went into the tens of thousands."[16] It was possible to discern the future political dimensions of the new mass medium. Faced with this outlook as well as with the increase of listenership (prompted by the lowering of fees from an annual fee of sixty reichsmark to a monthly two reichsmark in February 1924) to more than eight hundred thousand, of whom it is safe to assume that nine-tenths were working class, Communist members put continuing pressure on the Workers Radio Club (ARC) to transform itself from a leisure club into a political task force.

With the Exception of the Communists

Comrades, wherever you may be,
Have you not been struck
By the government's words:

"All parties are allowed
To broadcast their propaganda on air—
With the exception of the communists!"

They will all recite their ether trifles.
The liberal morons,
The militarist aspirants of the Third Reich,
The zealots of the Holy Inquisition,
The whining Social Democratic supplicants,
All who patch the bastion's rotting tower
With the grout of the people's community,
"All—with the exception of the communists!"

For they will all
Recommend exiting backwards,
They will all
Bear the order of oppression,
Some out of greed for power,
Others out of helplessness.
All of them—.
"All—with the exception of the communists!"

—Erich Weinert, "Mit Ausnahme der Kommmunisten," *Rote Fahne,* June 16, 1932, translated from German

The Reichs-Rundfunk-Gesellschaft (Reich Radio Society) was founded in April 1926 with the state as a majority stakeholder. Working in unison, federal and state governments (the former represented by the Reich Postal

FIGURE 13.7. Heading of the *Arbeiterfunk—Der neue Rundfunk*. From *Arbeiterfunk*, November 6, 1927.

Office and the Ministry of the Interior) appointed supervisory committees and cultural advisory councils from which representatives of the Workers Radio Movement and other progressive forces were excluded. In the course of rising media-political confrontations fueled by growing class antagonisms (which were evident in the 1926 plebiscite on the expropriation without compensation of the aristocracy), increasingly differentiated ideas about possible ways of achieving a democratic radio alternative came into being. The Workers Radio Press developed a comprehensive and meticulous critique of bourgeois radio—a critique that was related to the introduction by the Deutsche Welle of extraordinarily powerful German long-wave transmitters. Starting in 1926, the broadcasts of the Deutsche Welle pursued an imperialist agenda and targeted Germans living abroad. Closer to home, its so-called Arbeiterfunk (Workers Radio) no doubt addressed some of the radio-related interests of the working class, but ultimately it was a collaborationist endeavor.

The struggle for equal representation during parliamentary and presidential elections was a core item of the ARK's activities. Already during the first parliamentary election, bourgeois parties had been granted air time while the Communists had been excluded. Demands that the ARK be represented in the cultural advisory councils were part and parcel of the Workers Radio Movement; and although the Communist Party may have had a favorable view of the future fate of a truly democratic radio within the context of a fundamental change of society, it nonetheless advocated the establishment of radio councils to overview programming as a means of achieving the overall goal. But even while struggling for participation in decision-making processes rather than withdrawing into complete opposition, the Communists voiced demands for a radio station of their own. This demand was linked to the ultimate goal of abolishing capitalist conditions; it therefore became a question of power. The Social Democrats (SPD) distanced themselves from the demands of the KPD. The latter, in turn, unequivocally distanced themselves from all attempts to create an isolated radio domain for workers, as had happened in other capitalist countries.[17]

When, in early 1928, the number of listeners in Germany passed the two million mark, the KPD once again addressed the thorny issue of how to negotiate the means and ends of a democratic radio alternative: "We realize that we will only be able to broadcast truly proletarian programs once the proletariat has come to power. But we have to strive to achieve some of our goals today—under capitalist conditions."[18] Among these demands was the ongoing insistence that progressive artists, scientists, and politicians be granted access to bourgeois radio programs. But the guidelines of the fourth national conference of the Arbeiter-Radio-Bund (ARB) in September 1928 reduced the political struggle to the issue of participation in the ruling supervisory boards, which in the eyes of the right-wing leadership of the Social Democrats appeared satisfactory. Two examples illustrate the contradictory development of the Workers Radio Movement between 1927 and 1929. The first concerns the preparations undertaken by the SPD for the fourth national conference of the ARB, and the second is related to the postprocessing of that conference by the KPD. Talking about the so-called Workers Radio of the Deutsche Welle, a Social Democratic undersecretary and ARB board member offered an assessment of the programs addressed to workers:

> The lectures somehow touch upon questions of workers' lives.... With regard to working-class culture questions of good books and of a meaningful design of workers' living space were raised.... We want to show to the workers listening to these programs how they are to see themselves in their

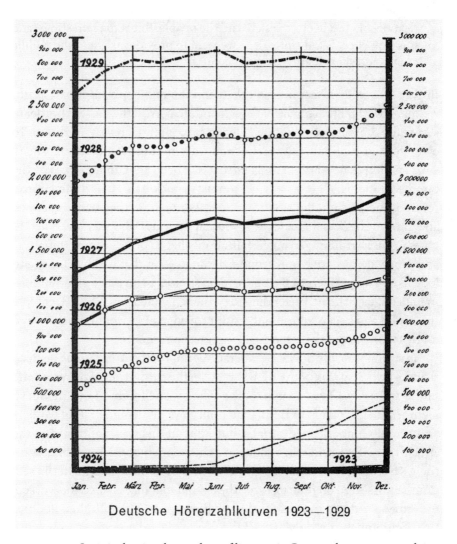

FIGURE 13.8. Statistic showing the numbers of listeners in Germany between 1923 and 1929. Reichs- und Rundfunk Gesellschaft Berlin, ed., *Rundfunk Jahrbuch* (Berlin: Union Deutsche Verlagsgesellschaft, 1930), 364.

relationship to society, the state and the people. But at the same time (and this side-effect is fully intended) we want to provide insight to the middle classes and show them what goes on in the depths.[19]

In 1928, as capital investments peaked and the military budget had mushroomed to 827 million reichsmark, demands to correct the Eastern border to secure the capitalist exploitation of living space grew increasingly candid. To enlighten the public about the referendum against the 80 million reichsmark construction of new armed cruisers that had been kept under wraps by the bourgeois media, a group of Communists snuck its way into the control room of the Berlin radio station on October 6, 1928. They had kidnapped an editor of the Social Democratic journal *Vorwärts* who was scheduled to give an officially approved on-air talk about alleged peacekeeping initiatives and proceeded to drive him around Berlin in a car that had been made to look like a guest vehicle of the Berlin radio station. In his place a Communist member of parliament, pretending to be the scheduled peace expert, entered the studio and began to read out the approved manuscript but then veered off by talking about the referendum against armaments.

In spring 1929, the Social Democratic editors of the *Arbeiterfunk* decreed that the principal task of the ARB should be the cultivation of tinkering and home assembly. On May 1, 1929, the infamous "Bloody May," the Social Democratic leadership in Berlin had police fire on demonstrating workers and beat up innocent pedestrians, but none of this was mentioned in the *Arbeiterfunk*. Following the exclusion of board members who did not want the ARB to devolve into an insignificant wireless association, its Berlin section was dissolved on June 16, 1929. The overall split that divided the workers movement had begun to affect the media domain. But with their actions, the Communists not only limited the future of workers radio to wire- and tube-based tinkering but also abandoned the airwaves to the far more serious and bloody encroachment by reactionary forces.

To quote the manifesto of the eleventh party congress of the KPD in early 1927, "Many Social Democratic leaders denounce and deny our struggle for a united proletarian front, . . . we communists will demand all the more urgently that all Social Democratic workers, all members of the free trade union, join hands to form a united proletarian front."[20] In spring 1929, as the radio corporations recorded peak sales and the fascist threat of a new imperialist war loomed, Germany had more than three million unemployed.

FIGURE 13.9. Cover introducing a device to count listeners. From *Funk-Stunde,* March 27, 1931.

FIGURE 13.10. The symbol of the German Workers Radio League ARBD (Arbeiter-Radio-Bund Deutschlands). From *Arbeiterfunk*, August 21, 1928.

"Fellow workers! Put an end to the party that betrays and murders workers! Put an end to the SPD! Drive off the agents of social fascism!"[21] proclaimed the manifesto of the twelfth party congress of the KPD of June 1928, which ended on the eve of the dissolution of the Berlin ARB and thus just before the split of the German Radio Workers Movement.

The Social Democratic ARB

The ARB pursued a problematic strategy particularly apparent in one of the principal claims that arose during the split—the demand for a "pure state monopoly."[22] Parliament was to ensure, strengthen, and legitimize the influence of the Workers Radio Movement. While certain key points of this demand differed from 1932 "Radio Reform" enacted by the Von Papen government with a view toward securing state authority,[23] it remains questionable to what extent the demand for a state monopoly represented an adequate response to the social dynamics and forces during the final phase of the Weimar Republic, especially when it served to obstruct more far-reaching goals.

This was already spelled out in Alfred Flathaus's 1929 article devoted to the project of a "Workers Radio": "From a political point of view any smart reactionary is well advised to accede to the seemingly radical demands of the worker's left-wing fraction. All at once, reactionaries would have a valid reason to refuse all demands working-class listeners submit to other radio stations by pointing out that they already have their own. . . . There is, really, no discernible advantage to a workers radio. We therefore wholeheartedly agree with the demands of the Arbeiter-Radio-Bund to strengthen the influence of the organized Workers Radio Movement rather than attack the monopoly of the Reich Postal Office."[24]

Even after the national congress of the ARB on June 25–26, 1932, at which the long-serving chairman Curt Baake was replaced by retired undersecretary Albert Falkenberg, the leader of the German Association of Public Servants, the bias against the Freier Radio-Bund (FRB) was apparent: "The National Congress of the ARB calls upon all freedom-loving 'Volkskreise' [ordinary people], especially the socialist ones, to fight for *equal radio rights*. Victory in this *cultural struggle* depends on uniting all the active forces of the working-class radio audience" (emphasis added).[25]

The *Volksfunk*

A conspicuous phenomenon of the ARB's increasing abandonment of all class-specific demands (as evident in the preceding quote) was the association's journal, which, starting in March 1932, was published under the title *Volksfunk—Illustrierte Wochenschrift für Berlin-Magdeburg-Stettin* (People's radio—illustrated weekly for Berlin-Magdeburg-Stettin). There were two distinct parts to this publication: an ample illustrated section full of tabloid-like reports on radio and film studios and almost completely devoid of any critical take on social events and processes, and a separate section dedicated to the ARB, in which the ARB itself had its say; its full title, which clearly separated it from the illustrated section, was *Rund um den Rundfunk—*

Kritische Programmbeilage des Volksfunk, Arbeiterfunk, Publikationsorgan der Arbeiter-Radio-Internationale und des Arbeiter-Radio-Bundes Deutschland e. V. und des FRB in der Tschechoslowakei (All about radio—critical supplement to the *Volksfunk, Arbeiterfunk,* the publication organ of the Workers-Radio-Internationale and the registered German Workers Radio League and the Czechoslovakian FRB). The feisty subtitle was abandoned in January 1933, and the supplement shrank to *Rund um den Rundfunk—Rundschau des Volksfunk-Arbeiterfunk* (All about radio—the people's and workers' radio review).

Excerpts from a report by Prussian secretary of culture Grimme in the first January issue of 1933 illustrate and explain the change of direction: "Politics as a whole has to orient itself to the cultural situation. Rather than engage in so many theoretical controversies, it seems more important *to connect with other large spiritual currents our time.* We have to see beyond fences, leave our towers and be more in tune with our times.... Socialism has to move beyond a class-based movement in order to unleash a *people's movement.*"[26]

In turn, the issue of January 20, 1933, contains the energetic words of ARB chairman Alfred Falkenberg: "We demand that radio put into practice what its new guidelines promise: That it serve all Germans and raise the global profile of German culture, which has as its noblest goal the spiritual elevation of the masses." With demands like these, there appeared to be no great need to take a stand against the looming threat of fascism!

The Communist-Led FRBD

One of the participants in the upswing phase of revolutionary culture between 1929 and 1933 was the Freier Radio-Bund Deutschlands (Free Radio League of Germany; FRBD). In the wake of the breakup of the Workers Radio Movement, it emerged as one of the principal organizers of the Communist forces. Following the KPD's strategy of centralizing all cultural mass organizations, the Free Radio League was affiliated to the Interest Group for Workers Culture, founded in 1929. Other affiliated organizations were the League of Proletarian-Revolutionary Writers, the Association of Revolutionary Visual Artists, the People's Film League, and the League of Worker-Photographers, among others.

When it came to radio, the Communist Party left no doubt as to its assessment that the Social Democratic Party was the "main social sup-

FIGURE 13.11. *Volksfunk* baby star. Cover of September 23, 1932.

port of the bourgeoisie,"[27] a stance that helped ensure that no effective antifascist front against Nazi radio was ever able to develop. A more promising hint of a broader struggle was contained in the declaration on the "fight against fascism" issued by the politburo of the central committee on June 4, 1930:

> The fascist movement aims to create a fascist state, to annihilate the entire workers movement, to install a regime of White Terror, of martial law and assassinations just as Mussolini has done in Italy.... The struggle against fascism must be fought as a political mass struggle on the broadest possible base.[28]

The *Arbeiter-Sender*

One decisive feature of the media-political strategy of the FRBD was to insist on the *class-bound* nature of bourgeois radio, as opposed to which the workers movement had to formulate a viable alternative. This is clearly

FIGURE 13.12. Advertising the *Arbeiter-Sender*: "because radio is the only medium for entertaining workers' families." From *Arbeiter-Sender*, January 29, 1932.

reflected in the many articles published in the *Arbeiter-Sender* (Workers-station) that attacked bourgeois broadcasts as well as the growing number of fascist programs by confronting them with objective interests of the working-class audience. One of the main targets of the critique was the so-called Erwerbslosenfunk (Radio for the Unemployed), designed to integrate the growing number of destitute unemployed and keep them from embracing revolutionary solutions to the problems caused by the global capitalist crisis. "Establish a Marxist education of workers, and make members of the council of the unemployed part of the Erwerbslosenfunk!"[29] were among the principal demands.

The struggle was primarily carried out on the extraparliamentary level. In the words of the last national convention of the FRBD, held in Berlin on December 27 and 28, 1932 (Figure 13.12):

> By introducing mass criticism of the Erwerbslosenfunk and demanding that discussions among the unemployed be broadcast, the FRB joins the campaign against hunger and cold. The struggle to reduce radio fees to 1 reichsmark and to exempt all unemployed, welfare recipients and pensioners

regardless of citizenship will be intensified. By establishing radio listener councils dedicated to abolishing fees and by waging a more intense campaign against all radio disturbances, the FRB commits itself to serving the interests of all listeners. By raising concrete programming demands addressed to all broadcasters and organizing conferences in all cities with radio stations, the FRB will mobilize millions of dissatisfied listeners to struggle against the fascist radio reactionaries.[30] .

The FRBD spent the last weeks of its legal existence preparing for its looming ban. Immediately following the fascist seizure of power, it initiated the creation of a Unity Board for the Working-Class Listeners of Greater Berlin (Groß-Berliner Einheitsausschuß werktätiger Rundfunkhörer):

> The unity board encompasses all the large proletarian mass organizations, members of trade-unionist opposition, works committee representatives from large enterprises such as the Oberspree cable-manufacturing plant, Bergmann-Rosenthal, Kristallate etc., representatives of the Greater Berlin Council of the Unemployed, the district committee of the Revolutionary Trade-Unionist Opposition, Red Help, the International Workers Support, and so on.[31]

Unfortunately, the *Arbeiter-Sender* was not able to keep its editorial promise that it "will return to the decisive step of founding a Greater Berlin Unity Board." It was to be the paper's last legal issue.

The Prohibition of the Workers Radio Movement

One of the first major actions of German fascism in the radio domain was to shatter the legal organizations of the Workers Radio Movement. The initial attack was directed at the FRBD. On February 24, the *Arbeiter-Sender* was banned, though at first only for a month. The FRBD was only able to publish a backdated issue consisting of nothing more than the title page announcing the ban: "The publication of the *Arbeiter-Sender* has been banned until March 22," signed by Magnus von Levetzow, the Berlin chief of police.[32] The *Arbeiter-Sender* was never published again. Two days after its ban, the entire organization of the FRBD was declared illegal. Many of its members were arrested, their facilities and offices either seized or destroyed. From this point on, the history of the Free Radio League has to be written as an antifascist struggle.

Ticke tacke ... ticke tacke

Achtung: Funkstunde Berlin,
angeschlossen Radio Wien
und alle deutschen Sender.

Sie hörten Deutsche Weihestunde
vom „Kyffhäuser" Reichskriegerbunde.
Ansprache: General von Horn.

Es folgt romantische Musik;
Lieder aus dem letzten Krieg.
Wir schalten um nach München.

Achtung, meine Herren und Damen,
Verzeih'n Sie, daß wir später kamen;
Sie hören jetzt Tagesberichte.

Wetterdienst: Im ganzen Reiche
bleibt das schlechte Wetter das gleiche.
Niederschläge sind zu erwarten.

Reuteramt meldet aus dem Osten:
Japan schickt an den Amur Posten.
Zur Atlantik wird uns gemeldet:

Erste Hilfe war ein deutsches Schiff.
Nun treibt der Dampfer gegen ein Riff.
Sanssouci wird neu angestrichen.

Im Hafen L'Havre wird gestreikt
für alten Lohn bei verkürzter Zeit.
Der Streik geht schon langsam zuende.

Eben trifft noch eine Meldung ein.
Die Streikfront soll noch verbreitert sein.
Der Nachrichtendienst ist beendet!

Hallo, Berlin, da sind wir wieder.
Sie hören nun alt-deutsche Lieder;
es singt der Britzer Jungfrauenchor:

Der du da bist direkt von Gott gesandt,
und: Was ist des deutschen Vaterland?
Danach für fünf Minuten Pause.

Berlin, sie hörten deutsche Lieder.
Kommen in fünf Minuten wieder
mit Börse, Wetter, Zeitansage.

Ticke-tacke, ticke-tacke,
ticke-tacke, ticke-tacke,
ticke-tacke...

In einer Minute wird's dreizehn sein...
Da hau'n die Proleten den Kasten klein,
und richten sich ihren Rundfunk ein!!

FIGURE 13.13. Poem by Eduard Schürer, published in the last issue of the *Arbeiter-Sender,* February 3, 1933.

The *Volksfunk* did not issue a direct statement on the banning of the FRBD, but it did not altogether ignore it either. The issue of February 24, 1933, features above the newspaper's name the caption "Now more than ever!" The lead article, "Where Is Our Radio Headed?," contains a strong attack on Nazi radio and concludes with the summons,

> Therefore, freedom-minded listeners, insist on your rights and demand that other listeners you know do the same. With your ongoing and passionate protests show the powers that be that you are on guard. Fight against Nazi propaganda on radio, do not cease to demand equal air time. Enlighten

FIGURE 13.14. Intermediality 1925: radio in film as the cover photo of a magazine for broadcasting. From *Der Deutsche Rundfunk,* November 1, 1925. Courtesy of the Zielinski Archive.

all like-minded listeners about the reactionary character of the broadcasts and see to it that on election day the Nazis get the response they deserve in order to pave the way for a truly free radio.

From Detector Receivers to Shortwave Transmitters

The Reich Postal Office's ridiculously low number of 467 officially registered radio customers in the inflation year 1923 vividly illustrates one of the overriding motives for the self-organization of interested workers in the early radio days: fascinated by the new medium, most workers were economically not in a position to purchase any of the devices available on the market. In 1923, Germany had five million unemployed;[33] in 1924, the relationship between weekly net income and the subsistence minimum was 26.50 to 41.20 reichsmark.[34]

German radio dealers had formed an association even before the system went on air, but their actual profits never met their high expectations. Sales figures left much to be desired—the majority of those afflicted by the monetary devaluation were forced to explore alternative options if they wanted to listen to radio. Enthusiasts met in Workers Radio Clubs (Arbeiter-Radio-Clubs): "The mass of working-class tinkerers sought to get together.... At first this took place without any political overtones, it merely served to provide a platform for the exchange of experiences gained while building their own devices and to receive instruction from more advanced electro-technicians and students."[35]

Guided by those who had acquired radio expertise, members together assembled crystal detectors, the most primitive devices capable of picking up electric signals and converting them into comprehensible acoustic waves. The detector receivers operated without a power source by means of a variable capacitor that could be adjusted to the wavelength of the local broadcaster. Associations and common workshops engaged in mutual assistance that enabled many working-class listeners to enjoy radio:

> Already at this early stage there was a noticeable difference to bourgeois radio associations. While the latter tended to train their members, especially the younger ones, to be operators and use Morse code in order to ensure there would be a sufficient number of radio operators, we of the Workers Radio Clubs put greater emphasis on assemblage, on modifying parts and wiring in order to build and try out new and differently constructed

FIGURE 13.15. "The future evening entertainment in the small town and in the countryside." From *Berliner Illustrierte Zeitung*, October 21, 1923. Reprint on the cover of Peter Dahl, *Arbeitersender und Volksempfänger: Proletarische Radio-Bewegung und bürgerlicher Rundfunk bis 1945* (Frankfurt am Main, Germany: Syndikat, 1978).

apparatuses. Well-functioning receivers, though of considerable size, were already available. There were also receivers the size of matchboxes, which were considered toys. At the time, though, we were only allowed to build detectors. Those eager to assemble tube units capable of receiving more than the local radio stations needed the so-called audio test permits, which involved passing an exam administered by the bourgeois radio association and paying a fee of 25 reichsmark.[36]

These procedures, in combination with the high costs of tube units, were the reason why the detector receivers dominated the radio scene during its first five years, despite the fact that the latter were still beyond the purchasing power of individual workers: including accessories (crystal detector plus headphones), a brand-name device still cost around seventy reichsmark. It was only during the second and third years that the series units dropped to half their initial price.

FIGURE 13.16. Illustration on the relationship between economy and broadcasting and its impact for the market. Stock market news and the announcement of the actual prizes had been a prominent part of early radio programs. Image from Peter Dahl, *Radio—Sozialgeschichte des Rundfunks für Sender und Empfänger* (Reinbek bei Hamburg, Germany: Rowohlt, 1983), 20.

As a result of the weak broadcasting power of the first German emitters (up to five hundred watts) and the limited range of the detector receivers, the first Workers Radio Clubs tended to be located in the vicinity of the broadcasting centers. Yet, once a more powerful network of program broadcasters emerged across Germany, club tinkerers faced new challenges. At the same time, more groups emerged in remoter areas.

Club members wishing to receive more than their local station had to assemble units more sensitive and technologically more complex than the early detectors. Tube units further required that self-made operators attain a greater command of the theoretical background knowledge. The simplest single-tube unit, as presented to the public at the first German radio show in Berlin in 1924, cost as much as detectors had in their early years, while a four-tube receiver could amount to 500 reichsmark, which did not include expenditures related to the power source—either an accumulator or an anode battery (power supply lines were not available until 1928–29). The fact that at the turn of 1925–26, almost 600,000 of the 1.02 million official permits issued by the Reich Postal Office were for self-built radios (or for those supplied by "wild" manufacturers) reveals the degree to which financial constraints and the tracking of illegal listeners limited overall participation in the new radioscape. The radio audience only began to increase in the wake of three changes: fees were lowered (a high-priority item on the list of the Workers Radio Movement's demands), the means of payment were modified, and better and cheaper radios were introduced (Figure 13.17). In 1925, the percentage of self-built tube units was still at 6.4 percent, but that of detectors was at 47.6 percent! It is safe to assume that it was the goal of every group to own at least one functioning tube unit, which, if listeners wanted to enjoy programs together, had to be equipped with speakers instead of the customary headphones.

On the technical side, this early stage of German broadcasting was accompanied by a flurry of innovations resulting in hundreds of patents. Unfortunately, it is not possible to document and measure in sufficient detail the contributions made by members of the Workers Radio Clubs. However, it is a fact that radio manufacturers built and patented numerous proposals tinkerers and club members had either presented at local promotional events or published in trade journals. This inventive piracy only receded when, under pressure from the major players—AEG, Siemens, Huth, Lorenz, and Philips—the number of smaller manufacturers declined and the market was

FIGURE 13.17. Leaflet demanding a lowering of the radio fees. From *Arbeiterfunk,* January 29, 1928.

under better control. The technical expertise of the organized listeners of the Workers Radio Movement kept pace with the ongoing expansion of the European network: the increasing crowding of the airwaves demanded the suppression of all noise sources interfering with reception as well as more effective ways to keep the broadcast frequencies from interfering with one another. The proliferation of domestic appliances in particular tended to spoil the "enjoyment." The postal office and big industry did their utmost to detect such sources of interference and track down the many illegal listeners. Especially after the onset of the 1929 economic crisis and the rise in unemployment, qualified members of the Workers Radio Movement were recruited as "radio aides." Many groups had constructed their own direction finders to pinpoint sources of interference; now tinkerers were hired by the postal office and paid premiums for locating those who either interfered with broadcasts or refused to pay fees.

In the eyes of the Workers Radio Movement, the utilization of these "radio aides" by the Committee on Broadcast Interference (Ausschuß für

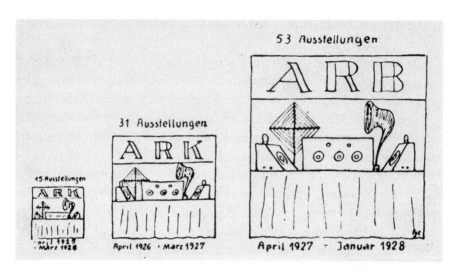

FIGURE 13.18. Growth of promotional events between April 1925 and January 1928—from fifteen to fifty-three exhibitions. From *Arbeiterfunk*, February 5, 1928.

Rundfunkstörungen), composed of representatives from the postal office, the electronics industry, and research institutions (the Heinrich Hertz Institute in Berlin), was an ambivalent undertaking. Widespread refusal to pay fees and mass cancellations of subscriptions were a source of discomfort to both the Reich Radio Society and the Reich Postal Office. (They rightly interpreted these actions as a critique of their programming.) The fact that the ARB had joined the commission and supplied many of 5,379 radio aides (the number as of September 1930) was viewed by the FRB as an act of treason against the interests of the working class.

The split of the Workers Radio Movement into two associations separated by differing party politics coincided with the inclusion of German-language broadcasts by Radio Moscow. (Starting November 7, 1929, the Soviet Trade Union broadcaster WZSPS, later followed by the Comintern Radio, broadcast an hour-long program for German listeners.) Once the problems of long-distance reception had been solved, the communist Free Radio League organized joint listenings. Long-distance reception, however, required special antennas. When, in early 1932, the Reich Postal Office started to interfere on a regular basis with these broadcasts by using the shrill whistle of a telegraph station, causing many listeners to complain to the ministry and the Reich Broadcasting Corporation, technicians of the *Arbeiter-Radio-Amateur* (Workers radio amateur), a technical supplement to the *Arbeiter-Sender,* proceeded to hand out advice on how to eliminate the interference (no. 7/1932).

As of the late 1920s, both the Social Democrats and the Communists pursued the construction of mobile transmission units that could be used for purposes of public agitation in the many election campaigns. These systems required extensive technical and material means, for they were either independently constructed self-contained, network-independent, high-power amplifiers with their own speakers (dynamic loudspeakers had replaced horn loudspeakers), or the repurposed amplifiers had been constructed in such a way as to access the power outlets of members who happened to live nearby, thus ensuring that proclamations, speeches, and music reached the most remote backyard. Recordings of work choirs and ensembles were used, such as those produced by the Versandhaus Arbeiterkult (Workers Culture Mail-Order Business). With very few exceptions, most of the numerous demands to allow proletarian cultural institutions access to microphones were denied. Such "courtyard events" also featured agitprop groups, for instance, the Rote Welle (Red Wave) of the BRPS member Pijet in northern Berlin. The marked and unmistakable contrast to the official radio programs was fully intended.

One of the most important recurring themes running through the history of the Workers Radio Movement was the demand for their own broadcasting station. It was already contained in the first 1924 bylaws of the Workers Radio Club (ARB).

In May 1928, the ARB published a set of demands, of which the twelfth read, "Release of trial transmitters also for serious amateur associations." State Secretary Bredow had in fact pledged in writing to the board of the ARK to authorize trial broadcasting facilities. But after the restructuring of German radio in 1926, there was no more mention of it. The participation in cultural advisory boards, which had advisory functions only, could not replace these demands.

While amateur operators in other Europeans countries as well as overseas had long been broadcasting shortwave Morse codes—according to contemporary press releases, the United States boasted no less than thirty-five thousand shortwave amateurs in 1927—German amateur operators had to wait until October 1, 1928, to receive a license. Once again, bourgeois associations received preferential treatment. The Prussian Ministry of the Interior handed out 800 CQ licenses to the Funktechnischer Verband and

FIGURE 13.19. Local promotional event, Berlin-Neukölln: "Radio should be the people's pulpit." From *Arbeiterfunk*, November 6, 1927.

150 to the Workers Radio Movement; this imbalance was subsequently further skewed to the disadvantage of the workers. Even the incorporation of the Drahtloser Amateur Sendedienst (Wireless Amateur Transmission Service; DASD), which enjoyed a close cooperation with the German Army, changed nothing. "Those are the benefits of collaborating with the bourgeoisie in the technological domain," the Rote Fahne (Red Flag) angrily commented.

Those interested in radio broadcasting but affiliated either with the Social Democratic Party or the Communist Party were rejected. A member of a Hamburg-based amateur shortwave radio group wrote to the *Arbeiterfunk*, "We are of the opinion that all technically interested Germans should receive a transmission license. Our shortwave group doesn't own a single approved shortwave transmitter" (Figure 13.19).[37] Nonetheless, amateurs proceeded to install and operate shortwave stations, which, upon discovery, incurred seizures and penalties.

In May 1932, the *Arbeiter-Sender* reported, "Because of the installation and operation of nonauthorized broadcast installations between the months of

January and March a total of 231 persons have been convicted, as opposed to 288 persons in the same period in the preceding year and 264 persons from October to December 1931."[38] Counterintelligence reports estimate that the SPD-affiliated ARB had about a hundred trained radio operators, while the FRBD, which prior to the fascist seizure of power had put special emphasis on their training, boasted no fewer than 350 "red operators." The latter managed to cause quite a stir during the ultraleft phase of the Communist Party by means of several spectacular actions carried out by illicit broadcasters that interfered with live public broadcasts (mainly by tapping into the cables connecting broadcast studios to radio masts). For instance, the 1932 New Year's Eve speech of Reich President Hindenburg was interrupted by another speaker, who demanded, among other things, "Fight against the hunger dictatorship! Help yourselves! Fight against every penny taken from unemployment benefits!"[39]

On November 4 and 6, 1932 (right before the national elections), astonished Berliners in Wedding and Neukölln heard on the Berlin Frequency (Berliner Welle) 450 a broadcaster distributing pamphlets and election slogans for the Communist Party. The mobile station returned in December 1932; the text is contained in police stenograms.[40]

At least one case in which broadcasts were interrupted because of severed cables is known to have occurred after the Nazi seizure of power, when the "Red Operators" had already been outlawed. It is worth recalling that

> on February 15, 1933 [the Communists] managed to thoroughly humiliate Hitler at his first public appearance as Chancellor in the Stuttgart town hall: Hitler's speech was to be broadcast throughout the land. But no sooner had he unleashed his torrent of words—"I will say this to the world's representatives: Our fight against Marxism . . ."—when the speakers emitted an audible crack and Hitler's voice vanished. Based on Kurt Hager's carefully prepared plan a feisty group . . . had cut the cable right next to the town hall with an axe. The fascists were fuming. While they railed against sabotage, the KPD was already circulating a flier: "We have denied Hitler the right to speak!"[41]

The Workers Radio Press

Owing to the widespread reluctance of the major broadcasting corporations to grant access, the press of the Workers Radio Movement was forced to step in as a substitute for the nonexistent workers radio. Since the publications in question were party-bound organs of the various associations (ARKD, ARBD, FRBD), these substitute functions were fulfilled in many different ways. Press organs had to organize the many activities of the associations and act as the main platform for the demands put to the bourgeois radio corporations. Aside from specific case studies (e.g., the *Volksfunk* and the *Arbeiter-Sender*), it is helpful to generalize in order to present some helpful remarks concerning the structure and content, functions and agitational format, of the newspapers in question. Each particular newspaper will be categorized under categories that persisted throughout the entire history of the Workers Radio Movement and their press.

Newspapers for the Working-Class Audience

Some of the determining features of the Workers Radio Movement were mirrored in their associated publication outlets, above all the critical engagement with middle-class radio programs and the elaboration and discussion of numerous alternatives. This was evident in many different rubrics, for instance, in program announcements. National editions (as opposed to regional editions) listed almost all European radio programs, including the many Soviet broadcasts that were accompanied by detailed comments, such as the Comintern station, the Stalin station, and of course the Trade Union station. Naturally, the German-language programs received special emphasis, and the newspapers were careful to highlight the wavelengths on which they were broadcast.

When it came to German stations, special selections were made with the specific interests of workers in mind. The selected programs were viewed either in "Critical Preview" (one of the categories informed by the guiding question "What do we want hear?") or as part of "Weekly Radio Plays and Events." A point of special importance that received a lot of space was the critique of broadcast programs. Under the heading "Critical Weekly Review (or Postviews) from the Reich," especially relevant broadcasts of the past week were extensively discussed.

The publication of alternative programs was considered especially important. For instance, manuscripts by workers or progressive artists censored by supervisory committees and cultural advisory boards were included. Program criticisms in the general sense also comprised the critical engagements with the politics of the broadcasting companies, frequently under their own heading (e.g., "Minor Broadcast" in the *Arbeiterfunk*).

Bastler-Zeitung (Tinkerers Newspaper)

The discussion and elaboration of technical issues and problems dominated the Workers Radio Press. In various phases (especially in the initial stage and, following the split, in the ARBD outlets), it was the dominant topic that overshadowed all other content. It was a characteristic feature of these technical discussions to be separated and appear in the shape of supplements associated with specific organizations. The *Bastlermeister* (Master tinkerer) was a supplement to the *Arbeiterfunk*; the *Arbeiter-Kurzwellen-Amateur* was billed as the *Mitteilungen des AED, Kurzwellen-Arbeitsgemeinschaft im Freien Radio-Bund* (Transactions of the AED, the Free Radio League's Shortwave Working Group). In terms of content, the topics under discussion ranged from organizational problems to the detailed analysis of technical problems all the way to straightforward instructions on how to build radios or improve reception.

Newspapers as Advertising Media

The technical portions of the Workers Radio Press were of obvious interest to the advertising industry. There were, however, significant differences with regard to content, form, and origin of the inserts in the various newspapers. The more politicized the Workers Radio Movement became, the more large electronics companies withdrew their ads (especially from the *Arbeiter-Sender*). The advertisement sections were abandoned to retailers. Advertisers were keen to connect with the articulated demands of the working-class audience. For instance, a drawing depicting a dissatisfied listener smashing his radio to pieces was adapted by a firm eager to hawk its durable tubes.

Finally, it is striking (and not only for the radio-related press) that workers were addressed as class-conscious consumers. The most striking example is an ad for a Solidarity Cigarette promoting its two grades of "proletarian cigarettes" (Figure 13.20).

Newspaper for Individual Reproduction

One important function of the Workers Radio Press was to provide entertainment. The ways in which this was accomplished were as diverse as the goals they strove toward. Fictional texts, serial novels, short stories, tales, poetry, and so on, were published on a regular basis. For a time, there were puzzle corners, humor sections, and especially caricatures, which primarily targeted middle-class radio. (One of the most famous caricaturists of the *Arbeiterfunk* was Georg Grosz.)

Belletristic texts ranged from "Bill of Fare Tips" (*Volksfunk*) to "How

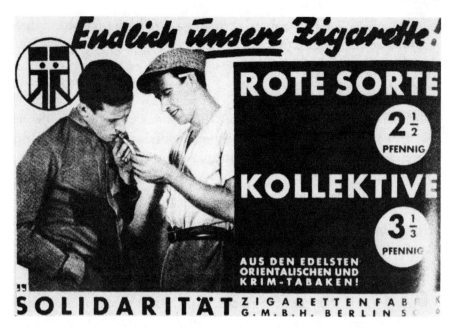

FIGURE 13.20. "At last, our cigarette!" Proletarian advertising in the Arbeiter-Radio-Presse. From *Arbeiter-Sender*, December 9, 1932.

Do I Behave While Sunbathing?" *(Arbeiterfunk)*. Artistic texts in particular revealed the breadth of the alliance that made up the Workers Radio Movement. Contributions by working-class writers found themselves next to those of progressive intellectuals and artists such as Kurt Tucholsky, Erich Weinert, Erich Mühsam, Ernst Toller, and Alfred Döblin, just to name the most famous. On the whole, however, artistic contributions noticeably declined after the split. In the Social Democratic *Volksfunk*, they were primarily replaced by glossy news, and in the Communist *Arbeiter-Sender* by direct political agitation (which was also the result of a stronger division of labor within revolutionary cultural organizations).

Medium of Organization

Another item of major importance for the overall development of the Workers Radio Movement were the many public announcements by various organizations. These rubrics, which disappeared from the *Volksfunk* following the fascist seizure of power, reflected the busy schedule of organizations like the ARKD, ARBD, and FRBD. Events of the many suborganizations were announced (public readings, joint listenings, discussions, exhibitions, etc.), and preparatory and postfactor reports of important organizational meetings were discussed. The newspapers' successes and failures were scrutinized, decisions and the constitution of new regional groups were announced, and there were far-reaching discussions of the movement itself.

The very multitude of names, many of which provide the sole access to hitherto undiscussed problems of the workers movement and its culture, is no doubt worthy of systematic research. However, given the bewildering diversity of organizations, it may be more helpful to list the newspapers once again in chronological order. The following overview will limit itself to those papers of national importance and exclude publications associated with regional organizations:

1. The *Arbeiterfunk,* the organ of the officially registered Arbeiter-Radio-Klub (Workers Radio Club), had a restricted circulation and temporarily ceased publication after the second issue. Date of first publication: August 1924.
2. *Der Neue Rundfunk* (The new radio), the new organ of the officially registered Workers Radio Club, began to appear on a weekly basis in April 1926. Place of publication: Berlin.
3. *Arbeiterfunk—Der Neue Rundfunk* (Workers radio—the new radio), associated with the officially registered Arbeiter-Radio-Bund Deutschland (Workers Radio League Germany), appeared weekly as of March 1927. Place of publication: Berlin. Publisher: Verlag der Neuen Gesellschaft, Berlin. Publisher and chief editor: Albert Baumeister.
4. The *Arbeiter-Sender* (literally, "Workers transmitter"), temporarily also known as *Unser Sender* (Our transmitter), was the organ of the Berlin section of the officially registered FRB/Freier Radio-Bund Deutschlands (Free Radio League), first published in January 1929, but on a regular weekly basis only as of January 1930. Place of publication: Berlin. Publisher: Orbis-Funk Verlag. Chief editor: Klaus Neukrantz.
5. The *Volksfunk* (People's radio) was the organ of the Workers Radio League that succeeded the *Arbeiterfunk* in March 1932. Place of publication: Berlin. Publisher: Volksfunk-Verlag. Chief editor: Albert Baumeister.

From Program Critique to Workers' Radio

In general, the right of working-class audiences to have their problems addressed on air was recognized—as can be glimpsed from comments by such diverse figures as Count von Arco, Albert Einstein, Berlin police commissioner Zörrgiebel, and Reich radio commissioner Bredow. Nonetheless, only a very small portion of broadcasts actually occurred:

> Broadcasts for workers. In the week from February 17 to 23, 1930, German broadcasters offered a total of 240 events that were of interest to workers.... On any given day of this arbitrarily chosen week, then, we have an average 1.4 pure and 2 partly mixed hours of events for workers.... No matter how you choose to twist and turn the numbers: 9–11% of programs for 80% of the listeners reveals an unacceptable discrimination against the interests of working class listeners.[42]

A review of the broadcast contents makes it easy to debunk the alleged impartiality behind which officials hid. In full agreement with the Workers Radio Movement that it amounted to no more than a mere semblance of neutrality, both left-wing and bourgeois authors minced no words in their contributions to the Workers Radio Press. "The privileged beneficiary of the capitalist order, the bourgeoisie, has with a complete lack of supplies—but also with undeniable skills—turned all innovations that aspired to further the promotion of culture and human well being into means and tools of exploitation of the working class."[43]

> **Berlin Radio Hour**
>
> Attention, Berlin radio stations!
> Still ahead of all the nations!
> Berlin is filled with strong intent:
> A coffee party for the Occident.
>
> Blessed are they who free of guilt and moral toll
> Preserve the German radio soul.
> Attention all, here comes
> Berlin Wave 505, the on-air band!
> Anthem of the Fatherland!
> Accompanied by muffled drums.

O the blissful sound of virgin prayers,
Wrapped in a ruby-colored tune
As radio wives submit to pious airs.
Melancholy sofas start to croon.
A soulful respite for those in power.
The sound of marching grenadiers:
Just a spring-time radio hour.

Household tips. An intermission.
Premature weakness?—Take Kukirol!
Attention, some bureaucrat or politician
Talks about occupied lands and alcohol.
Homey sounds and chimney dreams,
Marching bands for our marines.

The spirit that connects us all on air
Has settled for a little nap
Dressed in socks of camel hair
With cake and coffee in its lap.

As Germany's anthem enraptures all
Couch potatoes sigh and bawl,

Our radio communities are joined in rows
United against a world of foes.[44]

Given the total neglect in the dominant political circles of the needs of workers and their families, it comes as no surprise that the Workers Radio Movement voiced strong criticism. For proletarians living in miserable conditions, programs offering "Tips for Champagne Breakfasts" and "Engagement Receptions" were an insulting travesty. Apart from the countless easy listening programs and operettas, workers took a special dislike to the numerous edifying lectures on topics such as "Which Country Serves Our Lunch Table?," "The Magic of Gliding," "The Liberating Power of Nature," "Can Optimism Restore Our Health?," "The Development of Silesian Horse-Breeding and Horse-Shoeing Practices in Lower Silesia," "Men Who Mastered Their Lives," and "Saint Hubert, Patron Saint of Hunters." The proportionally overrepresented ecclesiastical celebration broadcasts were criticized in the following poem by Helmut Weiß in the *Arbeiter-Sender* in April 1932:

Black Frocks....
You see them standing in the studios
As long and frequently as possible!
And no matter how much you turn the dial
There is no escaping them,
Because day and night *they* are on air.

They have recipes for "anguished souls,"
All anguish, we know, is linked to "souls"!
They give you Sunday cantatas instead of bread
And each of them is a patriot
Of Protestant or Catholic persuasion.

Thinking—just leave that to the state,
To bishops and to businessmen.
They have advice for your worries.
For the hungry, an on-air sermon
Which is, of course, more convenient.

They stir the acoustic pulp
And tell you pretty stories
Between organ concerts and litanies
And they also keep you entertained
With reports of "Cheka atrocities."

They want to concoct a potent brew
And blind your eyes with radio.
You're not supposed to see what's going on.
They want to block your exits with phrases
No fear!!
He will be found.

Because of the strong and long-standing presence of freethinking in the workers movement, there were demands to broadcast ceremonies related to this particular tradition, such as Youth Dedications (Jugendweihen). However, only the reformist wing was successful in pursuing their demands, since these forces, too, had been affected by the split.[45]

The domineering presence of civil servants and right-wing party representatives on the supervisory committees installed in 1926 ensured time and again that broadcasts responding to workers' interests were censored. In cases in which such programs were approved, there were frequent "technical problems." For instance, during the broadcast of the dedication of a new workers' sport school, the microphones were simply turned off. On a different occasion, the broadcast of a mandolin orchestra was interrupted. For a long time, and contrary to demands by a large majority of workers, Labor Day parades were not broadcast. And when a broadcast finally did take place, it was interrupted by "technical problems" during the speech by left-wing Social Democrat and trade unionist Siegfried Aufhäuser. When, in 1929 and 1930, there once again were no Labor Day broadcasts (with officials citing the "no politics on air" guideline), the *Arbeiterfunk* was up in arms (Figure 13.22):

> In the eyes of German broadcasters Labor Day does not exist. Following orders from the new Reich government, the old Braun-Keudell agreement has been resurrected, which states that according to the dictates of political neutrality Labor Day parades are to be considered a partisan matter. There will be no mention of any mass parade. The supervisory committees are to follow cabinet resolutions and nothing else.—Attention! Attention! Radio has become reactionary![46]

The initially promised participation of the ARK and later the ARB in the cultural advisory committees turned out to be an illusion: "This whole mystification comes across as ridiculous once one realizes that according to official regulations these councils were hamstrung even before they started their work."[47]

The decision of the ARB to abandon certain programmatic demands of the Workers Radio Movement cannot be fully attributed to the introduction of special programs for working-class listeners, including

- "The Work Hour"
- "Chamber Hour for Workers and Employees"
- "Social Insurance Radio"
- "Radio for the Unemployed" and "Public Employees Radio"

Frequently these programs turned out to be nothing more than empty promises of better times to come, made by ministry officials and majority Social Democrats who had enforced the renunciation of the original goals of the Workers Radio Movement. The ARB's main press organ, the *Arbeiterfunk*, featured both "unpolitical" and integrative tendencies. The former is evident

FIGURE 13.21. The collaboration between church and military and its disproportional representation in radio programs was a constant target of attack for the Workers Radio Movement. From *Arbeiter-Sender*, July 1, 1932.

FIGURE 13.22. Labor Day demonstration with demands of the Workers Radio Movement. From *Arbeiter-Sender,* April 29, 1932.

in the mild criticism expressed in a letter to the editor in *Arbeiterfunk* 3 (1930): "I am employed, and like the majority of listeners I therefore can only enjoy radio in the evening. Like any other person I am worn down after an eight- to ten-hour work day and crave pleasure, comfort and quiet. Music... offers me this enjoyment.... Dear Postal Office, spare me, spare us your no doubt well-intentioned evening lectures and instructions." The integrative tendency, in turn, is revealed by the fact that sections of the *Arbeiterfunk* were allocated to members of the ministerial bureaucracy to allow them to justify their censorship—for instance, to Undersecretary Häntzschel's "The Political Surveillance of Radio."[48] It is also noticeable in celebrations of the Arbeiterfunktag (Workers Radio Day) featuring a speech by Prussian secretary of the interior Carl Severing that was broadcast on all channels.

For the First of May!

By Gerda Weyl

For 40 years workers have celebrated World Labor Day on May 1.—For 40 years the class-conscious proletariat has demonstrated for understanding and fraternization among nations, for the protection of human labor, for the eight-hour workday.—On May 1, 1930 red flags will fly over massive rallies....
So comrades, come rally
And the last fight let us face
The *Internationale* unites the human race!
Radio is the technological realization of the *Internationale* and the tool to unite peoples and nations. Broadcasts accompany current events and mediate of current affairs.—But for German radio May 1, 1930 will not exist. Agreements on the parity of political opinions (and thus on true neutrality), the fruit of recent negotiations, are ignored. The preparations by the Secretary of the Interior count for nothing. There was a resolution to acknowledge church holidays and respect religion by broadcasting appropriate programs as well as to respect the wishes of the working population on May 1—but only the first half of this resolution has been met. The social reactionary government shows its true colors.—For instance, instead of the lunchtime concert Berlin listeners will be offered something special: An original live concert and in the late afternoon a twenty-minute lecture on "The Significance of May 1."
There will be no broadcasts of any Labor Day celebrations. There will be no reports of any mass rallies. The supervisory committees are merely sticking to cabinet decisions.—Attention! Attention! Radio broadcasts reaction!

> You represent what is bound to fall!
> You're casts, no more, despite it all!
> We are the people, the human race,
> For now and ever, despite it all!
> Despite this and despite that:
> Come show us what we really face!
> You may constrain us, but that is all
> The world is ours, despite it all!
> Radio shall become the people's pulpit!⁴⁹
>
> —Editorial, "For the First of May," *Arbeiterfunk*, April 25, 1930

Another tendency was the commitment "to retain Marxist terminology in connection with the analysis of class struggle" and to further uphold "despite all enmities between the parties a common language [between Social Democrats and Communists]," as expressed in the editorial of the first issue of the *Arbeiterfunk* in 1930:

> [The progress of radio] will only be fully realized when it can also be used in the service of the masses. Currently it is still a monopoly of the ruling classes; it restricts itself almost wholly to programs catering to their needs. Unabashed or concealed, it touts the opinions of our class enemies, be they reactionary forces or the ruling bourgeoisie. All that agrees with our worldview, every free thought, every free word, is eradicated from texts and lectures by the bourgeoisie, the ruler of radio. G—r.⁵⁰

However, the ARB was unable to take the step from demanding critical listeners to the deliberate and organized action against correctly perceived problems. Propagating proletarian programs, as the FRB's *Arbeiter-Sender* did, could have met with success, especially if it had addressed followers of both parties.⁵¹

The *Arbeiter-Sender* (Workers transmitter) truly wanted to be the transmitter of the workers: during the final phase of the Weimar Republic, it carefully scrutinized the increasingly fascist content of all types of programs, for instance, the prevalent broadcasting of military music.⁵²

In addition, under the leadership of Klaus Neukrantz, new forms of expression were developed that made use of readings designed to evoke the radio experience: the "weekly radio tale"—written by authors like Neukrantz and Georg W. Pijet—usually dealt with hot-button issues that either had received insufficient attention in other media or had been subject to deliberate distortion, such as strikes and international events. Alternately, manuscripts were printed that had been rejected by broadcasters, for instance, by Rudolf Frank and Georg Lichey, Berta Lask, Ernst Toller, and Bruno Nelissen Haken.

The reports of the *Arbeiter-Sender* had the further function of guiding mass movements while at the same time exposing the current conditions. Reporters took their transmission system to the tent city of the evicted at the Müggelsee; as part of the report "Kuhle Wampe Listens to the Radio," the inhabitants of the tent city wrote an open letter to the Reich Broadcasting Company demanding the free usage of a radio station.

Suggestions for the Better Use of the Apparatuses

Following the publication of his article "Young Drama and Radio," written for the Berlin-based *Funkstunde,* the young Bertolt Brecht became the target of harsh and polemical criticism from the *Arbeiterfunk.* He appeared to have overestimated the progressive features of the new medium at the expense of theater:

> While he [Brecht] condemns the theater that performs his plays, he curries favor with a radio station under the spiritual regime of Alfred Braun.... Brecht's *Funkstunde* essay is either a case of gross negligence or, more seriously, an example of how time and again the well-honed ensnaring tactics of our radiocracy manage to put to sleep any "critical conscience." The question of whether or not the mess of pottage for which Brecht sold the birthright of his independence merely consisted of a honorarium for a new year's contribution, must await the further investigation of his case.[53]

Whatever became of this special "case," Brecht did derive a few lessons from his limited radio experiences. His insights into radio's concrete social embeddedness and the objective restrictions capitalism imposes on means of mass communication are in no small part the fruit of his first and only attempt to write and stage an original radio play, *The Flight of the Lindberghs,*[54] which premiered in Baden-Baden on February 27, 1929. In historical retrospect, Brecht's fragmentary theoretical comments—which his later publisher Suhrkamp, among others, falsely made out to be a "radio theory"—move him close to the essential claims and goals of the workers movement, even though he was never organizationally involved. This becomes evident in the brief sketch of his "pedagogical experiment"—its performance at the 1929 Baden-Baden music festival was designed to illustrate the notion of the *Lehrstück* (Figure 13.23):

> On the left side of the platform was the radio orchestra with its apparatuses and singers; on the right of the platform was the listener, who performed Lindbergh's role, i.e., the pedagogical part. He sang his part to the instrumental accompaniment supplied by the radio. He read the spoken sections without identifying his own feelings with that contained in the text, pausing at the end of each line of verse; in other words, in the spirit of an *exercise*. On the back wall of the platform was the theory being demonstrated in this way.[55]

In other words, the listener—who in Brecht's ideal conceptualization was a collective, for instance, a school class—was to be *activated,* to become part of the performance by means of speaking and signing along, thereby transcending the passivity of mere consumption. In its day, this was not an uncommon idea: for instance, quartets were broadcast that omitted the fourth instrument, thus inviting the music enthusiasts among listeners to play the missing instrument.

Yet none of the six radio performances of *The Flight of the Lindberghs* between 1929 and 1933 lived up to Brecht's ideas; each and every one was "a concert performance, i.e., a flawed one,"[56] that served to restrict the role of the receiver to that of mere listener and thereby jettisoned the pedagogical part of the experiment. To be sure, Brecht knew full well that one should not overestimate the consciousness-raising and activating impact of such a performance on the audience. "Its proper application, however, makes it so 'revolutionary' that the present-day state has no interest in sponsoring such exercises."[57]

The still necessary "rebellion by the listener, . . . his mobilization and redeployment as producer,"[58] cannot be brought about by his formal inclusion in the communication process alone. It is a matter of the masses at the receiving end gaining self-awareness as a social subject—a process that relies less on the pedagogical parts of the *Flight* (which in any case are tied to the specific reception circumstances of 1929) than on its content and ideology.

Brecht's material is an important part of the experienced reality of 1927, the year in which Charles Lindbergh flew across the Atlantic. The mass media celebrated the feat as an altogether incomprehensible sensation; Lindbergh

FIGURE 13.23. The Baden-Baden performance of *The Flight of the Lindberghs*, the listener (right), the radio (left). From Brecht, *Versuche*, booklets 1–4 (Frankfurt am Main: Suhrkamp, 1959).

himself was glorified as a singular individual hero. On both sides of the ocean, press and radio invited the public to reenact the adventure.

In Brecht's hands, however, the depiction of the spectacular event becomes an attempt to explain it, to render it comprehensible by aesthetic means. The individual heroization of Lindbergh is turned into the representation of an historical accomplishment. The first nonstop solo crossing of the ocean thus becomes a product of social labor, a historically necessary—and at this point historically enabled—victory of man over nature.

"In *The Flight of the Lindberghs* Brecht endeavors to break down the spectrum of the 'event' in order to extract the colors of 'experience': the experience which can only be drawn from Lindbergh's work (his flight), and which Brecht means to give back to 'the Lindberghs' (the workers)."[59] One means of realizing this aesthetic principle is the depersonalization of the "heroic" feat. The play replaces the individual (Lindbergh)—whose name is irrelevant—by the collective "aviators." The work of the airplane mechanics becomes a constitutive component of the flight; the crossing itself is an individual part of collective relation:

> Seven men built my machine in San Diego
> Often twenty-four hours without a break
> Using a few meters of steel tubing.
> What they have made must do for me
> They have done their work. I
> Carry on with mine, I am not alone, there are
> Eight of us flying here.[60]

Appearing in allegorical shape, the forces of nature insert themselves as mighty obstacles between the aviator and his destination. This is in turn countered by the conscious attempt of the aviator to intervene in nature, as formulated in one of the play's core passages:

> The steamship competed with the sailing ship
> Which had left the rowing boat far behind
> I am competing with the steamship
> In the struggle against what is backward.
> My airplane, weak and tremulous
> My equipment with all its defects
> Are better than their precursors, but
> In flying, I
> Struggle with my airplane and
> With what is backward.[61]

Brecht offers the listeners the adequate consequences to be drawn from the reflected experience of human abilities:

> Therefore take part
> In the battle against what is backward
> In the abolition of the other world and
> The scaring away of any kind of god, where-
> Ever he turns up.[62]

Involving the listener in the artistic performance serves as emotional and rational support to fully participate in the dynamics of reality. The listener is to be empowered to find his or her way in the various social fields of action—"first in the domain of aviation, later in the class struggle," as Walter

Benjamin characterized the stages of Brecht's *Lehrstück*. This is accompanied by an attack on the entire bourgeois culture industry and the ruling aesthetic standards cultivated by the mass media (then and now): "Epic theater attacks the basic view that art may do no more than lightly touch upon experience—the view which grants only to kitsch the right to encompass the whole range of experience, and then only for the lower classes of society."[63]

Keeping this distilled concentrate of *The Flight of the Lindberghs* in mind, it is possible to view Brecht's radio play as one of the alternatives that the workers movement was able to enforce in the interest of a socially progressive radio. In his 1932 speech on "The Radio as a Communications Apparatus," Brecht generalized on the objective commonalities with the revolutionary goals of the Workers Radio Movement, although he subjectively never realized them by collaborating with the movement: "It is simply not our task to renovate the ideological institutions on the basis of the existing social order through innovations. Instead our innovations must get them to abandon this basis. Hence for innovations, against renovation! *By means of constant, never-ending suggestions about better applications of the apparatuses in the interest of the many,* we must shake up the social basis of these apparatuses and discredit their application in the interest of the few."[64]

Appendix

> Unknown
> I'd like to have some time on air...
> I'd like to have some time on air
> To say what I want.—It's only fair.
> Just for once.—And just an hour.
> I'd "rouse the rabble"—spread hate and fire.—
> Just for once I'd like to "share"
> My slice of life with you on air.
> Just the plain facts.—And nothing more.—
> I think a miracle would occur.
> —I want to see the angry looks
> Of the bourgeois fat cats and greedy crooks
> Of the easy listeners and rumba slobs
> Of all of those who turn the knobs
> To listen to some humdrum presidential drone,
> When all they hear is: "Germans! Well,

You'll now hear a woman describe the hell
Of making jute for Grimme & Cohn.—
Don't touch that dial! We'll be right back!"
———————————. And all is quiet.
And then, a hiss, a wispy crack,
A sudden scratch, a raspy tone:
"It's Mrs Krause from Grimme & Cohn!"
Day in, day out, for thirteen years
I've toiled away at spindles and gears—
Surrounded by a motley crew of "knaves".—
Servants of mammon.—The spindle's slaves!
You'll want to know: What do they pay?
When all is well, sixteen a week.
There's no break for us.—Nine hours a day.
For thirteen years I've endured the reek
Of working in this rotten cage.
And now we're into the twentieth week
Just three days left.—At half the wage.
Messieurs Grimme and Cohn, I'd like to know
Will this year's dividend again be low?
Like, four percent instead of eight?
—No!—It's 8.7.—But wait:
The annual balance states it's 9.—
8.7—I guess that's fine
But, dear Sirs, it clearly seems
You've "written off" a tidy sum.—
For the new shed.—For new machines.—
That 8.7? We've been learning:
You wrote it off your so-called "earning."
We work for a pittance, yet your whips are such
We end up producing twice as much
In honor of the annual dividend.
Messieurs Grimme and Cohn, my hands are mangled.
Is that part of the share you robbers wangled?
Part of the blood money to expand your lair?
Now I am done. And beyond repair.
... Mrs. Krause! Really! Enough for now!
Words like "robbers" and "blood" we cannot allow!
We appreciate your working-class candor
But what you're saying is libel and slander.
———————————————. Zap!

—The power's cut. A final crack.
Mrs. Krause! Get out!—And that's a wrap!"
_____.

I'd like to spend some time on air.—
Just wait, dear Sirs—The time is near.—
Against your constant patriotic ruckus
We'll nourish the hate that grows within us.—
The revolution! It's in on our hands!
Stoke the flame, let it abound,
Let it intone the only sound
That can truly unite nations and lands.
When red flags wave from the radio tower,
The proletariat will be in power.

Translated by Geoffrey Winthrop-Young

Notes

1 "Funkerinnerungen eines alten Soldaten," *Der Deutsche Rundfunk,* March 9, 1924.
2 Hans Bredow, *Im Banne der Ätherwellen: Vol. 1. Der Daseinskampf des deutschen Funks* (Stuttgart, Germany: Mundus, 1954), 250–51.
3 Bredow, *Im Banne der Ätherwellen: Vol. 2. Funk im ersten Weltkriege* (Stuttgart, Germany: Mundus, 1954), 79–80.
4 Further, see E. Reiss and S. Zielinski, "Internationaler Medienzusammenhang: Am Beispiel der Entwicklung des Rundfunks in Deutschland, England und Frankreich," *Das Argument* 10 (1976): 173.
5 See Bredow, *Funk im ersten Weltkriege,* 82–87, 96–97.
6 Bredow, 97.
7 E. K. Fischer, *Dokumente zur Geschichte des deutschen Rundfunks* (Göttingen, Germany: Musterschmidt, 1957), 10.
8 Bredow, *Funk im ersten Weltkriege,* 170.
9 Bredow, 188.
10 *Radio-Kurier,* July 12, 1924.
11 See *Telefunken-Rundschau,* June/July 1924.
12 Bredow, *Funk im ersten Weltkriege,* 189.
13 *Telefunken-Rundschau,* June/July 1924.
14 *Der Deutsche Rundfunk,* December 23, 1923.
15 See program announcement in *Der Deutsche Rundfunk,* Spring 1924.
16 Bredow, *Funk im ersten Weltkriege,* 189.
17 Further, see Reiss and Zielinski, "Internationaler Medienzusammenhang," 175 and 187ff.
18 *Die Rote Fahne,* January 3, 1928, quoted in U. Brurein, *Zur Geschichte der Arbeiter-Radio-Bewegung in Deutschland: Vol. 1. Beiträge zur Geschichte des Rundfunks 1* (Berlin, 1968), 19.
19 *Arbeiterfunk—Der Neue Rundfunk* 20 (May 11, 1928).
20 *Arbeiterfunk—Der Neue Rundfunk* 20 (May 11, 1928), 265.
21 Marx-Engels-Lenin-Stalin-Institut, ed., *Zur Geschichte der Kommunistischen Partei Deutschlands* (Berlin, 1955), 227.
22 W. Bierbach, "Die Rundfunkreformvorschläge von Reichsminister Carl Severing: Anmerkungen zur Rundfunkpolitik der Weimarer SPD," in *Rundfunk und Politik 1923–1973,* ed. Winfried B. Lerg and Rolf Steininger (Berlin: Spiess, 1975), 49.
23 See Reiss and Zielinski, "Internationaler Medienzusammenhang," 177 and 186.
24 *Arbeiterfunk—Der neue Rundfunk,* May 31, 1929.
25 "Rund um den Rundfunk," supplement to the *Volksfunk-Arbeiterfunk,* July 1, 1932.
26 *Volksfunk,* January 1, 1933.

27 Bärbel Hebel-Kunze, *SPD und Faschismus: Zur politischen und organisatorischen Entwicklung der SPD 1932–1935* (Frankfurt am Main, Germany: Röderberg, 1977), 54.
28 Wolfgang Ruge and Wolfgang Schumann, eds., *Dokumente zur deutschen Geschichte* (Frankfurt am Main, Germany: Röderberg, 1975), 22.
29 *Arbeiter-Sender*, January 6, 1933.
30 *Arbeiter-Sender*, January 6, 1933.
31 *Arbeiter-Sender*, February 3, 1933.
32 See Brurein, *Geschichte der Arbeiter-Radio-Bewegung*, 2:49.
33 For the economic and sociohistoric material under discussion here, see H. Mattek et al., *Wirtschaftsgeschichte Deutschlands: Ein Grundriss*, vol. 3 (Berlin, 1974); *Geschichte der deutschen Arbeiterbewegung*, ed. Institut für Marxismus-Leninismus, chapters 6–9, 1917–33 (Berlin, 1967–68); Jürgen Kuczyinski, *Die Geschichte der Lage der Arbeiter*, vol. 5 (Berlin, 1964).
34 Cf. Kuczyinski, *Die Geschichte der Lage der Arbeiter*, 5:222.
35 R. Lehmann, "Vom Arbeiterbastler zum Funkkorrespondenten," *Beiträge zur Geschichte des Rundfunks* 2, no. 2 (1976): 82.
36 Lehmann.
37 *Arbeiterfunk* 1 (1930).
38 *Arbeiter-Sender* 24 (1932).
39 Horst Hanzl, *Rundfunk und Arbeiterklasse* (Leipzig, Germany: Karl-Marx-Universität, 1965), 26.
40 Brurein, *Geschichte der Arbeiter-Radio-Bewegung*, 2:16.
41 Hans Teubner, *Exilland Schweiz: Dokumentarischer Bericht über den Kampf emigrierter deutscher Kommunisten 1933–1945* (Berlin: Dietz, 1975), 331.
42 *Arbeiterfunk* 15 (1930).
43 *Der Neue Rundfunk* 1 (1926).
44 Erich Weinert, "Berliner Rundfunkstunde," *Der neue Rundfunk* 10 (July 1927), translated by Geoffrey Winthrop-Young.
45 This and other information regarding the connection between the Workers Radio Movement and freethinking are indebted to Friedrich Knilli's 1977 lecture at Technical University of Berlin on the Workers Radio Movement.
46 *Arbeiterfunk* 17 (1930).
47 *Der neue Rundfunk* 28 (1926).
48 *Arbeiterfunk* 14 (1930).
49 [Translator's note: The poem is the last stanza of Ferdinand Freiligrath's later variation (1848) of his famous poem "Trotz alledem!" (Despite it all!), first published in 1843. The poem was originally based on Robert Burns's "A Man's a Man for A' That."]
50 "What Radio Is to Us Today."
51 See Reinhard Kühnl, ed., *Der deutsche Faschismus in Quellen und Dokumenten* (Cologne, Germany: Pahl-Rugenstein, 1975), 94–95.
52 See *Arbeiter-Sender* 23 (1932).

53 *Der neue Rundfunk—Arbeiterfunk* 2 (1927).
54 Brecht later renamed the play to *Der Ozeanflug* to depersonalize it.
55 Bertolt Brecht, *Gesammelte Werke* [Collected works] (Frankfurt, Germany: Suhrkamp, 1967), 18:126.
56 Bertolt Brecht, "The Radio as a Communications Apparatus," *Brecht on Film and Radio*, edited and translated by Marc Sieberman (London: Bloomsbury, 2016), 40.
57 Brecht.
58 Brecht, 39.
59 Walter Benjamin, "What Is Epic Theatre?," in *Understanding Brecht* (London: Verso, 1998), 20.
60 Bertolt Brecht, "Lindbergh's Flight," in *Collected Plays*, vol. 3, part 2, ed. John Willett (London: Methuen, 1997), 7.
61 Brecht, 11.
62 Brecht, 11–12.
63 Benjamin, "What Is Epic Theatre?," 10.
64 Brecht, "Radio as a Communications Apparatus," 45.

14 Urban Music Box, Urban Hearing

Avraamov's Symphony of Sirens *in Baku and Moscow, 1922–1923—A Media-Archaeological Miniature*

In 2017, Arseny Avraamov's magnificent Symphony of Factory Sirens, first performed in the cities of Baku and Moscow in the 1920s, was performed once again in the Czech city of Brno. This latest performance featured FM Einheit (rhythm machine of the industrial band Einstürzende Neubauten) and other musicians from around the world, as well as cement mixing machines, cannons, a locomotive engine, several cars and Harley Davidsons, a two-hundred-person choir, and various other sonic sensations. In the role of the composer, I cited from Avraamov's texts, which had been written in the style of avant-garde poet and time-management researcher Aleksei Gastev. With this text, I first formulated my early discoveries of Gastev as well as Avraamov's spectacles, for an international audience in São Paolo, Brazil.

This essay was originally published in 2005.

※ ※ ※

"Factory sirens wailed. Industrial horns gave a concert. Music and songs were silent. Just people and flags and more people. The sounds of the *Internationale* coursed through the human tide."

> The instruments of this strange and extraordinary concert are widely dispersed: A crude structure erected in the Moges yard is equipped with fifty engine whistles and three sirens. On the other side of the Moskva, opposite the Palace of Labor, is the "percussion section" made up of artillery batteries that play the part of the drums. Red Army soldiers fire volleys. The conductor ... had to stand higher than usual on the roof of a four-storied building in order to be seen from both sides of the river. ... On your marks! The students of the conservatory, amongst them several children, rush to the wire levers connected to the sirens. Each siren equals one note. The conductor on the roof waves his flags. ... The percussion roars and a thundering echo rumbles through the Zamoskvorechye district. ... What followed could only be heard by those at a far distance. Those in closer proximity as well as the active participants could do no more than try to stuff their ears as tightly as possible to prevent their eardrums from bursting.[1]

These quotes come from contemporary reviews of a concert event that was only performed twice: on November 7, 1923, at noon in the center of Moscow, and exactly one year earlier in Baku, the capital of Azerbaijan. It was the most powerful city symphony ever staged. Its composer was the Donkosak Arseny Mikhailovich Krasnokutsky (1886–1944), who also worked under the pseudonym Avraamov. A music theoretician and acoustician, Avraamov designed several new musical instruments (among others, the string polychord), invented his own universal tone system made up of forty-eight tones for which he himself composed music, taught at the conservatories in Rostov and Moscow, and for a while held a number of high political offices in the young Soviet Union. He signed his manifestos and pamphlets with the three-letter word "Ars."

The two performances of the *Symphony of Sirens* in Moscow and Baku differed in many details. In the Soviet capital, the performance began at 12:30 P.M. with an artillery volley that signaled the start to all the city's inhabitants, followed by resounding fanfares whose piercing sound resembled the signals of minesweepers. Accompanied by gun and artillery volleys, the *Internationale* rang out, sung by a huge amateur choir of "Young Guards." Seasoned machine gunners not only imitated drum rolls but also wove intricate rhythmic figures. At the same time, twenty airplanes, used at various parts of the symphony, roared above Red Square. We know more about the first performance in Baku, the city on the Black Sea whose rich oil resources have in the course of history made it the object of desire for various European invaders and whose periphery has been turned into an eerie machine landscape by a crowded belt of giant metal oil pumps. In the *Baku Worker,* Avraamov published precise instructions for the performance that allow us to re-create it in greater detail.[2] The symphony consisted of three parts, each marked off by twenty-five cannon shots. I am indirectly quoting the composer's description to allow for additional explanations and clarifications on my part.

Part 1: "Alarm." The midday cannon that normally signals the beginning of the revolution's anniversary celebration is canceled. The first salvo at 12:00 P.M. sharp is followed by the foghorns of the ships at anchor in the harbor. The fifth salvo is followed by the sirens of the cargo transfer sites, the tenth by the second and third group of factory sirens. After the fifteenth salvo, the first group of factory sirens falls in accompanied by the sirens of the fleet. At the same time, the great brass orchestra starts up the *Warszawianka* (also

FIGURE 14.1. *Symphony of the Sirens*. Avraamov directing on the roof of a building.

known as *Whirlwinds of Danger*). After the eighteenth salvo, the airplanes add their deafening noise. The twentieth shot is the cue for the sirens of the train depots and the engine whistles. Led by the composer's flag signals, machine guns and steam orchestras join in. With the last five cannon shots, the first part reaches its climax, which ends with the twenty-fifth shot. Pause. The "magistral," an organ composed of steam boilers that function as the lead instrument of the performance, sounds the all-clear signal. Able to produce seventeen different tones, it plays a rudimentary version of the *Internationale*.

Part 2: "Struggle." Triple siren chord. The airplanes fly lower. From the harbor bursts an industrial "Hurray." The *Internationale* is played four times. In the middle of the second stanza, the brass orchestra intonates the *Marseillaise*. As the main melody of the *Internationale* is repeated, it is picked up by the masses assembled on the central square, who, acting as a chorus, finish all three stanzas. While the *Internationale* is being sung, the factory sirens in the depots and stations as well as the engine whistles fall silent.

Part 3: "Apotheosis of Victory." It begins with a festive chord accompanied by machine-gun salvos and the minute-long ringing of the city bells. The ceremonial march of the masses is underscored by another two renditions of the *Internationale*. The symphony ends with all the factory sirens of Baku and its districts.

The energetic stimulus for the *Symphony of Sirens* was the poetry of Aleksei Kapitonovich Gastev (1882–1939). A journalist, writer, tram driver, teacher, metalworker, and trade unionist from Suzdal, Gastev spent large parts the century's second decade in prisons, in penal camps, and on the run or in exile. He belonged to the radical futurist scene of St. Petersburg/Petrograd. Between 1913 and 1920, he developed alongside his political activities an extreme economy of language thoroughly imbued with the spirit of technology that he referred to as the "machinic."[3] In 1921 he published his final collection of poetry in Riga. It consisted of ten poems titled *Pack of Orders*. The formal-aesthetic apex of these ten poetic commandments of proletarian culture comprises verses composed of single-word lines containing machinic orders and instructions. Crossing the boundary that separates art from daily life, Gastev proceeded to build up institutes for the systematic analysis of work management, first in Moscow and then in other cities across the young Soviet Union. Similar to the biomechanics of Meyerhold and Eisenstein, Gastev sought to develop on the basis of a binary code of the machine (lift and thrust) an economy of work that fundamentally differed from the sluggish agrarian modes of production and was instead fully attuned to, and able to merge with, the rhythm of the machine. In line with the ideal of a proletarian human machine, the goal was to create an assembly of living expert systems. After an initial period in which he was courted and protected by the Communist Party leadership, Gastev found himself increasingly sidelined as a delusional activist of a new life and work world. Toward the end of the 1930s, he fell victim to the Stalinist purge and was murdered after one of the infamous show trails.

Krasnokutsky aka Avraamov and his many musical and technical aids were not interested in staging the *Symphony of Sirens* as a gigantic work of art that extended the closed concert space into the city itself. Much like Dziga Vertov's visual projects aspired to connect the camera to the new urban reality by having the cinematic eye invade the latter and be moved by its rhythm, the musicians and composers sought to design forms and practices designed to blend with the altered quotidian soundscape of Russian cities. The goal was to achieve virtuoso performances on the modern claviature of urban acoustic scenarios. There was nothing unique about the subject of *Symphony of Sirens*. The contemporary artistic context had been mapped out by the fairground scenes in Stravinsky's *Petrushka* and Kastalky's *Street Symphony,* a montage of photographs of street scenes with the dissonant sounds of car horns, or the experiments of Charles Ives in the United States, who, much like Avraamov, experimented with polytonality and staged the acoustic flows of North American cities. Horns and sirens became the voices of the new cities, whose character was shaped by the transformational work of industry. Joined into an ensemble, they functioned as mechanico-acoustic choirs designed to accustom the urban residents to new practices of hearing in tune with the world of labor.

"And what are we to dream of when the 'true' music sounds so pathetically sweet and sour and our technical deficiencies delay the arrival of that other, real music we desire?" Avraamov wrote in 1925, evaluating the *Siren Symphony*. He voiced far-reaching demands for a new "musical science," including, among other elements,

> 1. Construction of four radio-musical instruments with not only laboratory but also colossal social significance (unlimited increase of the strength of sound coupled with maximum precision of intonation and timbre).... 4. Topographical acoustics: Analysis of the conditions for the powerful tintinnabulation of musical apparatuses across entire cities.... 5. Design problems in the process of musical creation: the laws of composition under the condition of open-air artistic performances, the change of the soundscape of city life.

The Russian revolutionary government first used telecommunication as a mass medium on November 12, 1917, when Lenin's radio address "To All! To All!" announced that the Soviet Congress had accepted the Peace Decree. Ever since then, his formula that "communism equals Soviet power plus electrification" is connected to radio as well, although his address was in strict technological terms not yet radio but a message in Morse code. In the very same year in which the *Symphony of Sirens* was staged in Baku, Lev Sergeyevich Termen (Theremin) presented his "Thereminvox"—such was the original name of this early electronic instrument—in the Kremlin. It was played by the performer using his hands to manipulate the electromagnetic waves emitted by two antennas. One antenna determined pitch; the other controlled volume. Lev Termen had invented the instrument in 1920 in St. Petersburg, where he was in charge of the laboratory of the Physical Technical Institute. This, in turn, happened to be the same year in which Gastev wrote his last collection of poetry in Riga. Medial imaginations play as prominent a part as the manipulability of the natural and the constructed. "Electro-strings to the center of the Earth" (from the poem "Order 06"). "Switch off the Sun for half an hour / Write twenty kilometers of words across the night sky. / Extend consciousness to 30 latitudes. / Read 20 kilometers in 5 minutes. / Switch on the Sun again" (from "Order 07"). "Report: six hundred cities—passed test. / Twenty cities gone—rejected" (from "Order 10").[4]

In 1918 Gastev published a poem called "Factory Whistles" that contained the lines "When in the morning whistles blow in the factory districts, this is no call to slavery. It is the hymn of the future."[5] His demand for a *Symphony of Work Blows and Machine Noise* must be understood against the background of a society that, despite all revolutionary enthusiasm, experienced formidable difficulties when it came to reshaping modes of production and reproduction, which had been wholly determined by the slow and gradual processes of nature, into an industry able to supersede traditional agrarian modes of production. The utopian aspirations of these demands point toward an association of equally entitled systems of experts in which nobody would ever again feel the need to work because life, labor, and technology are, finally, united, and all alienation has been overcome.

Notes

1 Unless noted otherwise, all quotes are translations from S. Rumjantsev, "Communist Bells," *Soviet Music* 11 (1984): 54–76. It is the most precise account of the musical event under discussion here. I am indebted to Lioudmila Voropai for her translations from the Russian original and to Andrey Smirnov of the Moscow Theremin Center for providing the source. Around 2001–2, it was also the initiation for the research project leading to the exhibition and book *Sound in Z—Experiments in Sound and Electronic Music in Early 20th Century Russia,* ed. Andrey Smirnov (London: Koenig Books, 2013). Images are taken from René Fülöp-Miller, *Geist und Gesicht des Bolschewismus. Darstellung und Kritik des kulturellen Lebens in Sowjet-Russland* (Zürich: Amalthea, 1926).
2 For an English translation of the instructions, see Adrian Curtin, *Avant-Garde Theatre Sound: Staging Sonic Modernity* (London: Palgrave Macmillan, 2014), 188–89.
3 On Gastev and the St. Petersburg scene further, see Zielinski, *Deep Time of the Media: Toward an Archaeology of Hearing and Seeing by Technical Means* (Cambridge, Mass.: MIT Press, 2006), 227–54.
4 Aleksei Gastev, *Пачка ордеров* (Riga, 1921); translated from the German translation *Ein Packen von Ordern* (Oberwaldbehrungen: Peter Engstler, 1999).
5 Alexei Gastev, "Factory Whistles," in *Russian Poetry: An Anthology* (New York: International, 1927), 212.

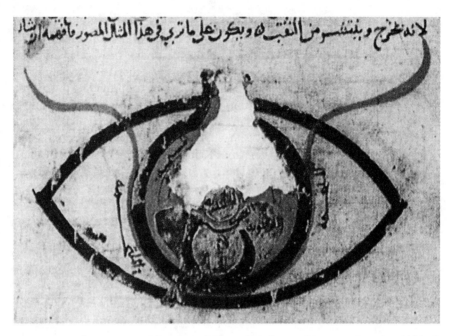

FIGURE 15.1. This anatomical drawing of the eye by Abū Zayd Ḥunayn ibn Isḥāq al-ʿIbādī (808–circa 873), who translated Galen's optical treatise into Arabic, is considered as probably the earliest in the Muslim tradition. Hunayn was a Christian Arab known in Latin as Johannitius who came from al-Hīrah near Baghdad. His ten books on the eye were described by Julius Hirschberg, *Die Arabischen Lehrbücher der Augenheilkunde* (Berlin, 1905), as "the first learned Arabic text book on ophthalmology with the name of the author that has come down to us" (16). The illustration used as a visual motto is taken from a later Arabic manuscript referring to Hunayn ibn Isḥāq. This manuscript dates from circa 1200 and is held at the National Library in Cairo. (Fuat Sezgin, *Wissenschaft und Technik im Islam*, 4:19, gives the signature as "Dar al-Kutub [National Library], [Signature:] Taimur, 319.")

15 How One Sees

> *Having formulated a neurologically grounded theory of visual perception, one that placed the human being at its center as active receiver and participant in the world, physicist and mathematician Ibn al-Haytham is among the leading figures in a southern modernity that was flourishing in Mesopotamia a good four hundred years before the European Renaissance began. This essay illustrates, for the field of optics, how the long-familiar knowledge of those Sun- and light-starved northerns was in effect generated through the eyes of Basra, Baghdad, and Cairo.*
>
> *This essay was written together with Franziska Latell and originally published in 2010.*

⁂

A Short Genealogy on the Variation of a Model

In intellectual history, there is the phenomenon of master thinkers. They are admired because they dared to reflect on matters in ways that nobody thought of before them. Or they combined existing thoughts in such a way that something extraordinarily original was the result. In the history of science and technology, we marvel at master models. From such models, many other particular concepts and construction principles are derived, which themselves can also be highly influential. The camera obscura is such a phenomenon. It appeared in the optical canon of the Chinese Mohists around the fourth century B.C.E. as a tool for studying the shadows of a person under different conditions of projection, including the throwing of two shadows by using two light sources. The inversion of the visual phenomenon, because the aperture of the "dark chamber" that faced the light source did not have a lens, was already a subject of discussion. Aristotle described projections of the Sun rather vaguely as natural phenomena, which is the reason why the invention of the apparatus is often—without hesitation but falsely—attributed to him.[1] After the turn of the first millennium C.E., two outstanding polymaths, mathematicians, and natural philosophers, who made high-profile studies of astronomy, seeing, and multifaceted visual phenomena, specified the model of projection of illuminated and glowing three- or four-dimensional objects in a dark room. They lived and worked in the Far East and Middle East. The Mohist Shen Kuo (also known as Shen Kua, 1031–95) came from a town in the Chinese province of Chekiang, which today is called Hangchow. The slightly older Muslim scholar Abū ʿAlī al-Ḥasan ibn al-Ḥasan ibn al-Haytham (965–1039/1040), who features prominently in Hans Belting's chapter of this volume, came from Mesopotamia and wrote the major part of his theory of optics in the Egyptian capital of Cairo. In Europe and the New World, he is known most frequently as Alhacen or Alhazen; Eilhard Wiedemann sometimes refers to him as Al Husen, and there are quite a few more variants of his name in the scientific literature.

Science historians maintain that both protagonists, from entirely different cultures, described the camera obscura more precisely as an instrument for making observations, particularly in astronomy, through exact study of the mathematical-geometrical laws that are obtained during the projection of rays of light through a tiny opening in a dark room. Additionally, both scholars worked on reflection and refraction, even on double refraction, which occurs, for example, in the prismatic light of rainbows and in crystalline objects. Both Shen Kuo and al-Haytham were early protagonists of geometrical optics, which proceeds mathematically and experimentally.

According to Abū Nasr al-Fārābī's (870–950)[2] *Book of the Enumeration of the Sciences*, written in the early tenth century and translated into Latin as *De scientiis*, "optics examines the same objects as geometry; namely, figures, dimensions, positions, order, equality, inequality, and so forth," although with the important "difference, that most of what occurs of necessity in geometry, because it corresponds to a certain order... when observed appears to be converse. Thus things that in reality are square seem round from a distance; very many things that are parallel one sees as converging, things that are equal appear as unequal, and unequal things equal.... *Therefore, with the help of this science, one can distinguish between that which appears different than it really is when seen, and that which appears as it really is*" (emphasis added). Optics teach "according to the true circumstances of what is looked at to find in the matter, the quantity, form, position, and order and the other things which of the things is where the gaze can be mistaken." In conclusion, al-Fārābī gives an incisive summary that became the program of perspectival representation in the European High Middle Ages: "Through this art, one can gain knowledge about the measure of a distant object from the magnitudes, in cases where it is difficult or impossible to reach it, and via the magnitudes of the distances [between the objects] and us; to this belongs the height of tall trees ... the width of streams and rivers, the height of mountains, and the depth of waters, ... as well as the distances of and magnitudes of celestial bodies.... That is done with an object via an instrument, which serves to direct the gaze in such a way that it does not err, and with another object without an instrument."[3] Al-Fārābī's concept of seeing in principle follows Euclid. Like the mathematician and geometer Ya'qūb al-Kindī (circa 801–73) before him, al-Fārābī assumes an active ray of sight; it carries light from inside the eye or human body and is beamed through the eye at external objects to scan them for perception. This divine scanner is one of the foundations of the Platonic worldview.

Ibn al-Haytham includes the aspects of perspectival representation in his theory but reverses the hypothesis of a ray of vision, which was valid from Euclid to Ptolemy. According to Eilhard Wiedemann, many of the Arab eye doctors and philosophers of that period adhered to a hypothesis that had been formulated, for example, in the tenth-century *Writings of the Brethren of Purity (Ikhwān al-Ṣafā)*: "Light emanates from bodies"; it "penetrates the transparent bodies"; it "assimilates their colors and takes them to the eye-balls.... The other view, that rays emanate from the eyes, is rejected as being misguided."[4] With his theory of intromission, or receiving instead of emission or transmitting, as the starting point of a worldview, Ibn al-Haytham formulated a further master model that integrated the existing knowledge of optics that had been generated so far. This model has been enormously influential in the history of natural philosophy and sciences. Even the cool modern physicist par excellence Ernst Mach, who was always anxious to exclude anything with even a taint of scholasticism or magic from his calibrated scientific universe, acknowledges the Arab scholar in his standard work on physical optics from 1921: "From Alhazen comes the first anatomical description of the eye, which the accepted nomenclature of today still follows."[5] The North American science historian and expert on optics David C. Lindberg, who introduced the reprint of Risner's *Opticae thesaurus* from 1572, became famous especially because of his work on medieval theories of vision and light, on Roger Bacon and Johann Kepler. He emphasizes in the summary of his description of Ibn al-Haytham's books on optics the independence of the Arab scholar from the ancient Greek authorities, whom Ibn al Haytham knew extremely well through his work as a translator: "Alhazen was neither

Euclidean, nor Galenist, nor Aristotelian—or else he was all of them.... Although [his theory] contained classical material everywhere, the resulting edifice was a fresh Islamic creation."[6] In essence, Lindberg's study deduces that the optical theory of Kepler derives from that of Ibn al-Haytham. He supports the hypothesis of continuity in the history of science: "The transition from medieval to modern optics was evolutionary, not revolutionary."[7] Lindberg even goes so far as to interpret Ibn al-Haytham as the author of core elements of the modern theory of vision, as it was formulated in detail in Kepler's theory of the retinal image and transferred by René Descartes into the model of seeing that became the icon of the scientific modern era of Europe. The intromission approach to vision had existed before Ibn al-Haytham; however, according to Lindberg, all of these were theories "of coherent images or forms.... Alhazen was the first to utilize the analysis of the visible object into point sources, each of which sends forth its ray, as a basis of an intromission theory of vision."[8]

As archaeologists who have been sensibilized methodologically through Nietzsche's and Foucault's concepts of genealogy, we are skeptical about any assertion that something or someone was the first. We do not accept that the history of anything has an ultimate stratum of basalt that is impossible to break through and discover even earlier candidates. Certain Presocratic concepts concerning visual perception only make sense if one does not assume complete objects, but smallest particles (atoms), which constantly move between what sees and what is seen, for example, the theory of pores by Empedocles (circa 490–430 B.C.E.).[9] One thousand five hundred years before Ibn al-Haytham and Shen Kua, this poet-philosopher from Agrigento, situated geographically right across from North Africa, pondered the question as to how a huge mountain could enter the tiny organ of perception, and his solution was to dissolve the natural phenomenon into smallest particles for the process of perception. Also the insight that between the points on the object—better, in the field of vision—and those inside the eye, a clear relationship must be established,[10] can in principle be recognized in the compatibility proposition laid out by Empedocles. The natural philosopher Democritus (circa 460–370 B.C.E.), who came a few years after Empedocles, was also an atomist. He expanded Empedocles' theory of pores in two ways: first by introducing a medium, the void, wherein the various configurations can arise, and second by suggesting a concrete—in his terms, material—in-between. The streams that emanate on the one side from the perceiver and on the other from what is perceived compress the air between them. The various constellations of atoms in motion are impregnated on the air and appear there as "idols" *(eidola),* images of real objects, which are identified by the sensory organs as different configurations. The Arabic natural philosopher Ya'qūb Ibn-Ishāq al-Kindī (801–73) refers to those concepts very clearly when he summarizes in his *De aspectibus* the different positions of the "Ancients" regarding vision and visual perception.[11] Without doubt, though, one must agree with Lindberg's view that Ibn al-Haytham was able to combine "the mathematical, anatomical, and physical traditions and created a single comprehensive theory."[12] This is also supported by Gérard Simon in his groundbreaking study *Le regard, l'être et l'apparence dans l'optique de l'Antiquité,*[13] which is quite critical of Lindberg.

All texts on the history of optics, including those by Lindberg and Simon, were written based on translations of Ibn al-Haytham's work that were incomplete to a greater or lesser extent. Already around 1200, Latin translations of the optical studies by the Arab scholar existed, possibly also under the title *De visu* (On seeing).[14] However, they contained many omissions, and some included intermixed interpretations of the adepts, as, for example, in the work by Witelo from Poland. The famous translation of Ibn al-Haytham by the Hersfeld mathematician Friedrich Risner (1520–80) of 1572, which was published together with Witelo's al-Haytham treatise in one volume as *Opticae thesaurus,* is incomplete and—according to Abdelhamid I. Sabra—not consistent with the original Arabic text. In numerous essays, Eilhard Wiedemann has translated passages from al-Haytham's work into German, and we have to thank Matthias Schramm for translating important passages from the books on optics already in the early 1960s, also into German. However, it was not until toward the end of 1989 that the Warburg Institute in London published the first two volumes in English by the Egyptian scholar Abdelhamid I. Sabra, who has devoted a large part of his life to translating and interpreting the seven books of the compendium on optics by al-Haytham.[15] In an essay published later, Sabra summarized in admirable clarity the special achievements of the eleventh-century polymath:

> Ibn al-Haytham's theory of vision was the only one circulating in Europe, up to the time of the Renaissance, that interposed between the centre of vision and the seen object a surface on which a configuration of illuminated points of color directly corresponded to their arrangement in the field of vision. That interposed surface was the slightly flattened spherical surface

of the crystalline humor, and the visually relevant class of points of light and color existing in it marked intersections of the surface with the straight rays proceeding from points in the field toward the centre of the eye, or vertex of the geometrically defined "visual cone." The theory maintained that perception/idrāk/comprehensio of any object in the field, and of all its visual properties (size, shape, distance, and the rest), consisted in a mental Reading of this color mosaic... after it has been transferred as a coherent whole through the humors of the eye and through the optic nerves, and after being ultimately presented to the brain where the final reading process was performed by a sense-faculty understood as a faculty of discrimination and judgement (tamyiz).[16]

Only now that there is access to Sabra's meticulous translation of books I to III of the *Kitāb al-Manāẓir*, the "treasure of optics," and his commentaries on the rest of the books in English and Arabic is it possible to formulate a more differentiated view. Thus, recently, A. Mark Smith has supported the view that Kepler's optics represents a larger revolutionary leap compared to the Arab one, because Kepler was supposedly less concerned with the analysis of vision than with the mathematical-geometrical description of light. "Keplerian optics is luminocentric rather than oculocentric."[17] Ibn al-Haytham's theory is clearly oculocentric.

The strong interest in the eye as the organ of sight corresponds to the physiological and medical focus that Arab-Islamic natural scientists undoubtedly had, from Abū Zayd Hunayn ibn Ishāq al-'Ibādī (808–73) from Baghdad up to the Persian Kamāl-Dīn (1267–1319/20). The endeavors to understand the visual abilities of the healthy eye and to heal the impaired organ of sight were at the forefront of a relationship to the natural world, which was interested in enjoying and improving it. Often, the optics researchers were also eye doctors.[18]

Alī ibn Īsā was one of them. He practiced as an ophthalmologist in Baghdad, presumably at the hospital there, which was founded as early as the eighth century. Alī ibn Īsā was a contemporary of Ibn al-Haytham, although a little older. His *Handbook of Ophthalmology* he completed around the turn of the first millennium, in circa 1004, according to Hirschberg and Lippert.[19] In the eighth chapter of book I, the author applies himself to the "foremost among the parts of the eye," which he considers is the "crystalline humor." He defines it as "colorless, clear, glowing, round," with the caveat, which is also very important to Ibn al-Haytham, that "incidentally, [it is] not a perfectly round shape, but a little oblate.... The moderate roundness exists so that it cannot be harmed so easily.... Oblateness, though, is possessed [by the crystalline humor], so that it can oppose many parts of the perceivable things.... An even shape touches more parts of the things, which it borders on, than a spherical one." After this very physical concept of perception, Alī ibn Īsā, whom the German translators at the beginning of the twentieth century still called "our Ali," emphasizes why, from the point of view of a doctor, he characterizes the crystalline humor as the element through which the process of seeing occurs: "When a star inserts itself between it and the object of sight, then seeing is suspended; but when the star is pushed away from it with an instrument, eyesight returns."[20]

Even masters do not think in isolation. The crystalline humor was also for Ibn al-Haytham the most noble part of the eye and the actual medium of sight. However, what possibly distinguishes Ibn al-Haytham especially as an extraordinary experimental thinker over and above his physiological, neurological, and mathematical-geometric insights repays a more thoroughgoing analysis. This concerns an interesting theological-philosophical implication, which Hans Belting also refers to in his essay, and which we would like to formulate from a media archaeological perspective in a different way. In Philip Wiener's *Dictionary of the History of Ideas*,[21] a trail is already laid out. The paradigmatic switch from a sender/transmitter to a receiver hypothesis, which Ibn al-Haytham initiated mathematically and geometrically in the eleventh century, which Witelo, John Peckham, and Roger Bacon understood and enlarged on in the thirteenth, and which led the Latinists among European scholars to the concept of *perspectiva*,[22] is linked to a dramatic change in the concept and evaluation of light, which remained significant at least until Athanasius Kircher's metaphysics of light in the seventeenth century.[23] The idea of an active ray of vision, which emanates from within the viewer's eye and falls on external objects, is connected to the notion of autonomous, divine light in classical theories of vision and Aristotelian–Christian conceptions and is called *lux* in the Latin tradition. Reflected light or the light emitted by luminescent objects is the profane *lumen*, which in the intromission theory by Ibn al-Haytham, astrophysicist and keen Moon-gazer, then advanced to become the decisive *illumination*. In this theory, the divine *lux* has become superfluous—like Lucifer as fallen angel and carrier of the light. However, that is a different subject altogether that we shall not go into here.

Ten Variants of a Master Model

Our minimal genealogy presented here in images and figure captions is not an attempt to write an instant form of a discursive subhistory or archaeology of optics or the concept of vision. Like the example of the imaginary cut through the organ of sight, which Ibn al-Haytham performed to describe its inner structure and explain the way it functioned, our experiment seeks only to demonstrate with examples how the idea of the Arab scholar inscribed itself in the European history of science and culture of the early modern period.

Ibn al-Haytham (circa 965–1039/1040)

The two sketches, which have been published and described many times as visualizations of the insights of Ibn al-Haytham, are quite different regarding their visual composition. The sketch with the double eye (Figure 15.2) is based upon the idea of a transversal cut through the center of both eyes, which reaches up to the region where sensory event and cerebral activity are adjacent. The depiction is intertwined, delicate, and focuses on describing the connection of the two "distinct phases," which, according to Simon, distinguish the theory of sight by the Arab scholar: "an external optical phase, which led to the creation of the image on the crystalline lens, and an internal phase of transmission and progressive sensory processing up to the cerebral locus of cognition."[24] On the neurobiological side, this is an early intimation of how the two hollow visual nerves, through which the visual spirit flows, are connected via an optic chiasma. This sketch is not included in the various Latin adaptations that are known to us, including that of Risner, and therefore it is only known with Arabic labeling. Of this, too, several variants exist, which, interestingly, are always traced back to the same source, namely, to a manuscript with the number 3212, fol. 81b in the Istanbul Fatih library that Sabra dates to the year 1083. He comments on some of the variants in the depiction with the nice remark that the commentators wanted to set their own different accents as to whether the sketch refers to their own eyes or those of the person opposite.[25]

For Figure 15.3, we reproduced the schematic description from Sabra's book and made a negative of the two diagrams to make it easily distinguishable from the second model. Sabra uses the following terms to translate the Arabic version into modern English:

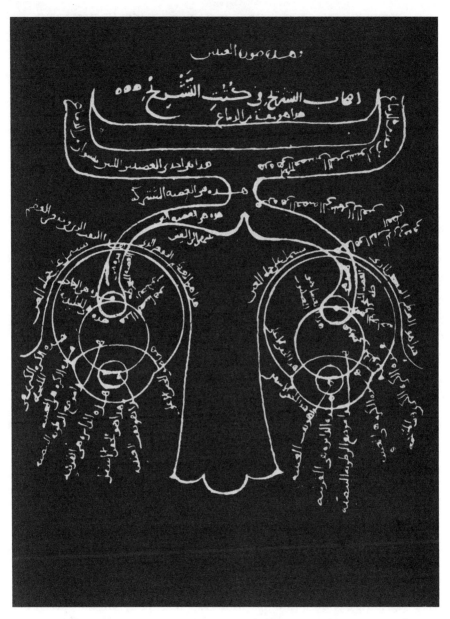

FIGURE 15.2. A. I. Sabra, *The Optics of Ibn al-Haytham* (London: Warburg Institute, 1989), 2:42a.

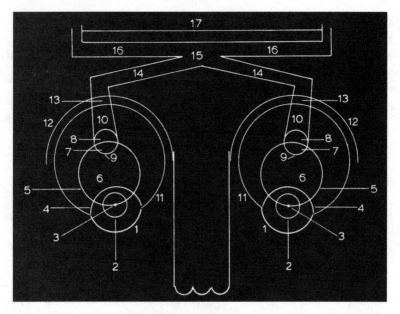

FIGURE 15.3. Sabra, *Optics of Ibn al-Haytham*, 1:63.

1. Lower eyelid—2. Cornea—3. Uveal aperture (pupil)—4. Upper eyelid—5. Uveal sphere—6. Albugineous humor—7. Crystalline humor—8. Vitreous humor—9. Web-like membrane (arachnoid), encircling the crystalline-vitreous body—10. Conically shaped nerve—11. Conjunctival sphere, containing the eyeball—12. Concave bone in which the eye is set (orbit)—13. Aperture in the concave bone—14. Nerve attached to one of the eyes—15. Common nerve (optic chiasma)—16. Optic nerve originating from the brain—17. Front of the brain.[26]

Genealogically, however, we are mainly concerned with the career of the second model. The imaginary sagittal cut vertically through the eye possesses a strong defining power. In Riesner's diagrammatic form (Figure 15.10), it seems as though the eye has been torn out of the skull together with the nerve strands that lead to the brain, or the other way around: as though it were possible to push the analyzed organ back into the skull so that it would function there like the author described it.[27] The widely laid out optic nerve seems like a sword, a dramatic coupling. In the plane of regard is located the core of Ibn al-Haytham's theory, the crystalline humor (Latin: *humor crystallinum*), which, through the membranes or the borders to the other two humors, the vitreous humor (Latin: *humor vitreus*) and the albugineous humor (Latin: *humor aqueus*), is compressed into a convex lens, with the *aranea* (which is best translated as "cobweb") as moveable interface. Ibn al-Haytham needed this construction to salvage his idea of an upright-standing inner universe of points as the equivalent to the universe of external objects. In modern optics, a convex lens is also known as a positive lens.

Kamāl al-Dīn al-Fārisī (1267–1319/1320)

It took almost three hundred years before the work of Ibn al-Haytham was acknowledged properly in an Arabic treatise. The author was Kamāl al-Dīn al-Fārisī, and his work was published at the end of the thirteenth century. In essence, the treatise describes and comments on Ibn al-Haytham's theory of vision and the eye but does not expand on his insights (Figure 15.4 and 15.5). The observations on the anatomy of the eye applied the findings of experiments on the eyes of animals to humans, for in that period, only anatomical investigations performed on animals were allowed. It is reported that Kamāl al-Dīn al-Fārisī dissected a sheep, and this helped him to his most important findings regarding the structure of the eye. "Through his deliberations and experiments Kamāl al-Dīn al-Fārisī arrived at a result that was not taken up again until 1823—by Johannes Evangelista Purkyně.[28] Kamāl al-Dīn al-Fārisī was the first to establish perfectly the reflection on the front surface of the lens and substantiate it excellently within the framework of his theory,"[29] as Schramm has described Kamāl al-Dīn al-Fārisī's achievement.

Roger Bacon (1214–92/94)

"I shall draw, therefore, a figure in which all these matters are made clear as far as is possible on a surface, but the full demonstration would require a body fashioned like the eye in all the particulars aforesaid. The eye of a cow, pig, and other animals can be used for illustration, if any wishes to experiment. I consider this figure better than the one that follows, although the following one is that of the ancients. For it is impossible that the centre of the vitreous humor should be below the sphere of the anterior glacialis, because in that case right will appear left and the reverse, as will be shown below: not yet on surface of its body, because in that case an impression on the right would go too far to the right, and the one on the left to the left, and they would never meet in the common nerve, wherefore the centre will lie outside toward the anterior part of the eye.... The size of the opening is determined by the boundary lines of the visual pyramid, which is *abl*. For let *al* be the base of the pyramid, which is the visible object, the impression

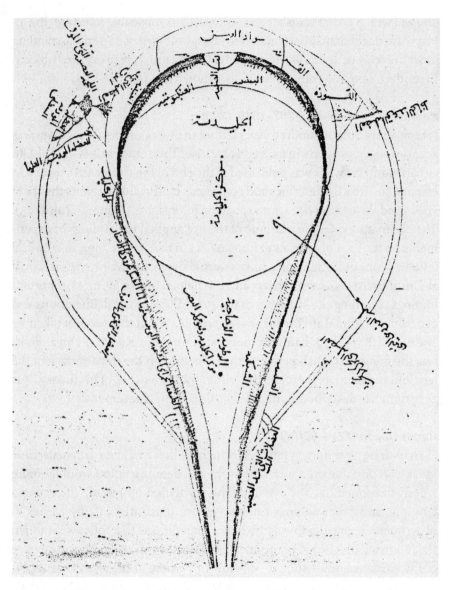

FIGURE 15.4. The sagittal cut through the eye after Kamāl al-Dīn al-Fārisī (circa 1300); Istanbul, MS Ahmet III, 3340, fol. 24b. Sabra, *Optics of Ibn al-Haytham*, 2:49.

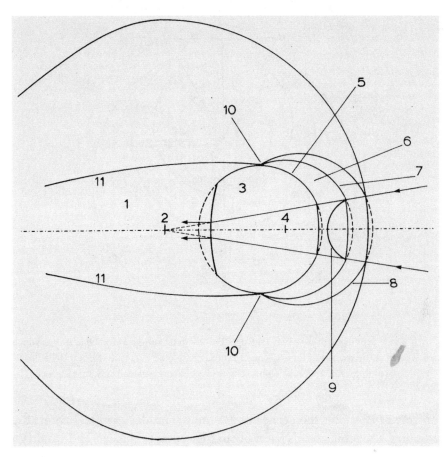

FIGURE 15.5. Schematic of the human eye according to Kamāl al-Dīn al-Fārisī with A. I. Sabra's simplified labeling: "1. Vitreous humor—2. Centre of the eye—3. Crystalline humour—4. Centre of the uvea—5. Web-like tunic, encircling the crystalline humor—6. Albugineous humor—7. Uveal sphere—8. Cornea—9. Uveal aperture (pupil)—10. Wreath—11. Retina." Sabra, *Optics of Ibn al-Haytham,* 2:49.

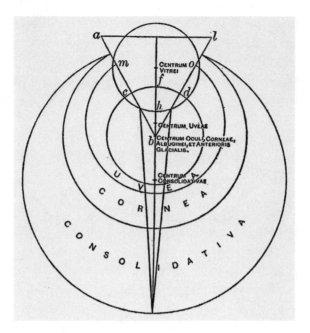

FIGURE 15.6. Schematic of the eye from Roger Bacon's most famous work, *Opus Majus*, which was written at the request of Pope Clement IV and completed in 1267. In Robert Belle Burke, *Opus Majus of Roger Bacon* (Philadelphia: University of Pennsylvania Press, 1928), 2:442.

of which penetrates the cornea under the pyramidal form and enters the opening, and which tends naturally to the centre of the eye, and would go to it if it were not met first by a denser body which it is bent, namely, the vitreous humor, *chd*. For this reason, therefore, I have so placed the centre of the vitreous humor and have so drawn its sphere.... The uvea, moreover, is drawn in a complete sphere, except in its opening, *mo*, and yet it is not in the eye completely spherical, as has been stated, but is so in its anterior part, in which is the opening, because sphericity is not required for vision except in that place; elsewhere it is of irregular form, in order that a void may not exist between the uvea and the cornea; and the lines *c* and *d*, which are drawn toward the interior of the eye, are in the sides of the nerve of the uvea, and between these lines is the opening uvea, which accordingly is toward the interior of the eye. Above this opening the vitreous humor is formed, as is apparent; for the aperture of these lines terminates at the extremities of the portion of the vitreous humor, and its distance is between *c* and *d*, and this distance between the sides of the nerve is filled with the vitreous humor as far as the common nerve in the surface of the brain. This nerve, however, in which is this path of the vitreous humor, spreads and expands in the circuit of the humor vitreus, glacialis, and albugineus, as far as the anterior opening of the uvea, *mo,* which is opposite to its own opening, which is *cd.* Then follows the cornea, and then the consolidativa, as indicated in the figure."[30]

Erazmus Ciolek Witelo (circa 1230/35–circa 1280/90)

The Polish monk, theologian, and natural philosopher Erazmus Ciolek Witelo translated and commented on Ibn al-Haytham's *Book of Optics* approximately at the same time as Kamāl al-Dīn was engaged on his work. Witelo published his variant in 1270. Similar to his Arab contemporary, Witelo tried to stay close to Ibn al-Haytham's text and considered it particularly in the light of the mental dimension of visual perception. Witelo's *Perspectiva* became for centuries a standard reference for optics in the West. The diagram of the anatomy of the eye in Risner's edition of the book became a general model and was frequently identified as authored by Ibn al-Haytham. Anatomically, Witelo did not add anything essential to Ibn al-Haytham's elaborations either regarding the composition of the three humors and four coats or tunics of the eye or in connection with the central importance of the glacial transparent mass, which he regarded as "the organ of vision proper." Witelo accentuates the Arab scholar's clearly medical perspective: "Its transparency alone is able to assimilate the visible forms; it is in the center of all fluids [*humors*] and all web-like coats [*tunics*]; if some other tunic or humor is affected but the glacial humor remains intact, the administration of medicine to the eye will always effect healing and recovery and sight will be restored. However, if the glacial humor is damaged, sight will be impaired and there will be no hope of healing through medical treatment. Particularly the crystalline, or glacial humor is therefore the organ of sight; for this reason it is especially protected by nature."[31]

In the second part of his explanation, Witelo gives an exact description of the structure and position of the optic nerve: "The humors and the tunics of the eye originate from the brain substance, for in the frontal area of the brain two optic nerves [*nervi optici*] grow in two places. They are curved [*concavi*] and both have two tunics, which are attached to two tissues [*telae*] of the brain. These nerves project to the center of the forebrain, where they combine into a single optic nerve. Further on, this nerve divides again into two optic nerves of the same kind, which exchange their positions so that

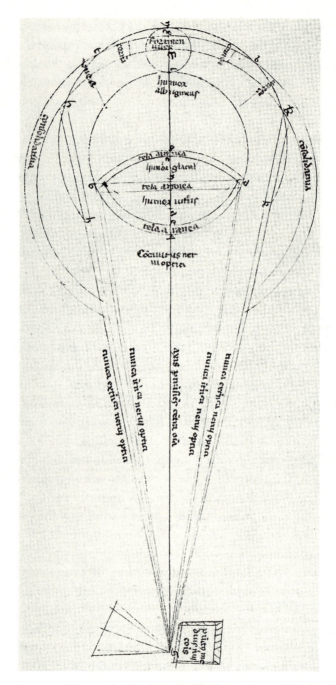

FIGURE 15.7. Anatomy of the eye after Witelo, Oxford Bodleian Library, MS Ashmole 424, after Fuat Sezgin, *Wissenschaft und Technik im Islam*, vol. 4, 26.

the right-hand nerve takes the left position and the left-hand nerve takes the position on the right."[32]

Witelo places the position of the *nervus opticus* on an axis with the lens—a determination that did not change until the work of Christoph Scheiner,[33] a Jesuit astronomer and one-time assistant of Athanasius Kircher: "The aperture, which is situated in the front area of the uvea, lies opposite the aperture of the concave nerve for the optic nerve cuts through the conjunctiva and the vascular tunic [*tunica coniunctiva et uvea*] and passes through all tunics of the eye before it reaches the crystalline sphere, which cuts through the pyramids of the nerve as well as the vitreous humor, which is located in the pyramid-like concavity [*pyramidiale concavum*] of the optic nerve. This cut, therefore, is common to the optic nerve and the crystalline sphere.... It is this concave nerve that conveys the spirit responsible for the visible [*spiritus visibilis*] from the brain to the eye. Through its fine veins [*venae parvae*] nutrients [*nutrimentum*] reach the eye and circulate through it via the supply pathways [*per vias nutrimenti*]. At the point of intersection of this nerve in the frontal part of the brain is the faculty of vision [*virtus visiva*], which perceives and distinguishes all things."[34]

Andreas Vesalius (1514–64)

In 1543, the Flemish doctor Andreas Vesalius (Andries van Wesel) published *De humani corporis fabrica (The Fabric of the Human Body)*, which is regarded as founding modern human anatomy. This 650-page anatomy textbook is Vesalius's most famous publication; he was only twenty-nine when he wrote it and had just graduated as a surgeon and anatomist. The book was instrumental in Vesalius obtaining professorships at the universities of Padua, Pisa, Bologna, and Basel. In it Vesalius explains the structure of the complete human body: skeleton, muscles, veins, arteries, organs, and also some parts of the brain. The voluminous seven folio volumes of *De humani corporis fabrica* contain 83 plates and a total of 420 illustrations depicting dead and dissected human bodies in allegorical poses and everyday situations, which are assumed to be the work of Jan Steven van Calcar (1499–1546). The illustrations of the human eye, however, are not highly original (Figure 15.8). Notwithstanding, Vesalius provided the impetus for ascribing the receptor role to the retina and not to the flattened lens, and—unlike his predecessors—Vesalius does not depict the optic nerve as a hollow tube, an empty channel.

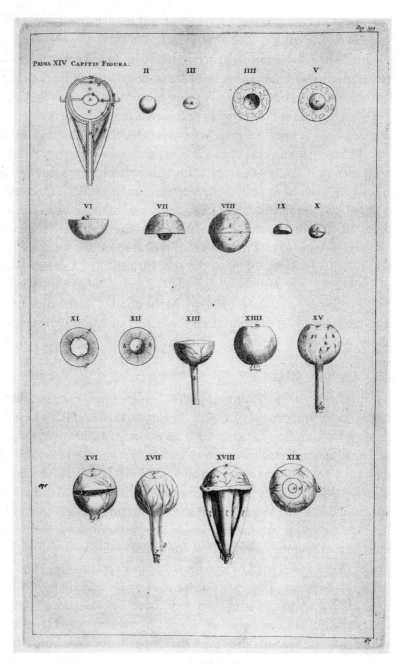

FIGURE 15.8. This illustration from a plate in *Opera omnia anatomica et chirurgica* by Andreas Vesalius (1725, Plate 67, 554–55) shows the structure of the human eye (I) as well as the structure of the optic nerve and the muscles controlling eye movement (XVIII).

CAPUT XIV.
DE OCVLO, VISVS INSTRVMENTO.

PRIMAE DECIMIQUARTI CAPITIS FIGURAE,
ejusdemque characterum Index.

Haec figura, earum quae praesenti Capiti praeponuntur prima, alteram oculi partem exprimit, una sectione ab anteriori sede per posteriorem, atque ita per nervum visorium divisi: perinde scilicet, ac si quis divisae secundùm longitudinem cepae alteram partem ea superficie delinearet, qua alteri parti connata continuaque fuerat. Atque hoc etiam modo coelos, & quatuor elementa, in plano depingere solemus.

A Humor cryſtallinus.
B Tunica anteriori humoris cryſtallini ſedi obducta, tenuiſſimaeque ceparum pelliculae inſtar pellucida.
C Humor vitreus.
D Nervi viſorii ſubſtantia.
E Tunica, quam reti aſſimilamus, quamque reſoluta dilatataue viſorii nervi efficit ſubſtantia.
F Tenuis cerebri membranae portio, quae nervo obducitur viſorio.
G Vvea tunica, in quam tenuis viſorium nervum induens membrana degenerat, ac expanditur.
H Hac ſede uvea tunica in poſteriora comprimitur, & corneam illi obductam non contingit tunicam.
I Pupilla, ſeu foramen, quo uvea eſt pervia.
K Tunica ab uvea initium ducens, & ciliis ſeu palpebrarum pilis imagine correſpondens, ac interſtitium pariter vitrei humoris ab aqueo.
L Durae cerebri membranae portio, nervum viſorium obvolvens.
M Dura oculi tunica, quam dura cerebri membrana conſtituit.
N Durae oculi tunicae pars, quae cornu inſtar pellucida viſitur.
O,O Humor aqueus.
P,P Muſculi oculum moventes.
Q Adhaerens, albave oculi tunica.

FIGURE 15.9. The figure caption of the "Instrument of Vision" lists the principal parts of the eye. Vesalius, *Opera omnia anatomica et chirurgica*, 554.

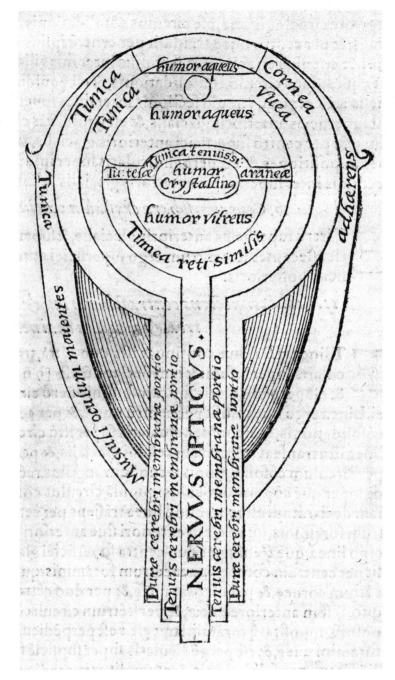

FIGURE 15.10. Friedrich Risner, *Opticae thesaurus: Alhazeni Arabis libri septem, nunc primum editi; eiusdem liber de Crepusculis et nubium ascensionibus,* Item Vitellonis Thuringopoloni libri X (1572).

Friedrich Risner (circa 1533–80)

The most well known European model of the eye's anatomy in the Middle Ages is by Friedrich Risner from 1572 and shows the *humor crystallinus* at the center of the diagram. Risner's *Opticae thesaurus* contains Witelo's treatise on optics and parts of all seven books of Ibn al-Haytham's *Optics*, of which the first treats the theory of vision and contains the diagram reproduced here as Figure 15.10.[35] This illustration is not contained in the Arabic manuscript. Hirschberg claims that the information inside the illustration was taken "from Vesal's *Anatomy*, VII c. 14, Basil. 1555."[36] Risner used exactly the same illustration for both authors (Alhazen: Book I., 6, Witelo: Book III., 87).

Francesco Maurolico (1494–1575)

This schematic (Figure 15.11a) appeared in *Photismi de lumine, et umbra ad perspectiuam, et radiorum incidentiam facientes*, which was written by the mathematician, physicist, architect, and natural philosopher Francesco Maurolico in 1521, published posthumously in 1611. Maurolico came of a Greek family from Messina in Sicily and was a Benedictine monk in a monastery in Santa Maria del Parto à Castelbuono and an eminent professor at the University of Messina. The long figure caption for the diagram details the parts of the eye and their functions:

Ocular Terms

- A Crystalline or glacial humor, pupil
- B Web-like tunic, surrounding the glacial tunic, transparent or onion skin [*caeparum pellis*]
- C Vitreous humor, nutrient of the crystalline humor
- D [*Nervus opticus*] optic nerve
- E Web or web-like tunic, originating from the optic nerve
- F Tunic of the pia mater or thin cerebral tunic, coating the aforementioned nerve
- G Uvea, following [*secundina*], originating from the aforementioned tunic
- H End of the uvea, optically opaque villosity darkens the humors to improve seeing
- I Opening of the uvea, lets light rays through
- K Coarse tunic [*tunica villosa*], originating from the uvea, separating the vitreous from the albugineous humor

FRANC. MAVROL.
Visualis organi theoria.

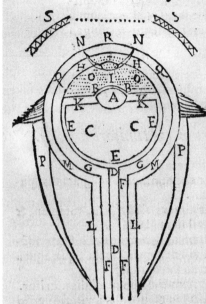

A Humor chryftallinus, glacialis, pupilla.
B Aranea pellucida, glacialem veftiens, pellucida ficut cæparum pellis.
C Humor vitreus, chryftallini nutrimentum.
D Neruus opticus, viforius.
E Retina, retiformis pellicula à viforio neruo procedens.
F Piæ matris, feu tenuis meningis pellis dictum neruum veftiens.
G Vueæ tunica, fecundina à dicta pelle procedens.
H Finis vueæ adumbrantis opaca villofitate humores ad perfectiorem vifum.
I Foramen vueæ radios admittens.
K Tunica villofa, ab vuea deriuata, vitrei &, aquei humoris difcrimen.
L Duræ matris, feu craffæ meningis pellis fecundo veftiens neruum opticum, fimilis palpebrarum pelli.
M Sclerotica, pofterior corneæ pars, à dicta pelle procedens, dura.
N Cornea tunica ex quatuor pelliculis perfpicuis, ac tenuibus ad tutandos humores compacta; cornu inftar pellucida.
O Humor aqueus fiue albugineus, tutamen & excrementum glacialis.
P Mufculi oculū mouentes, è diuerfo foramine quàm opticus.
Q Confolidatiua, tunica alba, denfa, ex pericranio, feu pellicranio progenita, oculum totum veftiens.
R Foramen confolidatiuæ, tranfitum vifui ad corneam, reliquafq. tunicas præbens.
S Palpebræ cum cilijs, claufura oculum complentes.

FIGURE 15.11. Maurolico, *Photismi de lumine* (Naples: Longus, 1611), 72.

ARBOR distinctionis humorum, & tunicarum oculi.

In oculo quædā pertinent.
- Ad visionem, aut
  - Ad recipiendū. — Opticus neruus idest visorius. Humor chrystallinus, vel glacialis, pupilla.
  - Ad transmittendū. — Humor aqueus, albugineus, tutamen, & superfluitas glacialis.
  - Ad adumbrandum. — Vuea tunica, quæ opaca villositate adūbrat visum cum pellicula, quæ aqueū à vitreo discriminat.
- Ad visus conseruationem, siue quo
  - Ad nutrimentum. — Vitreus humor, à quo glacialis nutritur. Retina seu retiformis à visorio neruo progenita, vitreo alimentum præstās. Secundina vueæ pars posterior à tenui meninge vueam nutriens. Sclerotica corneæ pars posterior à crassa meninge corneam nutriens.
  - Tutamen.
    - intrinsecum. — Aranea pellicula glacialem cohibens. Retiformis vitreum complectēs. Cornea tunica ex quatuor pelliculis, generale humorum propugnaculum.
    - extrinsecum. — Consolidatiua alba, & pinguis à pericranio totum contextum vestiens. Palpebræ à cute cranij ad clausuram. Cilia & supercilia ad arcenda noxia.
  - Seruitium motus. — Musculi seu nerui ex cerebro oculum mouentes.

L Tunic of the *dura matris* or thick cerebral tunic, second coating of the optic nerve, similar to the skin of the eyelids
M *Sclerotica*, posterior part of the cornea, originating from the aforementioned tunic, hard
N Cornea, composed of four transparent and thin tunics together, to protect the humors, like transparent horn
O Albugineous or whitish humor, protection and effluent [*excrementum*] of the glacial humor
P Muscles, which move the eye, from a different aperture than the optic nerve
Q Stabilisers [*consolidativa*], white and dense tunic, originating from the pericranium or the scalp, encasing the entire eye
R Opening in the *consolidativa*, allowing access to the cornea and the other tunics during the process of seeing
S Eyelids with eyebrows, which close the eyes.[37]

Maurolico also acknowledges the source of his anatomical knowledge: "We have taken this from the treatise on anatomy by Andreas Vesalius from Brussels, an exceptionally gifted scholar of our times, in order that what is described here may be better understood. We also did this in connection with the explication of the treatises on optics [*perspectivae*] by Roger Bacon and Johannes Petsans."[38]

Maurolico's branching diagram aims to distinguish *(distinctio)* the humors (Figure 15.11b) and the tunics of the eye:

In the eye certain components serve (1) the function of seeing and others serve (2) to preserve the faculty of sight, as follows:

(1.1) Reception: the *nervus opticus*, the optic nerve [*visorius*], the crystalline or glacial humor and the pupil.
(1.2) Transmission: the albugineous, whitish humor, protection and superfluity of the glacial humor.
(1.3) Darkening: the uvea, that through its optically opaque villosity [*opaca villositas*] darkens the sight together with the tunic that separates the albugineous from the vitreous humor.
(2.1) Nourishment: the vitreous humor, that is nourished by the glacial humor; the web or web-like tunic, which originates from the optic nerve and nourishes the vitreous humors; die secundina, that is, the posterior part of the uvea, which originates from the thin

cerebral tunic and nourishes the cornea; the sclerotica [*sclera*], that is, the posterior part of the cornea, which originates from the dense cerebral tunic and nourishes the cornea.

(2.2) Protection (a) inner: the cobweb-like tunic, that surrounds the glacial humor; the web-like tunic, that surrounds the vitreous humor; the cornea, that originates from the four tunics and functions as a general protective Wall for the humors; and (b) outer: the stabilizing [*consolidativa*], white, thick skin, that originates from the scalp [*pericranium*] and envelops everything; the eyelids, that originate from the skin of the skull [*cutis cranii*] and serve to close the eyes; the eyelashes and eyebrows that serve to ward off damage.

(2.3) Movericaement: the muscles or nerves [*nervi*], that originate from the brain and move the eyes.[39]

Robert Fludd (1574–1637)

At the beginning of the seventeenth century, the English doctor and natural philosopher Robert Fludd published in two large folio volumes his eminent work on the history of the macrocosm and the microcosm. Obsessively and in numerous variations, he describes analogies between his imagined cosmic reality and the human body as a microcosmic reflection. These two engravings (Figures 15.12 and 15.13) present the eye from a surgical perspective.

Johannes Kepler (1571–1630)

Ad Vitellionem paralipomena, quibus astronomiae pars optica traditur (*Optics: Paralipomena to Witelo and Optical Part of Astronomy*) is the title of the famous work by Johannes Kepler, in which he—via Witelo's *Perspectiva*—refers to Ibn al-Haytham's theory of optics. Kepler stated the laws of refraction of light in transparent bodies mathematically and geometrically and as a consequence determined that the back wall of the eyeball, the retina, was the place where the visual object was projected before being relayed to the brain for further processing. Thus the retina became the actual organ of perception upon which an image is seen inverted. This modern theory of the retinal image was put forward by an astrophysicist and mathematician. The form of Kepler's diagram (Figure 15.14a, I) is very similar to the master model, except that it focuses more on aspects that are important for the geometry of vision. Ibn al-Haytham's crystalline humor is replaced with the concept of the lens.

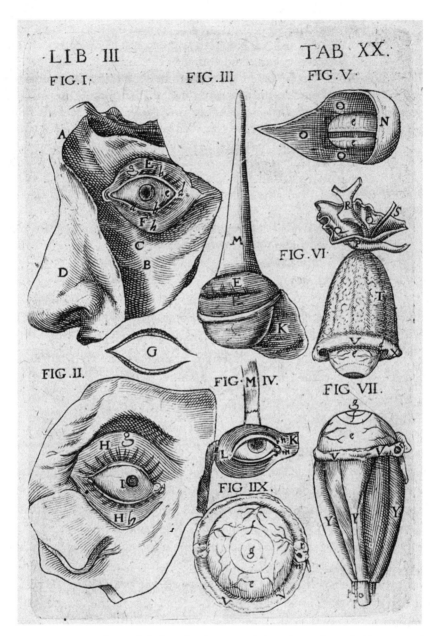

FIGURE 15.12. Here the English doctor and natural philosopher Robert Fludd presents the dissection of an eye in various stages and from various perspectives. The graphic quality of the illustrations is reminiscent of the work of the surgeon. From Robert Fludd, *Utriusque cosmi maioris scilicet et minoris metaphysica, physica atque technica historia* (Oppenheim, Germany: Theodor De Bry, 1617), 186.

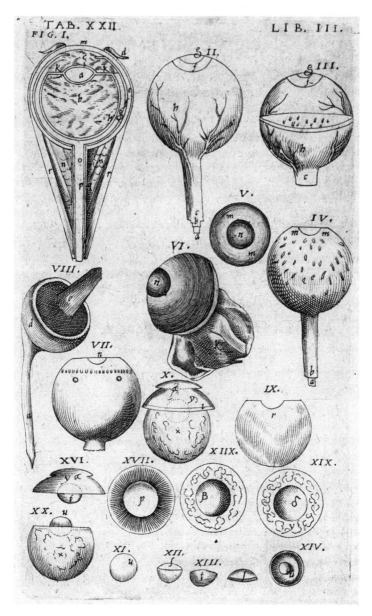

FIGURE 15.13. A simplified schematic of cross sections of the human eye by Robert Fludd, taken from page 195 of the same book: "a—Cristallinus humor, b—Vitreus, c—Aqueus, d—Est tunica adnata, e—Cornea tunicae pars optica, f—Vuea tunica, g—Retisormis tunica, h—Hyaloides tunica, i—Christalloides tunica, k—Pro cessus Ciliares, l—Impressio Vueae a crassa tunica abscedebs, m—Cornea pars crassae tunicae, n—Adepsinter musculos, o—Nervus opticus, p—Crassa meninx, q—Tennis meninx, r—Musculi."

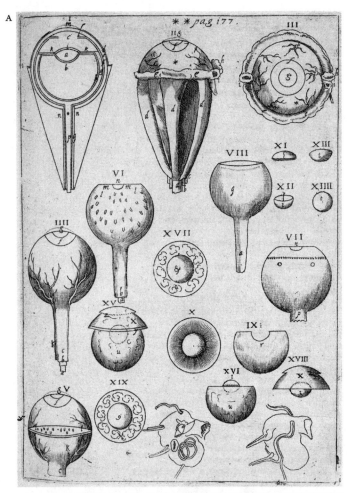

FIGURE 15.14. Johannes Kepler, *Ad Vitellionem paralipomena, quibus astronomiae pars optica traditur; potissimum de artificiosa observatione et aestimatione diametrorum delinquiorumque solis & lunae. Cum exemplis insignium eclipsium. Habes hoc libro, lector, inter alia multa nova, tractatum luculentum de modo visionis, & humorum oculi usu, contra opticos & anatomicos* (Frankfurt am Main, Germany: Claudius Marnius & Johann Aubrius, 1604), text 176.

Here is Kepler's description of the parts of the eye (Figure 15.14b) translated by W. H. Donahue: "I. Portrayal, through a line drawing, of the membranes and humors of the eye, in imitation of the real eye. In which A is the cristayline; B the vitreous; C the aqueous, humors; D the adnata tunica (i.e., the conjunctiva); E the opaque part of the thick tunic (crassa tunica, i.e., the sclera); F the uvea; G the retinal; H the hyaloid; I the crystalloid, tunics; K the ciliary processes of the uvea tunic; L the indentation of the uvea separating off from the thick tunic; M the corneal part of the thick tunica, whose protruding convexity, noted by others, is indicated by dots; N the muscles of the eye; O the visual nerve; P the thin membrane of the nerve; Q the thick membrane of the nerve."[40]

René Descartes (1596–1650)

Some years after Kepler's death, in 1637, a work was published that claimed to provide a solid basis upon which sciences could develop and that became one of the most influential works in the history of modern science: René Descartes's *Discours de la méthode pour bien conduire sa raison et chercher la vérité dans les sciences* (*Discourse on the Method of Rightly Conducting the Reason, and Searching for Truth in the Sciences*, commonly abbreviated to *Discourse on Method*). The philosophical and mathematical treatise was published together with three appendixes, which the book's cover states are essays in the method: meteorology, geometry, and *dioptrique*. In the latter, Descartes published his formulation of the laws of reflection and refraction in various bodies. He took the projection of an image on the retina further than Kepler and others before him, and his depiction of the process attained iconic status for the modern theory of vision. Substantiating his proposition purely by his own reason, Descartes effectively made a clean sweep of past knowledge of optics and dispensed with all references to his scholarly predecessors, although of course he was intimately acquainted with their work: "I do not pride myself that I was the first to discover any [of these ideas], yet I do pride myself that I have never adopted any [ideas] because they were asserted by others, or because they had never been put forward before; I have only ever adopted ideas because my reason convinced me to."[41]

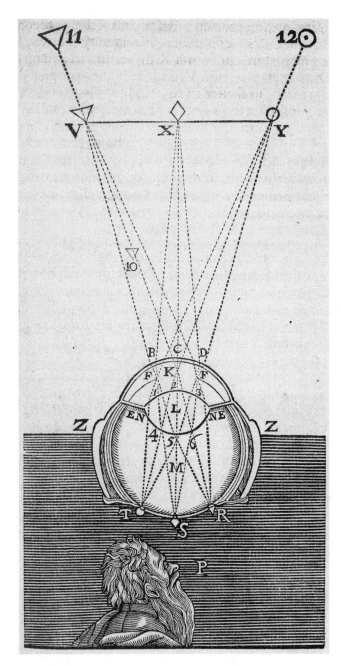

FIGURE 15.15. René Descartes, *Philosophiae seu dissertatio de methodo recte regendae rationis, et veritatis in scientiis investigandae: Dioptice, et Meteora* (Amsterdam: Apud Ludovicum Elzevirium, 1656), 93.

Résumé: The Arabs in the Middle

From a genealogical perspective, here we decline for now to differentiate between concepts of seeing, optics, and visual perception, whose contemporary acutely psychological emphasis appears to us to be the gaze. The various epistemological interests are too interwoven, even within a single research approach. Our brief genealogy, however, does support the positioning of Ibn al-Haytham's master model almost exactly at the midpoint in the line-up of model constructions up to now, with regard to both chronology and the model's special genealogical significance. This becomes clearer if we divide the development of knowledge about vision/optics/visual perception into five periods:

1. The concepts of the Presocratics and Atomists among the natural philosophers of the ancient world (Empedocles, Democritus, Lucretius) conceived of the act of seeing as sensory and reciprocal, as a form of constant exchange of energy between what is seen and the seer. Thus it is not useful to label these concepts "materialist," because their possible antipodes, namely, decidedly idealistic concepts, had not yet developed, nor had the division of the world into that which suffers and that which is active (object and subject).

2. Idealism was a prominent development of the second period of Hellenistic thought. In the *Timaeus,* Plato linked seeing to the cognitive faculty, which is formulated in an exaggerated idealistic form in his philosophical-political parable of the cave. Aristotle opposed both Empedocles' theory of reciprocity and the sensory-based explanation of the Atomists' concepts; he defined the eye as the foremost sensory organ for interpreting the world. Aristotle understood vision as the only active sense faculty, and this combined very well with Euclid's theory of ex-transmission—rays emanating from the eye are the cause of vision. This theoretical coupling became the dominant concept of vision in Western and Christian culture for the following twelve to thirteen hundred years.

3. Arab eye doctors, following Galen's work, then reinstated the close connection between vision and perceived object, initially without abandoning the Euclidean concept. Ibn al-Haytham's receiving-oriented concept of vision became established together with the theory of the projection of three-dimensional bodies as atomized punctiform light into the dark chamber of the eye—understood scientifically in the course of research in

astrophysics (astronomy)—and was a significant inspiration for the late medieval perspectivists in Europe and their revitalization of Aristotle.

4. The protagonists of the medieval perspectivists, Bacon, Peckham, and Witelo, studied al-Haytham with the intention of saving the divine, idealized view of Plato and of aiding the development of perspectival representation. The mathematical and geometrical expression of this was linear perspective, which is essentially concerned with mastering and (re-)constructing the world.

5. With Maurolico, Kepler, and Descartes, natural philosophy has logically come full circle and, with its diverse and differentiated way of contemplating visual phenomena, opens up to modern science. With the exactly calculated image on the retina, the inverted image of the world can be corrected in such a way that it is rendered fit for productions and reproductions of all kinds.

Two of the preeminent concepts of vision were developed with a view to the cave and actually from the perspective of the cave: Plato's political parable in the *Republic* and al-Haytham's seminal book that innovated optics, which he is reputed to have written confined to a dark chamber while under house arrest in Cairo.

The next findings concerning visual perception that signify a pushing back of the boundaries of established knowledge emerge one thousand years later, in research on how sight can be restored in blind people. Here the organ of vision is no longer required as a medium. Machines that simulate optical stimuli for the brain replace it. Visual perception functions by direct connection to the visual capabilities of the brain. Physically, we are capable of getting out of the cave, but are we capable of leaving it metaphysically, without divine assistance?

Translated by Gloria Custance

Notes

1 See Zielinski's short genealogy of the camera obscura in *Deep Time of the Media: Toward an Archaeology of Hearing and Seeing by Technical Means* (Cambridge, Mass.: MIT Press, 2006), 87–88. A detailed bibliography on the camera obscura, which includes the book on optics by Ibn al-Haytham, is provided by Christina Candito, "La camera oscura portatile," in Lo Sardo's catalog *Athanasius Kircher: Il Museo del Mondo*, 248–49.
2 He probably came from Afghanistan but went to Baghdad as a student and died in Damascus.
3 Eilhard Wiedemann on al-Fārābi's "Aufzählung der Wissenschaften De scientiis" [Enumeration of the sciences], *Beiträge zur Geschichte der Naturwissenschaften* 11 (1907–8): 87–88.
4 Eilhard Wiedemann, "Zur Geschichte der Lehre vom Sehen," *Jahrbuch der Photographie* 7 (1893): 318.
5 Ernst Mach, *Die Prinzipien der physikalischen Optik* (Leipzig, Germany: Barth, 1921), 60.
6 David C. Lindberg, *Theories of Vision from al-Kindi to Kepler* (Chicago: University of Chicago Press, 1976), 85. The reprint of Risner's Latin edition of al-Haytham and Witelo was published by the Department of the History of Science at the University of Wisconsin.
7 A. Mark Smith, "What Is the History of Medieval Optics Really About?," *Proceedings of the American Philosophical Society* 148 (2004): 180.
8 Lindberg, *Theories of Vision*, 59–60.
9 See the Empedocles chapter in Zielinski's *Deep Time of the Media*, 39–56.
10 See also Lindberg, *Theories of Vision*, 86.
11 "Quemadmodum plures antiquorum extimauerunt." See Al-Kindi, *De Aspectibus*, reedited in Latin by Axel Anthon Björnbo and Sebastian Vogl (Leipzig, Germany: Teubner, 1912), 9.
12 Lindberg, *Theories of Vision*, 85.
13 The French original, *Le regard, l'être et l'apparence dans l'Optique de l'Antiquité*, was published already in 1988; here we use the German edition: Gérard Simon, *Der Blick, das Sein und die Erscheinung in der antiken Optik* (Munich, Germany: Fink, 1992), 227.
14 Candito uses this title in a somewhat cryptic manner in connection with Alhacen; cf. note 4.
15 To date, only books I–III have been translated into English, which Sabra published in 1989. In 2002, he reissued books IV and V of the optics compendium by Ibn al-Haytham, *On Reflection and Images Seen by Reflection*, in a critical Arabic edition in Kuwait, introduced by a short summary in English. Sabra is currently preparing an English edition of books IV and V.
16 Abdelhamid I. Sabra, "Ibn al-Haytham's revolutionary project in optics," in

The Enterprise of Science in Islam: New Perspectives, ed. J. P. Hogendijk and Abdelhamid I. Sabra (Cambridge, Mass.: MIT Press, 2003), 96.
17 Smith, "What Is the History of Medieval Optics Really About?," 194.
18 See also Eilhard Wiedemann, "Beschreibung des Auges nach al Qazwīnī," *Jahrbuch der Photographie* 26 (1912): 67–73.
19 Ali ibn Isa, *Erinnerungsbuch für Augenärzte* (Leipzig, Germany: Veit, 1904), 23–24. In 1996, Fuat Sezgin published a reprint of this book as volume 44 of the series *Islamic Medicine*.
20 This strongly Galenic argument runs across pages 13 and 14 of the quoted text.
21 Philip P. Wiener, ed., *Dictionary of the History of Ideas*, 3:409–10.
22 Of course, these positions from the thirteenth century were not coherent but rather heterogeneous. Also, one should not forget that most of the ancient Greek texts (including Euclid's treatises on optics) were not translated into Latin from Arabic before the twelfth century.
23 See S. Zielinski and E. Fürlus, "Ars brevis umbrae et lucis," introduction to *Variantology 3* (Cologne, Germany: Walther König, 2008), 9–11.
24 Simon, *Der Blick, das Sein und die Erscheinung*, 228.
25 "The orientation of the diagram doesn't matter—are the eyes mine or yours?" Abdelhamid Sabra, pers. comm., March 14, 2009.
26 Sabra, *Optics of Ibn al-Haytham*, vol. 1, book 1 (London: Warburg Institute, 1989), 63.
27 Investigations of the anatomy of the eye in the Arab Islamic and European Christian tradition were performed by dissecting the eyes of animals (usually cows or oxen) at least up to the time of Descartes. It is tempting to analogize this diagram with the famous filmic cut through the eye of a cow with which in 1928 Buñuel and Dali in *Un chien andalou* sought to redefine seeing in film. However, that would be an impermissible historicization.
28 See the Purkyně chapter in Zielinski's *Deep Time of the Media*, 192–203.
29 Matthias Schramm, *Ibn al-Haythams Weg zur Physik* (Wiesbaden, Germany: Steiner, 1963).
30 Robert Burke, *The Opus Majus of Roger Bacon* (Philadelphia: University of Pennsylvania Press, 1928), 441–43.
31 Vitelloni, *Thuringopoloni opticae libri decem*, ed. Friedrich Risner (Basel, 1572), 85.
32 Vitelloni, 86.
33 See E. H. Schmitz, *Handbuch zur Geschichte der Optik* (Oostende, Flanders: Wayenborgh, 1981), 271.
34 Vitelloni, *Thuringopoloni opticae libri decem*, 86–87.
35 See Friedrich Risner, *Opticae thesaurus: Alhazeni Arabis libri septem, nunc primum editi; eiusdem liber de Crepusculis et nubium ascensionibus, item Vitellonis Thuringopoloni libri X* (1572), 6.

36 Cf. Julius Hirschberg, *Die arabischen Lehrbücher der Augenheilkunde: Ein Capitel zur arabischen Litteraturgeschichte* (Berlin: Reimer, 1905), 112.
37 Maurolico, *Photismi de lumine*, 72.
38 Maurolico, 72–73. Io. Petsan is mentioned in the nineteen-volume thematic bibliography of the Swiss naturalist Conrad Gessner (1516–65) under the entries on optics ("Titulus IV") in connection with Maurolico and a treatise on optics from 1542.
39 See Maurolico, *Photismi de lumine*, 71.
40 Cf. Johannes Kepler, *Optics: Paralipomena to Witelo & Optical Part of Astronomy* (Santa Fe, N.M.: Green Lion Press, 2000), 189.
41 Descartes's *Dioptrique*, quoted here after the essay by Michel Authier, "Zur Geschichte der Brechung und Descartes' 'vergessene' Quellen," in *Elemente einer Geschichte der Wissenschaften*, ed. Michel Serres (Frankfurt am Main, Germany: Suhrkamp, 1998), 473. Descartes makes one important exception: he mentions the Benedictine monk Francesco Maurolico. Within the framework of our research on an Institute for Southern Modernities (ISM), we shall take a closer look at this Sicilian scholar in a special publication.

16 Lüology, Techno-souls, Artificial Paradises

Fragments of an An-archaeology of Sound Arts

Despite its title, my book Audiovisions *(1989; English translation, 1999) was still very strongly image oriented. This was characteristic of much early media thinking about aesthetic production. I was nevertheless keenly aware of this pervasive oversight of the sonically technologically mediated world. In fact, my first forays into media-archaeological research had been with a view to the sounds of radio. This chapter outlines some of the high points in a deep-time reflection on the complex dimensions of the theory and practice of the sonic as such.*

This essay was originally written in 2012 and is published here for the first time.

※※※

Sound art relates to phenomena, events, processes, and installations that produce acoustic material for intellectual and sensual enjoyment with the aid of or through constructed artifacts. Genealogies of sound art that conform to this definition, as τέχνη *(technē)* in the broadest sense of the Greek term, which includes craftsmanship, are frequently written in such a way that they have virtually nothing to do with the complex genesis of artificial worlds of sound. Analogous to the visual arts, in which the mathematization of the image and the geometrization of the gaze began during the second, that is, European renaissance (with occasional reference to the mathematically unsophisticated Greek and Hellenistic tradition),[1] the mechanization and automatization of the arts of sound commenced in the High Middle Ages or early modern period, respectively (with occasional reference to the Pythagorean, that is, Greek and Hellenistic tradition). Curtis Roads, for example, in his reconstruction of the automation of music, which he refers to as "informed music," naturally includes some of the Greek concepts dating from the second century B.C.E. but then he jumps immediately to the thirteenth-century mechanical carillons from the Netherlands.[2] In his fascinating conceptual discussion of composing experimental music, Christoph von Blumröder does stress the fundamental link between composition and scientific research. However, for the author, its history only starts to get interesting when the word *experimentum* appears in Latin medieval treatises in Europe and only becomes truly relevant for practice and theory with Francis Bacon's operational definition of the word *experiment* as "sought-after experience" in *Novum Organum*, published in 1620.[3] And the legendary 1980 exhibition of sound art objects curated by René Block and organized by the Berlin Academy of the Arts in collaboration with the DAAD artists' program and the Berlin Musical Instruments Museum of the State Institute for Music Research (SIMPK) prefaced its fascinating displays with a genealogical diagram that featured only Western dates and names (Figure 16.1).

The introduction to the catalog did at least carry an illustration of an automaton for telling the time, which also produced sounds, attributed to the twelfth-century Kurdish engineer al-Jazarī.[4] The automaton pictured had in fact been designed and built three hundred years before that, in the early ninth century, at the Baghdad court of the ruling dynasty; it was subsequently updated in al-Jazarī's mechanical encyclopedia. In 806, the magnificent horologium of Caliph Hārūn al-Rashīd was loaded on board a ship at the port of Sidon (now in Lebanon) that narrowly escaped being attacked by a Byzantine fleet while sailing from the eastern Mediterranean to the Adriatic; the gift clock was offloaded at Venice and then taken via Treviso over the Alps to its destination, Aachen, the court of Charlemagne, where it arrived in 807. Four elephants were sent along with the clock, but unfortunately only one survived the journey: Abul Abas. Legend has it that his tusks were later used in Aachen to make the keys of keyboard instruments. Baghdad did not really expect any gifts in return from the northern barbarians that would be anywhere near as magnificent and accurate as the Caliph's present.[5] And it was naturally a water clock that Hārūn al-Rashīd sent to Aachen and not a solar-powered one, because they assumed in Baghdad that the sun would not shine enough in the northwest for the horologium to work properly.

The following fragments are connected with three genealogically oriented projects that intersect variously with the sound arts. (1) With the concept of "expanded animation," I seek to develop a concept of animation for the present days of the past and future that is entirely focused on movement, ensoulment via technology, and thus on the time-based dimension of this

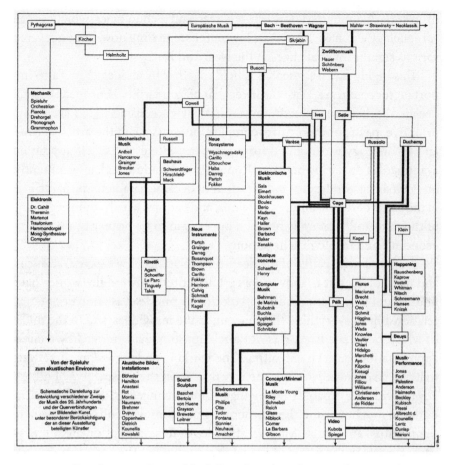

FIGURE 16.1. Development scheme of the different branches of music in the twentieth century and interconnections to visual arts. Opening diagram from *Für Augen und Ohren—Von der Spieluhr zum akustischen Environment: Objekte Installationen Performances,* exhibition catalog (Berlin: Akademie der Künste, 1980), 6.

cultural technique. In this effort, sound- and image-producing machines play a preeminent role. (2) The first renaissance of the knowledge of European classical antiquity did not take place in Italy and Europe but in Persia and the Arab region. The translation of ancient Greek manuscripts and further development of the ideas they contained by Muslim scholars, particularly in the half-millennium between the eighth and thirteenth centuries—also referred to as the Golden Age of Islamic culture—gave us access to important worlds of knowledge about technology and the arts. (3) Study of the deep-time relations between arts, sciences, and technologies, which we call *variantology* and have been working on for some time now, has uncovered various significant challenges in the Near and Far East.

This highly diverse research is held together by an idea that is also important for sound art. With the invention of "modernity," European culture declared itself the center of the universe. What was considered old-fashioned, primitive, passé, or innovative, complex, progressive, in the arts, sciences, and technology was defined from the perspective of the self-appointed avant-gardes operating between Rome, Paris, London, and, intermittently, St. Petersburg. Modernity and the modern became paradigmatic postulates in the battle between cultures, religions, and nations, which at the beginning of the third millennium is revealed once again to be primarily a battle for economic and ideological hegemony.

From the perspective of the deep-time relations between arts, sciences, and technologies, it is above all a certain attitude toward the world that I understand as *modern*. The hallmark of this mind-set is an experimental relationship to everything that surrounds the individual, and in the midst of which he or she lives; it is not a relationship characterized by testing, appropriating, or even exploiting. The world is understood as something that can be changed, in the same way that, from the perspective of others, any individual is capable of changing or being changed. In an ideal scenario, both the individual and the surrounding environment will be changed, to their mutual advantage. Modern attitudes of this description are found in various parts of the world in very different epochs. They are not confined to European countries in the period from the fifteenth to the eighteenth centuries. And in the vanishing point of the diverse inquiries, a space of possibilities opens for which we potentially do not need the hegemonial concept of modernity any longer.

Lüology

The school of the Pythagoreans was a hard one and had more prohibitions than wishes or desires. It is said that after rising in the morning, members were required to smooth the sheets of their beds so that all traces of their bodies from the time they were sleeping and dreaming—and thus time that was not subject to discipline or mathematization—were no longer visible. For the Pythagoreans, music expressed cosmic and divine harmony, which earthlings must attempt to translate into harmonious proportions perceptible to the senses. Symbolic values do not produce sounds. The medium of sound was understood as numbers, *numerus sonorus,* as it was later termed in the Baroque theory of harmony: the musical tones of numbers, or rather numbers that have been made to produce sounds. The same numerical ratios that determine the distances of the planets to the central fire of the Earth, and the organs of the body to the *anima mundi,* the soul, also applied to the discrete notes of the diatonic scale, and vice versa. The freemason Robert Fludd, philosopher and physician, as such responsible for the health of the body, charged Theodor De Bry in Oppenheim near Frankfurt with the production of impressive engravings of this genesis of the macrocosm and microcosm (Figure 16.2).

Neoplatonism, which resonates in this unity of the great and the small, we shall encounter again in the course of this text in connection with automata in the Muslim tradition. Until the early modern period, music as a discipline was subordinate to mathematics. The faculty for music theory and practice was disdainfully referred to as *scientia subordinata.* The vibrations of tones and the intervals were subordinated to the geometry of ratios and arithmetic. The Atomists, such as the Greek natural philosopher Democritus, who described vibration as the only real form of movement, were largely ignored for a very long time by the discipline of music theory. Among others, Marin Mersenne took up their ideas again in the seventeenth century; he understood frequency not as measures of length but as oscillations, that is, as a temporal phenomenon.

The history of Greek–Hellenistic hegemony in founding European arts of sound is well known and has been described by many commentators, from Adorno and Xenakis ("we are all Pythagoreans more or less") to contemporary media theorists (Friedrich Kittler). A part of this history—in the tradition of the Pythagoreans and the preeminence of the *mathematikoi*—is the school of Aristoxenus. A pupil of Aristotle, he wrote *Elements of Harmony* in the fourth century B.C.E., the first great musical treatise. Aristoxenus rejects the idea that numbers are the determining element in harmony. Only in the second instance can they formulate the harmony produced by experience, by the hearing and musical intuition: "using calculations to create intervals that no voice and no instrument could produce and that the ear could not identify"[6] makes no sense at all. Archaeologies of science and technology offer other topics and stories. But for these we must leave the sedate regions

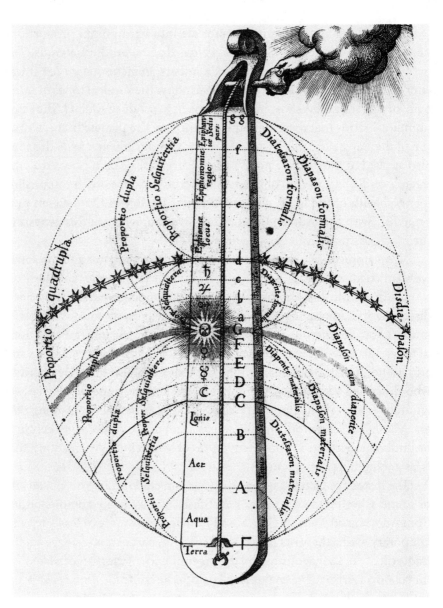

FIGURE 16.2. The great cosmos (above) as monochord, carefully tuned by the hand of God, the small cosmos on Earth (right) with the same proportions/intervals—a completely harmoniously tempered relationship. Illustrations from Robert Fludd, *Utriusque Cosmi Maioris scilicet et Minoris, Metap(h)ysica, Physica atque Technica Historia*, book 1, tract. 1 (Oppenheim, Germany: Theodor De Bry, 1617), 90 (left), and Tom. 2, Tract. 1 (Oppenheim, Germany: Theodor De Bry, 1619), 275 (right). Sächsische Landesbibliothek—Staats- und Universitätsbibliothek Dresden, Sig. 1.B.3237-1 and 1.B.3237-2.

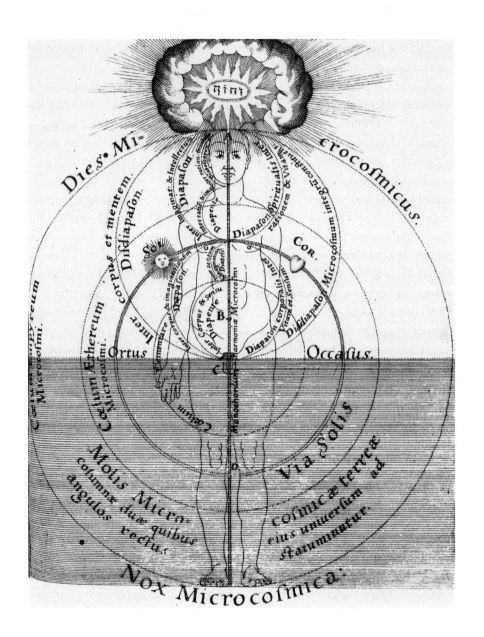

around the *mare internum,* the Mediterranean, and cross the ocean or the great deserts of Asia to China.

In his magisterial study of science and civilization in China, Joseph Needham emphasizes that in ancient China, the cultural technique of tuning was primarily oriented on physical sensations and phenomena of movement—activity.[7] Analogous to taste and color, sound is understood as an active experience. In ancient China, various high tones and their ratios to each other were investigated, for example, using the sounds produced by bubbling cooking pots. According to how deep and wide the pots were, and how much water was brought to the boil in them, they produced different sounds. The steam, the energy that was released during boiling and that pushed up the lid with a hissing noise, was termed onomatopoetically by the Chinese as *Qi,* which is undoubtedly one of the most important categories in Chinese natural philosophy. It is safe to say that *Qi* was and is extremely complex.[8] *Qi* stands for a variety of very heterogeneous concepts, such as breathing, temperament, and the principle of life, as well as for offering a meal. Applied to music, according to Needham, a particular form of pneumatic action is articulated in *Qi,* which later played a very important role in early music automata that utilized air pressure.[9]

Singing requires pneumatics-based actions. Raising the voice, the specific way a note is intoned, is the most active part of using the voice and expresses most clearly the uniqueness and thus the *kairos* nature of the voice. *Kairos* means the auspicious moment, the time we use, or rather how we use the time, how we decide to use it. Articulating a vowel, for example, pulls apart the vocal folds in the larynx and generates a short, explosive sound. This is clearly to be avoided at all costs when using a sensitive microphone. To articulate a vowel softly, the vocal folds are only pushed apart to the extent necessary to make the desired sound, and then they bounce back. In this case, the vocal sound is preceded by a very soft sound, which the Czech experimental physiologist Jan Evangelista Purkinjě discovered in his Wrocław laboratory in 1836, and which he called a *lekki chuch* (gentle whisper).[10]

With regard to the steaming cooking pots, an analogy occurs to me from media archaeology that concerns an archaic instrument featured in the Bible. When Jacob flees after betraying his twin brother and his blind father, he lies down at night to rest with his head on a stone and sees a ladder to heaven with angels ascending—like the steam from a cooking pot. This image of Jacob's ladder obviously can also be interpreted as an acoustic scale. In the parable about a possible ascent to the higher spheres, Jacob did not see the ladder in his dream; he heard it (or perhaps both). Led Zeppelin memorialized this idea impressively in rock music with their "Stairway to Heaven."

Xu Fei, eminent physicist, archaeologist, and musicologist of the deep time of Chinese knowledge cultures, views the development of standard pitch instruments for tuning as an important indication of the success and degree of humans combining arts, science, and technology. In his essay "Exploration and Achievements of the Acoustics of Yuelü,"[11] he refers particularly to the brilliant studies on musical tuning methods by the mathematician and acoustics researcher Chen Cheng-Yih, who in his book *Early Chinese Work in Natural Science* (1996) retains the lovely Chinese word *lüxue* (律学), which literally means the study of *lü* (律, "pitches"), that is, the study of the ritual tone system, and which is often translated as "temperament." As Chen considers this an inadequate translation, he suggests better not to translate the term and instead speak of "Lüology" to refer to the ancient Chinese methods of tuning as an important "branch of early science in Chinese civilization."[12]

From surviving ancient documents and recent archaeological discoveries, Xu Fei and his team have determined that the Chinese began to manufacture standardized wind-pipe music instruments around 6,000–8,000 years ago. The large number of artifacts evidences the existence of established musical practices and methods, which must have been shared by many people. The earliest yet found, the Gudi (骨笛) flutes carved out of animal bone and excavated in large numbers from the Jiahu Neolithic site, were clearly very popular instruments. Xu Fei has studied their technical and acoustic qualities and found that when they are played while held vertically at an angle, the results are comparable to today's musical instruments. "Most of the 6- and 7-hole Gudis can perform two true octaves when played with the old technique of vertical and inclined end-blowing. This result . . . affirms that the Jiahu bone flutes actually served as functioning musical instruments. Furthermore, . . . even at that time, humans had obviously already mastered suitable technologies for making musical instruments."[13] Not later than the end of the Yin Dynasty and the beginning of the Zhou Dynasty, the ancient Chinese invented pitch pipes (*lüguan*, 律管) "to regulate the pitches of bell chimes. During the Spring and Autumn Period (722–476 B.C.E.) and the Warring States Period (475–221 B.C.E.), the pitch-pipe had already become a

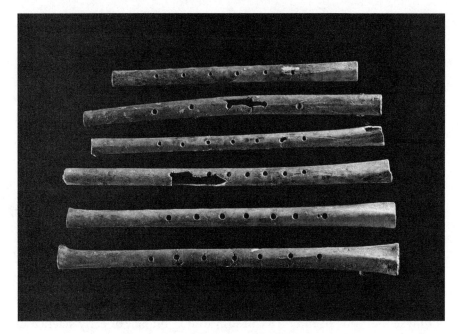

FIGURE 16.3. The six Gudi bone flutes (M511:4) excavated from the Neolithic site Jiahu, China.

commonly used tuning instrument. The widespread utilization of pipe-tuning instruments in ancient times met its own social demand" (Figure 16.3).[14]

The preceding sentence is succinct, but on closer inspection, there is a lot behind it philosophically. The thinking of scientist Xu Fei, grounded in dialectical and historical materialism, makes reference here to the connection between social practice and the cultural technique of tuning, which still echoes today in our expression "in tune with." The Pythagorean school also understood its practice as political and ethical. Precise tuning of instruments becomes necessary in situations where several instruments are played together and must harmonize. For individual players to play together without descending into cacophony or chaos, they must agree on and set a common pitch. "Ritualized official ceremonies with music and dance ... were a national institution."[15] This was the main reason why flute-like instruments were used for standardization. And it remained the most important reason why research on instruments and study of the tone system never stopped in ancient China, indeed, were actively fostered by the rulers of the various dynasties.

Techno-souls

The genealogy of automata that produce sounds I shall discuss in the context of a concept of expanded animation.[16] *Anima* in Latin means in a physiological sense a breath of air, the breath of the body, the breath of God, the *pneuma* that flows through the subconscious of a man and directly motivates, that is, moves, intentions and actions. The soul, *anima*, also has female connotations in Latin. Its male counterpart is *animus*. *Animus* is energy, the current of air that activates the vital spirit of the man and penetrates the female principle completely. As in Chinese natural philosophy, this does not refer primarily to biological gender differences and their potential compatibility. Yin and yang represent opposite principles that are actually perfectly compatible in the way they relate to each other. In animation, they work together cooperatively. In a wider sense, *anima*, as well as *animus*, stands for the principle of life. This dimension of meaning also resonates in the concept of animality. On one hand, the body and everything to do with it is quite alien to the *anima* and *animus*, but at the same time, they ensure the vitality of the body and are a constituent part of its corporeality. The Latin verb *animare* not only means "to animate," "to ensoul"; in Baroque music theory, it still meant the act of composing.

No kinetic device can function without motive force, without energy. "A machine may be defined as *a system of interruptions* or breaks *(coupures),*"[17] according to Deleuze and Guattari in *Anti-Oedipus*. "Every machine, in the first place, is related to a continual material flow *(hylē)* that it cuts into."[18] In order that an artifact moves, there must be energy and also organic or technical power transmission that carries the energy to the artifact. Such technical installations take over the position formerly occupied by the soul, or spirit, as the motive force in the human body. The earliest material forms of such artificial souls were water and air, frequently combinations of hydraulics and pneumatics. Rotating elements of automata were driven directly by falling water, or vessels produced air pressure by filling with water and then emptying again, which drove the mechanical movements of the artifact. In the traditions of Hellenistic Alexandria, Byzantium, Persia, and Mesopotamia, a vast number of these technical solutions was developed and over the centuries—up to the first renaissance in the Arab-Islamic countries from the ninth to the thirteenth centuries—became increasingly elaborate and sophisticated (Figure 16.4).[19]

FIGURE 16.4. Many magnificent and ornate sound installations are described in *Tales from the Arabian Nights,* automata that seven hundred years later will grace the Renaissance interiors of the aristocracy, albeit in forms that are far more modest. Here the metal birds are supplied with the pneuma not by water power but by a servant hidden behind the table. The engraving is from Capitano Agostino Ramelli's *Le Diverse et Artificiose Machine del Capitano Agostino Ramelli* (1588). From Annette Beyer, *Faszinierende Welt der Automaten* (Munich, Germany: Callwey, 1983), 32.

Beginning at the latest in the fifteenth century, explosive substances and fire were used in Europe to set immobile matter in motion. Pyrotechnics, in turn, is very close to the form of energy that has become the soul of all modern machines and automata, namely, electricity and electronics. To begin with, in connection with the concept of magnetism, electricity was celebrated as the mover of everything and was even the object of religious veneration. In the tradition of physicotheology, the *theologi electrici* developed in Germany and Switzerland, as well as in England—a movement that viewed electrical phenomena as representing the presence of God in the world.[20]

John Freke (1688–1756), for example, was a theologically well-versed English surgeon who worked at St. Bartholomew's Hospital in Smithfield, London. Freke specialized in diseases of the eye, invented many medical instruments, and caused a stir with an early natural philosophical-theological essay on electricity, "A Treatise on Electricity," which he sent to the Royal Society in 1746.[21]

For Freke, electricity is the noblest phenomenon in the sublunary world of the earthlings, the "First Principle in Nature," and is clearly responsible for all movement. Freke places special importance on stating that the "electrical fire" does not originate in the "apparatuses" or in any of their physical components. Instead, electricity resides in the air, which he calls "paebulum vitae," nourishment or food of life. Thus, for Freke, electricity is inextricably bound up with all things living, and it permeates the animal as well as the vegetable and the mineral. It makes the blood red ("rubefying the blood") and is therefore "flamma vitalis," the very flame or fire of life. Analogous to Robert Fludd's designation for the sun, Freke refers to electricity as "anima mundi," the soul or mover of the world.

In the technical section of his treatise, Freke describes exactly what happens when an electrified body meets a nonelectrified body—when the electricity jumps over from one body to the other and the electrical fire discharges acoustically with a crackling sound and optically with a spark: electrical audiovision in its most archaic form.

One of the most striking and mysterious formulations for a theology of electricity is found in Freke's preface. With electricity, one becomes directly acquainted with the "*Officer* of God Almighty." This term not only connotes the meaning of executing a function but also includes the notion of a visual indicator: electricity is a display of the actions of God, a medium of His ensoulments.[22]

Divine Automata

If we project the idea of one almighty god onto our semantic field, what we get is the prime mover. "Deus est semper movens immobilis" (God: the ever-moving immobility) is the nineteenth answer given by the twenty-four philosophers to the question "What Is God?" in the famous *Book of 24 Philosophers*.[23] Expressed in technological terms, God is the eternally moving—and thus mover of all things—perfect meta-automaton.

At the same time, what moves is what motivates (from *motus*, "movement") and what creates everything. (For Marcel Duchamp, movement was the most important marker of modernity.)[24] Out of water and earth, God created the creatures who brazenly call themselves *Homo sapiens* because He gave them a soul and gave them dominion over all other living organisms on Earth. This created the primal split which cannot be overcome in reality, only in the imaginary where cinema and other media reside. The ponderous and lethargic physical material is moved by the metaphysical soul, becomes flesh, has desires and hunger, struggles, suffers, loves, and dies. The soul is inside humans. Thus it appears to be automotive, a mover of the self, and this touches upon a highly complex theological question. It concerns the issue of free will and the way in which free will fits in (or not) with the concept of the omnipotence of the mover of all things.[25] In technological parlance, we might refer to humans as God's automata.

Over the last two and a half thousand years, this automaton of God's has consistently sought to praise God's creation—that is, first and foremost itself—by building presentations of the world-machine that produced harmonious sounds. This applies equally to Christians and Muslims. All of the automata, for example, designed or actually built by the Kurdish engineer Ibn al-Razzāz al-Jazarī (1136–circa 1206) from the Al-Jazira region[26] were created to glorify the prophet Mohammed and to honor al-Jazarī's powerful patrons, who gave him commissions and paid him royally for his work. Al-Jazarī designed and described his devices as though they were mechanical objectifications of the cosmic spirit: Allah's automata.

Such installations and devices are familiar from *Tales from the Arabian Nights*, which reached the Arabian Peninsula from Persia in the fifth century and was translated into Arabic in the eighth century. In *The Adventure of Djaudar the Fisherman from Kahirah and His Meeting with Sultan Baibars*, there is a particularly impressive description:

> In the middle of the hall was a fountain with four golden lions and four peacocks adorned with pearls and precious stones above them, and opposite the lions were four brass statues that each held a trumpet to its mouth. Next to the statues were four Greek female slaves with tambourines covered in the skin of gazelles and four female Franconians with lutes. The girls looked so fresh that one thought they were alive and could talk. . . . The brass statues bowed and the slave girls began to play on their instruments. I remained seated and listened to them, until Heifa said to me, "Even if you sit on that seat for a thousand years, the girls won't stop playing; they do not tire because they are not alive and they move by the power of magic."[27]

In addition to cultic functions and their entertainment value, many of the automata fulfilled practical purposes, which competed with the deity's claim to omnipotence. Superficially, a great many of the hydraulic and pneumatic automata served the purpose of getting the guests at orgies or revels drunk as quickly as possible. However, as large mechanical apparatuses, they also intervened in the given natural environment and changed it greatly to the benefit of its inhabitants by turning extremely dry and barren swathes of land into fertile green landscapes. This entailed transporting water from deep underground to the surface. In this way, pulleys, pumps, wheelworks, and other instruments produced instances of animation in the direct sense of the word: ensouling as bringing to life. The origins of these instruments can be traced far back in time to the cultures of ancient Egypt, Alexandria in the Hellenistic period, and Byzantium, where the goal was also to eliminate hard physical work, particularly for female slaves, as illustrated by a poem from the first century B.C.E., which celebrates the waterwheel:

> Let your hands now rest, you milling girls, and sleep
> Long; may your slumbers be undisturbed by the morning cockcrow.
> Ceres has decreed that from now on the nymphs will do your work,
> They leap and rush over the rolling wheel
> With many spokes which turns on its axis
> And drives four heavy, grinding millstones.
> Now we enjoy again the golden days of yore
> And eat the goddess's fruit without burdensome toil.[28]

The masterpieces of these automata were the clocks; they could be read, and they were also audible from a considerable distance. They were horological apparatuses, which one can also designate as archaic visual installations, complex structures in which the passage of time was inscribed so that it was visible, but in which time also resounded. The man-made artifacts counted the hours (or smaller temporal units) and thus intoned the structure of time within which people acted—the function of the timepieces was comparable to that of the muezzin who calls the hours of prayer from a minaret at regular intervals and in this way endows the urban or rural environment with a sacral structure. Etymologically, *minaret* derives from the Arabic word for lighthouse, *manāra,* which actually does not match this practice. For the minaret is first and foremost an acoustic installation, comparable to the bell towers of Christian churches.

In a long chapter subdivided into ten sections with densely interwoven texts and illustrations, al-Jazarī dismantles once again the legendary ninth-century clock of Caliph Hārūn al-Rashīd. He describes its components so precisely that, with the requisite skills, it is possible to build the timepiece and reconstruct it today.[29] This is a further notable aspect of a genealogy of technology-based arts: in the beginning (avant-garde), the engineers opened up their machines and even wrote manuals so that their customers could construct replicas. It was only with industrialization that the apparatuses increasingly mutated into black boxes, which is what the vast majority of them are today. When the first and second avant-gardes of the twentieth century again opened them and exposed their mechanical hearts to demonstrate that the machines themselves were producing reality, it was actually a throwback to the early protagonists of technology-based sound art. It is difficult to imagine futures without going back to the sources.

What becomes apparent is the extent to which the Arab engineers and geometers availed themselves of the knowledge of various ancient Greek authors, for example, when al-Jazarī meticulously describes a semicircular disc on which the twelve hours of daylight are represented as signs of the zodiac. In the third century B.C.E., Archimedes of Syracuse (circa 287–212) and Ctesibius of Alexandria (circa 285–222) were already working on mechanical solutions for regular timekeeping based on the principle of the clepsydra. And like his pupil Philo of Byzantium, Ctesibius worked on the construction of water-powered clocks and organs.[30] Both genealogies, of the horological devices and of musical instruments, are very closely connected as time-based arts of sound in the archaeology of music.

In his writing, the engineer al-Jazarī devotes a great deal of attention to perfecting the hydraulic drive of the water clock *(binkām)* so that the timepiece operates without interruption, and he also describes various mechanical and pneumatic principles for producing air pressure to activate models of birds, flutes, and trumpets. The hydraulic mechanism is based on the *clepsydra* (water thief). These simple water clocks with consistently spaced markings were developed in ancient Egypt and dated by Arab historians of technology back to 1500 B.C.E. Analogous to the hourglass, or sand clock, except that in the "water thieves," the water flows from one vessel into another, clepsydrae were constructed to measure predetermined time periods, for example, time limits for speakers during judicial proceedings or political debates. Via China, Hellenistic Alexandria, Byzantium, and Iran, this specialist knowledge was refined and then spread to the Arabian Peninsula and, from there, to Andalusia in Europe. It is reported that in the eleventh century, there were two handsome water clocks on public squares in Toledo.

The most impressive water-powered horologium, also regarding the sound, was built at the end of the eleventh century by the Chinese engineer Su Song in the palace gardens at Kaifeng, approximately at the same time as Ken Shuo's experiments with the camera obscura. Su Song's installation, which could be heard from far away, was ten meters high, driven by a waterwheel with a diameter of four meters, and used an escapement mechanism, the key invention which made the construction of all-mechanical clockwork in Europe in the High Middle Ages possible[31] (as well as the mechanism of the film camera later). Su Song's chiming timekeeper is reported to have featured kinetic mannequins with cymbals, trumpets, and gongs. In the case of the famous weight-powered water clock in Fez, Morocco, completed in 1357, it is fortunately still possible to experience the imposing physical presence of such public horological devices. The clock's engineer was Abou al-Hassan ibn Ali Ahmed Tlemsani. Its twelve windows can still be seen in a photo of 1913 in the legendary book *Les automates* by Alfred Chapuis and Edmond Droz (Figure 16.5).

The chimes from the Dar al-Magana clock house synchronized the sacred and profane time periods of everyday life for Moroccans, dating back to the time when Fez was their capital city.

In the chapter that describes the functions of the device—designed as twice the height of a man and therefore an impressive size—al-Jazarī focuses on the production and performance of sounds. The sounds were intoned by a group of five female musicians: two trumpeters, two players of different drums, and a cymbalist. At certain hours of the day and night, the figurines

FIGURE 16.5. The Dar al-Magana (clock house) in the Medina district of Fez, commissioned by Sultan Abū 'Inān Fāris, 1357. A hydraulically powered mechanism opened one window per hour and released an iron ball that fell into a brass vessel below and made a loud clanging sound like the ringing of a bell. Opposite the clock tower is a famous madrasa, the Islamic college of Bou Inania, whose diurnal rhythms the sound installations also structured. From Alfred Chapuis and Edmond Droz, *Les automates: Figures artificielles d'hommes et d'animaux. Histoire et technique* (Neuchâtel: Édition du Griffon, 1949), 40.

were set in motion via a mechanism involving two model falcons; the falcon is an important symbol in Islamic culture, denoting virtuality and virility. From the wide open beaks of the birds, small brass balls drop into beakers positioned below them, which sets in motion a system of hoists, connections, and pulleys that keeps the musicians animated for a while before they finally come to rest and then wait for their next mechanical appearance. At the beginning of the night, all of the twelve windows above the musicians' heads are closed and in darkness; at the end of the night's twelve hours, they are all open and brightly illuminated. This is a powerful image for Allah's automaton: the divine light, which causes day to break; the Almighty, who is all-knowing. And with the words "God is all-knowing," al-Jazarī ends this subchapter of his manual.[32]

At the turn of the first millennium, automata that produced sounds, which al-Jazarī developed for feasts and orgies, were particularly popular at the courts of the rich and powerful in Mesopotamia, Baghdad, and Basra. Made of valuable metals like silver or copper, these artifacts played out what the big water clocks achieved on a larger scale. Mechanical ducks or other birds squawk, or mechanical musicians play, as long as enough wine remains in the vessels to exert hydraulic pressure on the movable mechanisms. In al-Jazarī's descriptions, an overt social and gender-specific component appears time and again that is familiar from classical Greece, in the *Tales from the Arabian Nights,* and even more drastically in the compendium *101 Nacht.*[33] These technical devices were operated by female slaves (*jāriya*), and the musicians as well are referred to as slave girls (Figure 16.6).

The geographic and political hot spot of the knowledge culture of the Golden Age of Arab-Islamic sciences was the Bayt al-ḥikma, the "House of Wisdom," founded in Baghdad in the early ninth century. The initiator of this research and teaching facility, Caliph Al-Ma'mūn, who ruled for twenty years (813–33), not only commissioned many translations of classical Greek texts in natural philosophy but encouraged young scholars who were eager to learn to think independently and pursue an experimental approach to the world. This made the House of Wisdom an early modern laboratory.

Among those who profited from this institution were the three brothers Muḥammad, Aḥmad, and al-Ḥasan, the sons of Mūsā ibn Shākir, who as a mini-cooperative combined a whole universe of scientific qualifications in their brotherhood: from mathematics and geometry to astronomy, natural philosophy, and medicine, as well as music and engineering. They have gone

FIGURE 16.6. Four facsimile pages with various drinking automata from al-Jazarī's *Compendium on the Theory and Practice of the Mechanical Arts* (1206; repr., Frankfurt am Main, Germany: Institute for the History of Arabic-Islamic Science, Goethe University Frankfurt, 2002).

down in the history of science and technology as the Banū Mūsā. Their achievements include rescuing books V–VII of Apollonius of Perga's treatise on conic sections and saving them for posterity. They edited the translation of this classical Greek work on geometry and mathematics, which did not survive in the original.[34] Particularly prominent in their work are traces of the great Alexandrian automaton constructor Heron, whose texts they carefully translated from the Greek into Arabic. (Some of Hero's treatises on mechanics have only come down to us in Arabic translations, which were then translated directly into German.)[35]

Aḥmad is considered to be the greatest engineer of the three brothers. He is thought to be the main author of their *Kitāb al-ḥiyal (The Book of Ingenious Mechanical Devices)* from the mid-ninth century.[36] The book is a compendium filled entirely with sketches and exact instructions for building more than one hundred mechanical models *(shakl)*. These include a variety of artifacts, mechanical components, kinetic sculptures, and automata, the latter in the direct sense of devices that move of their own accord: scooping and drinking devices operated hydraulically and mechanically, pneumatically activated animals that make noises, oil lamps that not only refill automatically, but also have draft shields that self-adjust to keep the flame protected and eternally lit—in short, items that functioned as a special kind of *perpetuum mobile*.

That things can be perpetually in motion, without interruption, was evidently very important to the Banū Mūsā in regard to music, for one of their masterpieces exhibits this characteristic. This particular automaton is not included in the three surviving copies of the *Book of Ingenious Devices*. Three hundred and fifty years later, al-Jazarī must have had access to the manuscript describing the Banū Mūsā's automaton, because in his *Compendium on the Theory and Practice of the Mechanical Arts*, he makes explicit reference to the three brothers in the introduction to the chapter in which he describes mechanical devices with fountains and perpetually playing flautists.[37]

The Instrument Which Plays by Itself is the title of a manuscript that Henry George Farmer describes in his famous work *The Organ of the Ancients: From Eastern Sources* as a "solitary exemplar" that is now considered lost.[38] The only known copy was held for many years in the library of the Three Moons Monastery, today a college, of the Greek Orthodox Church in Beirut. With the aid of the Lebanese science historian George Saliba, we learned that it was stolen from the college in the nineteenth century and is now possibly in the library of the Greek Orthodox patriarch in Damascus or was taken to Deir El-Balamand in the El-Koura region in northern Lebanon. At the present time, it is difficult to visit either of these places.[39] However, around 1900, a photographic negative must have been made of the manuscript, which is now kept in the Bibliothèque Orientale of Saint Joseph University in Beirut. Louis Cheikho, the founder of the journal of Arabic studies *al-Mashriq* (The East), described the instrument of the Bānū Mūsā in an issue published in 1906 (vol. 9). George Saliba arranged for us to get access to the negative. We were able to establish that the manuscript in the negative is identical to the text in *al-Mashriq* of 1906, so it is highly likely that we now possess a photographic copy of the master text.[40]

The Instrument Which Plays by Itself from mid-ninth-century Baghdad is well worth the research effort of an investigation that was almost criminological. For the manuscript contains not only a description of an artificial, perpetually playing flautist but also an early programmable universal music machine. The title evidences that the Bānū Mūsā were quite aware of the significance of their technology and that it could be generalized. Clearly they wanted their invention to be understood independently of any specific form in which it was realized, such as the flute player. They describe a complete music automaton, which can vary the rhythm of the music and has the potential to be primed with different melodies. At the very beginning of the text, the Bānū Mūsā state their intention: "We wish to explain how an instrument... is made which plays by itself continuously in whatever melody... we wish, sometimes in a slow rhythm... and sometimes in a quick rhythm... and also that we may change from melody to melody when we so desire."[41]

Supplied with air pressure via hydraulic systems, animated songbirds or flute players have their place in ancient Chinese literature just as in sources from the traditions of Hellenistic Alexandria and Byzantium. Both Ctesibius and Philo of Byzantium described such apparatuses in connection with their chiming clocks and organs (Figure 16.7).

In the first century, the great experimenter Hero of Alexandria elaborated, embellished, and evolved these inventions into a vast phylum of theatrical machines, sometimes at the cost of their technical functions.[42] The technologically most advanced solutions—with regard to design of the driving mechanism and system—are attributed in the Greek tradition to Apollonius.[43] He had already developed a hydraulic-pneumatic mechanism

FIGURE 16.7. Terra-cotta model of a water organ modeled on Ctesibius's invention, National Museum of Denmark in Copenhagen. From A. G. Drachmann, *Ktesibios, Philon and Heron* (Copenhagen: Munksgaard, 1948), 11.

of such complexity that his anthropomorphic synthetic figure could play the flute endlessly—provided it was supplied with a constant flow of water. As a result of his circular design for an air chamber system, whereby a second water container filled up while the first was emptying and pressed out air for the flautist, the automaton had a constant inflow of energy in the most literal sense. The Arab concept of sound was physical (one meets with more references to Aristoxenus in the literature than to Pythagoras): "The musical tone is a simple sound that during a perceptible period of time remains in the body where it is produced,"[44] as Wiedemann quotes the universal natural philosopher al-Fārābī (circa 872–950) (Figure 16.8).

FIGURE 16.8. Depiction of a hydraulically animated flautist based on the treatise by Apollonius. Figures from Eilhard Wiedemann's text on musical automata in *Sitzungsberichte der physikalisch-medizinischen Sozietät zu Erlangen*, no. 46 (1915): 17–26; translated into English by Baruch Gottlieb in *Allah's Automata: Artifacts of the Arab-Islamic Renaissance (800–1200)*, ed. Siegfried Zielinski and Peter Weibel, 49–55, exhibition catalog (Ostfildern, Germany: Hatje Cantz, 2015).

The mechanical heart of the automaton is a hydraulically driven cylinder that rotates on its own axis, a *barbach*, as it is called in the Arabic manuscript, a cylinder carved out of wood. It has roughly the same length as the instrument to be played and is positioned parallel to the instrument. In the Bānū Mūsā's description, the instrument is a "sornay," a wind instrument that is played with a double reed, similar to the European oboe. The sornay, too, reached the Arab region via Persia. On the surface of the cylinder are bands made of wood or metal, which can carry small protruding pins of different lengths.

Prographein means "to prescribe." The way the pins are arranged on the cylinder formulates the musical prescriptions or instructions, the *program* of the instrument: the pins and bands are notation translated into hard material. Depending on how the pins are positioned on the bands and how fast the cylinder revolves, a mechanical translation opens or closes the valve over one of the sornay's holes, a single organ pipe, or moves another sound-producing element. If the diameter of the cylinder is increased, it is possible to put several melody fragments on one band, which enables changing from one melody mode to another without a break. The brothers also gave thought to the speed of rotation and its effects on the progression of the sound performance.

The Bānū Mūsā attach particular importance to the possibility of controlling any instrument with their automaton, and this underlines the universal-

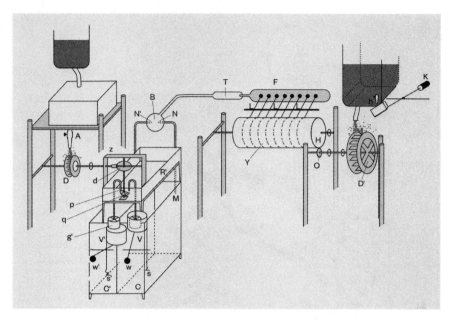

FIGURE 16.9. The basic construction of the programmable, universal music automaton of the Bānū Mūsā. *Left*, the hydraulic-pneumatic ensemble for the permanent supply of energy. *Center*, the apparatus of the air pump. *Right*, the programmable mechanical heart and the rotating cylinder with discrete pins and bridges, which in this example controls the valves of organ pipes. The depiction is taken from Zielinski and Weibel, *Allah's Automata*, 88. Visualization by Clemens Jahn, 2015.

ity of their aims. Using the same method that they describe in detail for the sornay, it is "sometimes proper that we should make an image which plays (*lit.* beats) on the lute [*'ūd*], or on an instrument of strings."[45] It sounds as though the Bānū Mūsā want to create an entire band of mechanical musicians (Figure 16.9):

> Then each of the two images conforms to the other, for the "organ" (zamr) conforms to the string [instrument], and the string [instrument] conforms to the "organ." And it is also possible that we should make figures of images which dance and follow this "organ" and these strings. And the contrivance in all this is like the contrivance of the "organ," so that every note of the strings corresponds with every note of the "organ" to the end of the piece of music (nauba).[46]

Thus there existed around 850 the concept of a master machine for automatically controlling mechanical bodies that generate sounds and

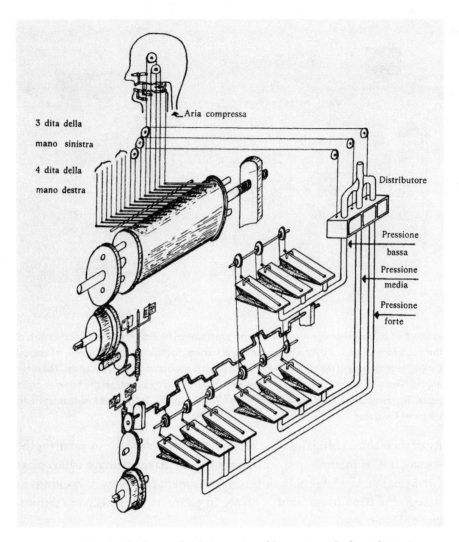

FIGURE 16.10. Nine hundred years after the invention of the rotating cylinder with pins, Jacques de Vaucanson (1709–82) built an android whose command administration functioned in a similar way. The figure shows a reconstruction by Edmond Droz, the legendary conservator of the automata museum in Neuchâtel (Switzerland). From *Guida introduttiva agli strumenti musicali meccanici, Collezione Marino Marini* (Ravenna, Italy: Tipografia Musiani, n.d.), 25.

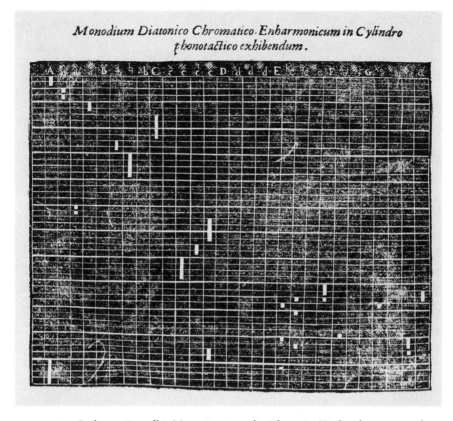

FIGURE 16.11. In this section of his *Musurgia universalis,* Athanasius Kircher demonstrates how, by projecting a phonotactic cylinder onto a two-dimensional surface, the outline of a program emerges that we will encounter two hundred years later in the punch cards used to control machines. Athanasius Kircher, *Musurgia universalis* (Rome, 1650), 520.

movements. The description is couched in awkward language, but from an engineering point of view, it is entirely comprehensible. The hardware of the Banū Mūsā is virtually identical to the programmable cylinders with pins that were used five hundred years later in the heavy European glockenspiels of the late Middle Ages, and again even later in the mechanical organs of the Renaissance, as well as for writing automata and automatic music instruments in the Age of Enlightenment. Giambattista della Porta, Robert Fludd, Athanasius Kircher, Jacques de Vaucanson, and many others not only drew on the work of the ancient Alexandrians and Byzantines but also systematically exploited the knowledge of the Muslims (Figure 16.10).

One of the only three extant manuscripts of *Kitāb al-Hiyal* (Book of

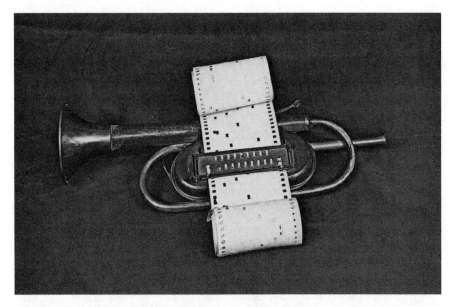

FIGURE 16.12. A punch card–controlled trumpet from the eighteenth century. From *Guida introduttiva agli strumenti musicali meccanici*.

ingenious devices) is held in the Oriental Department of the Biblioteca Apostolica in the Vatican.

With Kircher, we can see how a further step was taken in the technological genesis of automata. In *Musurgia universalis*, his main work on the theory and practice of the arts of sound of 1650, one not only finds hydraulically powered automata that use cylinders with pins. A two-dimensional projection of the three-dimensional information on the cylinders (Figure 16.11) gives rise to the technical solution of the punch card, which was used to control the music instruments of the Age of Enlightenment as well as nineteenth-century looms (Figure 16.12).

The step forward to abstraction, to the program, can be discerned more clearly from a projection onto a flat surface than with the three-dimensional revolving cylinder of the three brothers from Baghdad circa 850. They constructed the mechanical heart that powered nearly all the music automata of (early) modern times in Europe.

Universalization and automatic control of mechanical ensembles, including sound, are features that we define as of modern origin. Our brief excursion into the deep time of the automata of Islamic provenance has attempted to show that the European variant is not the starting point of modernity but instead must be viewed as a combination and concentration of knowledge and cultural techniques that were in circulation many centuries before. They were widespread, moreover, in regions that were excluded from modernity after the establishment of European hegemony.

Translated by Gloria Custance

Notes

1. On the much older Egyptian theory of harmonious proportions and the corresponding tools, see the works by Albert Presas i Puig, esp. *Numbers, Proportions, Harmonies, and Practical Geometry in Ancient Art*, preprint 262 (Berlin: Max Planck Institute for the History of Science, 2004).
2. Curtis Roads, "Research in Music and Artificial Intelligence," *Computing Surveys* 17, no. 2 (1985): 163–90.
3. Christoph von Blumröder, "Experiment—Experimentelle Musik," *Feedback Papers* 29 (July 1982): 1–2. Here translated from the German.
4. René Block et al., *Für Augen und Ohren: Von der Spieluhr zum akustischen Environment—Objekte Installationen Performances*, Musikinstrumenten-Museum des Instituts für Musikforschung Preussischer Kulturbesitz, exhibition catalog (Berlin: Akademie der Künste, 1980), 6–7.
5. Dzevad Karahasan, "Hārūn al-Rashīd schenkt Karl dem Großen einen Elefanten," *Trajekte—Zeitschrift des Zentrums für Literatur- und Kulturforschung Berlin* 19 (2009): 36–40. On the clock, see particularly Ulrich Alertz, "The Horologium of Hārūn al-Rashīd Presented to Charlemagne: An Attempt to Identify and Reconstruct the Clock Using the Instructions Given by al-Jazarī," in *Variantology 4: On Deep Time Relations of Arts, Sciences and Technologies in the Arabic-Islamic World and Beyond*, ed. Siegfried Zielinski and Eckhard Fürlus, 19–42 (Cologne, Germany: Walther König, 2010).
6. See the excellent essay on "Harmonics" by Annie Bélis in Jacques Brunschwig and Geoffrey E. R. Lloyd, eds., *The Greek Pursuit of Knowledge* (Cambridge, Mass.: Harvard University Press, 2003), 170.
7. Remarkably, J. W. Ritter, the Silesian physicochemist, used this idea and the same term when he developed his operational anthropology of the arts around 1800. Human activity for Ritter is the essential reference for the four stages within which the relationship between art and science develops. Sound as art within time is for him the ultimate perfection of activity. See my short essay "Im Zustand der Schwingung kann es keine Ruhe geben: Ein kurzes Portrait des Physikochemikers Johann Wilhelm Ritter (1776–1810). Peter Weibel zum Geburtstag," in *05–03–04: Liebesgrüsse aus Odessa—Für/*

For/à Peter Weibel, ed. Ecke Bonk, Peter Gente, and Margit Rosen, 8–12 (Berlin: Merve, 2004).

8   On *Qi,* see, e.g., Chen Cheng-Yih, "Cultural Diversities: Complementarity in Opposites," in *Variantology 3: On Deep Time Relations of Arts, Sciences and Technologies in China and Elsewhere,* ed. Siegfried Zielinski and Eckhard Fürlus, in cooperation with Nadine Minkwitz, 153–88 (Cologne, Germany: Walther König, 2008); Dagmar Schäfer, "*Ganying*—Resonance in Seventeenth-Century China: The Examples of Wang Fuzhi 王夫之 (1609–1696) and Song Yingxing 宋應星 (1589–ca. 1666)," in Zielinski and Fürlus, *Variantology 3,* 225–54.

9   See Joseph Needham, *Science and Civilisation in China,* vol. 4, *Physics and Physical Technology: Part I. Physics* (Cambridge: Cambridge University Press, 1962), 132–33. For a comparison of the Greek and Chinese traditions, see also Paul Feigelfeld, "Chinese Whispers—Übertragung und Übersetzung: Die stille Post des Wissens zwischen China und Europa 1500–1700," MA thesis, Faculty of Humanities and Social Sciences, Humboldt University Berlin, 2008, esp. chapter 4, 80–89.

10 On Purkinyě's experimental work, see my *Deep Time of the Media: Toward an Archaeology of Hearing and Seeing by Technical Means* (Cambridge, Mass.: MIT Press, 2006), 196–203.

11 See Xu Fei, "Exploration and Achievements of the Acoustics of *Yuelü*: The Theory of the Tone System in Ancient China," in Zielinski and Fürlus, *Variantology 3,* 257–94.

12 Chen Cheng-Yih, *Early Chinese Work in Natural Science: A Re-examination of the Physics of Motion, Acoustics, Astronomy and Scientific Thoughts* (Hong Kong: Hong Kong University Press, 1996), xxxii. Chen Cheng-Yih's most important study on this thematic complex is *Two-Tone Set-Bells of Marquis Yí* (1988), published by the Húbei Provincial Museum Wuhàn and the University of California.

13 Xu, "Exploration and Achievements of the Acoustics of *Yuelü,*" 260. During a workshop held in 2005 at the Academy of Media Arts Cologne, Xu Fei played a replica of a Gudi flute. The flute is held vertically and slightly tilted and produces a range of tones from powerful to muted. That it is possible to play music on an eight-thousand-year-old instrument that we recognize as modern was for me one of the most impressive experiences of researching the deep time of media archaeology.

14 Xu, 261.

15 Xu, 258.

16 See Siegfried Zielinski, "Expanded Animation: A Short Genealogy in Text and Images," in *Pervasive Animation,* ed. Suzanne Buchan, 25–51 (London: Routledge, 2013).

17 Gilles Deleuze and Félix Guattari, *Anti-Oedipus: Capitalism and Schizophrenia* (London: Continuum, 2004), 38.

18 Deleuze and Guattari, 38–39.
19 On the European tradition, see the chapter "Die Anfänge der Hydromechanik" in *Geschichte der mechanischen Prinzipien und ihrer wichtigsten Anwendungen*, ed. István Szabó, 143–56 (Basel, Switzerland: Birkhäuser, 1977), but which unfortunately leaves out the Persian and Arab contexts entirely.
20 See my essay "Theologi electrici" (chapter 11 in this volume) and with reference to the work of a contemporary sound artist, *God Is Electric, My Soul Is Electric, Nature Is Electric—For Paul DeMarinis*, exhibition catalog (Berlin: DAAD, 2010).
21 See John Freke, "II. On Electricity," in *A Treatise on the Nature and Property of Fire: In Three Essays* (London: W. Innys and J. Richardson, 1752).
22 John Freke, *A Treatise on Electricity, an Essay to show the Cause of Electricity and why some things are Non-Electricable. In which is also Considered Its Influence in the Blasts on Human Bodies, in the Blights on Trees, in the Damps in Mines; and as it may affect the Sensitive Plant*, 2nd ed. (London: W. Innys, 1746), 3–59.
23 *Liber XXIV philosophorum* is an anonymous medieval Latin text first cited in the twelfth century, of which until today no printed English translation exists. The translation cited here is provided online by the Matheson Trust for the Study of Comparative Religion at http://themathesontrust.org/papers/metaphysics/XXIV-A4.pdf.
24 See, e.g., Calvin Tomkins, *Marcel Duchamp: The Afternoon Interviews* (Cologne, Germany: Walther König, 2013), 67–93.
25 Henning Schmidgen describes the way in which physiopsychology has attempted to measure this problem for more than 150 years in his book *Hirn & Zeit: Die Geschichte eines Experiments 1800–1950* (Berlin: Matthes & Seitz, 2014).
26 See Alertz, "Horologium of Hārūn al-Rashīd."
27 Quoted from Gustav Weil's version of 1838: *Tausend und eine Nacht*, vol. 4, chapter 63, http://gutenberg.spiegel.de/buch/tausend-und-eine-nacht-vierter-band-3447/63. Translated here from the German.
28 Poem quoted from Conrad Matschoss, *Die Entwicklung der Dampfmaschine: Eine Geschichte der ortsfesten Dampfmaschinen und der Lokomobile, der Schiffsmaschine und Lokomotive* (Berlin: Julius Springer, 1908), 1:13–14n1, http://www.digitalis.uni-koeln.de/Matschossd/matschossd_index.html. Translated here from the German.
29 See especially Alertz, "Horologium of Hārūn al-Rashīd."
30 Several Arabic manuscripts make reference to Archimedes as the constructor of a large chiming water clock; classical scholars and archaeologists, however, doubt his authorship. See A. G. Drachmann, *Ktesibios, Philon and Heron: A Study in Ancient Pneumatics* (Copenhagen: Enjar Munksgaard, 1948), particularly 16–40. On the genealogy of the ancient water organ, see particularly the outstanding collection of source materials in Michael

Markovits, *Die Orgel im Altertum* (Leiden, Netherlands: Brill, 2003), 401–14.
31 See Ahmad Y. al-Hassan and Donald R. Hill, *Islamic Technology: An Illustrated History* (Paris: UNESCO/Cambridge University Press, 1986), 55–56.
32 The Institute of the History of Arabic Sciences at the University of Aleppo in Syria has conducted a major research project to reconstruct this valuable knowledge in which al-Jazarī's manuscript, completed probably in early 1206, plays a central role. Ahmad Y. al-Hasan has reedited it under the title *A Compendium on the Theory and Practice of the Mechanical Arts* (Aleppo: University of Aleppo, Institute for the History of Arabic Science, 1979). Recently I had the good fortune to receive a copy via a middleman in Andalusia. For German-speaking scholars studying al-Jazarī, the early works of Eilhard Wiedemann from the beginning of the twentieth century, and for English-speaking scholars, the meticulous translations and commentaries by the English engineer and historian of science and technology Donald R. Hill, particularly in Ibn al-Razzāz al-Jazarī, *The Book of Knowledge of Ingenious Mechanical Devices* (Dordrecht, Netherlands: D. Reidel, 1974), here 19, are invaluable. In 2002, the Turkish-born Orientalist Fuat Sezgin presented a completely reconstructed version of the manuscript, of which only a few copies are extant. Sezgin's reconstruction integrates single pages held in various libraries all over the world into the master manuscript held in the Topkapı Sarayı Library in Istanbul. I am very grateful to Fuat Sezgin for this exquisite gift.
33 *101 Nacht* (Zurich: Manesse, 2012).
34 Apollonius, *Conics, Books V to VII: The Arabic Translation of the Lost Greek Original in the Version of the Banū Mūsā* (New York: Springer, 1990). This text machine also begins with the invocation of Allah: "In the name of God, the merciful, the forgiving. I have no success, except through God" (preface of the Banū Mūsā, 2:620).
35 E.g., *Herons von Alexandria Mechanik und Katoptrik* (Leipzig, Germany: Teubner, 1900).
36 See the new edition of *Kitāb al-Hiyal* by Ahmad Y. al-Hasan (Aleppo: University of Aleppo, Institute for the History of Arabic Science, 1981). He translates the title as *The Book of Ingenious Devices*.
37 See "Category IV: On Fountains Which Change Their Shapes *(tabaddala)* at Known Intervals, and on Perpetual Flutes," in al-Jazarī, *Book of Knowledge of Ingenious Mechanical Devices*, 157–58.
38 Henry George Farmer, *The Organ of the Ancients: From Eastern Sources (Hebrew, Syriac and Arabic)* (London: William Reeves, 1931), 88–89.
39 George Saliba, pers. comm., May 12, 2014.
40 The first translation into German was by the science historian Eilhard Wiedemann using the text published in *al-Mashriq*. "Über Musikautomaten bei den Arabern," was published in *Centenario della nascita di Michele Amari*

(Palermo, Italy: Stabilimento Tipografico Virzì, 1910), 2:164–85, and also distributed as an offprint. Cf. the reprint in vol. 1 of *Gesammelte Schriften zur arabisch-islamischen Wissenschaftsgeschichte*, revised by Dorothea Girke for *Veröffentlichungen des Instituts für Geschichte der Arabisch-Islamischen Wissenschaftsgeschichte*, ed. Fuat Sezgin (Frankfurt am Main, Germany, 1984). For our exhibition project *Allah's Automata* (2015–16) at the ZKM (Karlsruhe), Imad Samir has recently produced a new translation into the German. Compared with the earlier translation, Samir's is complete and more exact in many details. It was published in Siegfried Zielinski and Peter Weibel, eds., *Allah's Automata: Artifacts of the Arab-Islamic Renaissance (800–1200)* (Ostfildern, Germany: Hatje Cantz, 2015), 69–86.

41 Farmer, *Organ of the Ancients*, 88.

42 Drachmann, *Ktesibios, Philon and Heron*, 114–19.

43 In the ancient literature, this flute player is incorrectly attributed to a text by Archimedes, which is actually by Apollonius. On the issue of its provenance, see also E. Wiedemann and F. Hauser, "Byzantinische und arabische akustische Instrumente," in *Archiv für die Geschichte der Naturwissenschaften und der Technik*, 140–66 (Leipzig, Germany: F. C. W. Vogel, 1918).

44 Quoted from Eilhard Wiedemann, "Ueber Musikautomaten bei den Arabern," 169.

45 Quoted from the new German translation by Imad Samir of the Banū Mūsā manuscript (see note 40), 82. Translated here from the German.

46 See note 45.

17 Designing and Revealing

Some Aspects of a Genealogy of Projection

> *Traditional media histories have long been written along the lines of either concrete individual technical systems (newspaper, computer, video) or institutionalized apparatuses (cinema, television, internet). With this chapter, I've begun paradigmatically to make interdiscursive fields into a topic of deep-time discussion. In the phenomenon of projection, we find art, science, alchemy, philosophy, psychoanalysis, and media research all converging to generate a fascinating dialogue. This text is the preliminary study for a larger genealogy of projection.*
> *The essay was originally published in 2010.*

⁂

Projection is a multifarious[1] concept with many origins. It is equally at home in the sciences as it is in philosophy and psychoanalysis, painting and architecture, or media technologies, including cartography. In the second part of his great Suprematism manifesto, which Kazimir Malevich wrote in early 1922, the master of modern abstract art pointed out how impossible a task it is to really get to grips with this phenomenon:

> The human skull represents infinity for the movement of ideas. It resembles the universe, for it also has no roof or floor and has space for a projection apparatus that makes shiny dots appear like stars in space. However big what is imagined may be, there is room for it inside the skull just as there is in the universe, although the space inside the skull is enclosed by a wall of bone. But what, then, do space, size, and weight mean when everything can fit into such a small container.[2]

In this impressive image, at once both very concrete and mysterious, Malevich not only formulates the infinite ambiguity of the phenomenon. From a contemporary perspective and in the context that interests us here, he also designates the exact point between Vilém Flusser's idea of the traditional imagination and his notion of the new imagination that is necessary under the conditions prevailing in an advanced technological civilization. Flusser, the philosopher of culture and technology from Prague, understood classical imagination as a process of imprinting something external on our perception.[3] This should be understood as an action of receiving. Interestingly, this notion corresponds in principle to the theory of visual perception developed by the Arab natural philosopher Ibn al-Haytham in the eleventh century.[4]

By contrast, Flusser saw the new imagination as beginning anthropologically with the technically aided gaze, and he understood it as a process whereby something that is inside us is impregnated with a perception of the external world. In principle, this matches—to stay with the history of science—the Euclidean and Aristotelian concept of vision. Of culture-historical significance is how technical artifacts are understood in this concept, namely, as "Organprojektion" (the projection of an organ), as Ernst Kapp termed it in 1877, summing up his efforts to write the "history of the origins of culture from new points of view" as "Basic Outlines for a Philosophy of Technique": "Projection is in all ... cases more or less the throwing onto or out, the putting forth, relocation, and displacement of something internal to the outside." Projection and imagination are in essence not very different "insofar as the innermost act of imagining is not free of the object that is in front of the eyes of the imagining subject." With reference to contemporary findings of psychology (a long time before Freud's work), Kapp insisted that projection should be understood as "the soul apparently stepping out of the body in the form of a sending-out of mental qualities" into the world of artifacts.[5]

With Flusser, the impregnation of the external with the inner world is not effected through the agency of a divine force that dwells within us but through abstraction via the intermediate step of a (re-)calculation of the world, that is, by means of traversing the symbolic world of numbers, Flusser's zero-dimension, computation. "The power to institute an image!" he said emphatically, suddenly switching to German in an interview in English with László Béke and Miklós Péternák in 1990 in Budapest.[6] And in a text that also dates from this period, Flusser explains, "It is what distinguishes this gesture from that of the other, image-making gesture referred to before: it is not abstract, retrograde, but the opposite—it concretizes, projects."[7] Here Flusser is getting close to the world of the engineer. "To design is to invent"—thus begins a seminal section of Eugene S. Ferguson's book *Engineering and the Mind's Eye*.[8]

The tension between the two gestures—the gesture of abstraction (from the lived-in world to the image) and the gesture of concretization (from numbers to sensibly perceptible visual object)—can be utilized as a moveable framework to sketch some of the origins and developments of projection. Hereby we shall also allow Ludwig Wittgenstein's notion of the "method of projection" as the "thinking of the sense of the proposition" "of the possible state of affairs"[9] to resonate.

The meaning of the word *projection* oscillates between two poles that are intimately connected with the two aforementioned gestures that run in opposite directions. On one hand, projection is about the spectacular (in the literal sense) proof that something, which is sent through an image machine constructed according to mathematical laws, was or is like that which we can see in the half-space (Johann Wilhelm Ritter) of the projection screen. On the other hand, projection includes the production of a reality as an image, which only exists in the way that we see it within the area of the created image.

This gives rise to a basic problem that is shared by all the avant-gardes, and particularly the artists who use film and other advanced technologies to generate images. One works with efficient techniques to produce affects and illusions, and at the same time one wants the pretty outward appearance that is generated to remain tangibly and intelligibly created by technology. For the spectators and listeners, this means being offered two realities at the same time, sojourning simultaneously in parallel worlds: in the world created specially for cinema and in the world that exists without cinema, which, however, becomes a different world with each film that is produced. "On the screen we can sit inside and outside ourselves at the same time," wrote the film buff Henry Miller to introduce the 1947 festival and symposium Art in the Cinema at the San Francisco Museum of Modern Art, describing the fundamental tension that results from this circumstance, the debates and productions also of the second avant-garde, after the disaster that modernism had experienced in fascism.[10] For the artists, this meant they had to abandon once and for all the concept of duplicating the world through using technology. "The experimental film, called such only because it dares to lie to the mirror."[11] The art of mirroring without reflecting, or to use Walter Benjamin's terminology with reference to the specific work of the artist, to be both magician and surgeon in the same person,[12] laying on of hands and cutting into the body.

In minimal an-archaeological explorations, I shall present remarkable events from the many-faceted genealogy of projection; the goal is not to seek a solution by simply deciding in favor of one or another of the outlined dualisms. In science as in art, the point is to develop the potential of projection within the polarity of "illusionization" and orientation in such a way that ensures recognition, with intellectual and aesthetic enjoyment, of the fabricated and artificial nature of the world thrown onto the screen. This implies understanding projection as something that enables one to look into a space containing possibilities. In a Flusserian sense, we can understand this possibility-space as a numerically determined, algorithmically organized space. I also see it as close to the "potential space" described by English pediatrician and psychoanalyst Donald W. Winnicott, which can certainly be filled pleasurably.[13] The semantic proximity in meaning of the words *number* and *game* (*Zahl* and *Spiel*) never failed to fascinate Vilém Flusser.

The Semantic Field and Some Fundamental Dichotomies

The word *projection* denotes a polyvalent field of concepts, artifacts, technical systems, psychological mechanisms, and above all visual practices. In all of these a paramount concern is to answer the question of how sections of lived-in or imagined worlds can be put onto a two-dimensional flat surface. In the twentieth century, this developed into a common and widespread cultural practice.

To understand the historical dimensions of the heterogeneous fields of theory and practice of projection up to the beginning of the modern age, a classification formulated by the Jesuit mathematician Zacharias Traber in 1675 is helpful. Traber took it over from the writings of the ancients. His treatise on the optic nerve, *Nervus opticus sive tractatus theoricus*, he divided into three books: optics, catoptrics, and dioptrics. Optics is the science of light and vision, which is subdivided into the study of biological and physical phenomena. Since classical antiquity, dioptrics has dealt with the phenomena of reflected light in transparent bodies, later also the mathematical and technical properties of lenses. Catoptrics is the study of the refraction and reflection of light produced by flat reflective surfaces. In the early modern period in Europe, catoptrics and dioptrics were described and taught together as catadioptrics. It became possible to put the two together after optical systems were developed in which refraction and reflection were combined using lenses, or combinations of mirrors and lenses, to project visual objects.

This is not about techno-ontological differences. Different focuses of

The trinity of classical OPTICS

DIOPTICS   CATOPTRICS

CATADIOPTRICS

The two scopic regimes : seeing by technical means

LOOKING THROUGH   *LOOKING ON*
[PERSPICERE]   [PROICERE]
[DURCHSICHT]   [AUFSICHT]

and the corresponding techno-aesthetical concepts of

VISUAL/SIGHT MACHINES   *IMAGE MACHINES*

MICROSCOPE   CAMERA OBSCURA
TELESCOPE   MAGIC LANTERN
TELE-VISION   CINEMA

FIGURE 17.1. Overhead projector transparency. S. Zielinski.

interest characterize these two subfields of optics, which can be defined as follows from a media-archaeological point of view: the dioptricians, with Johannes Kepler, Galileo Galilei, René Descartes, and Isaac Newton as the greatest scientific protagonists of a "physics of the visible" in the seventeenth century, were more interested in problems of looking through, whereas the catoptricians were fascinated by and probed phenomena of looking at. This juxtapositioning continues to have consequences for image technologies today. In his International Flusser Lecture from 2010, Alexander Galloway projected this onto the interface of computer games.[14] It would seem that here in some cases the *perspicere,* which classically belongs to electronic display technologies, intermixes with the *perspicere* of looking through reality as enlightenment.

With regard to the early modern era in Europe, one could also formulate the difference as follows: catoptric theater *(theatrum catoptricum)* existed as a media form. However, no praxis existed that could have been called dioptric theater, despite that the telescope and microscope, which were gradually becoming established at the time, provoked highly dramatic commentaries.

As for the artifacts and technical systems, possibly also with reference to perception, one can distinguish between image machines and visual machines/sight machines. Projection apparatuses, such as the lens-less camera obscura, the flat mirrors for simple reflection, or convex/concave mirrors for distortion, concentrate an observer's attention on the projected artificial images, whereas the dioptric instruments can be seen as artificial eyes. The practices associated with catoptrics were connected more with magic, showing the nonvisible, the organization of astonishing effects, tricks, and illusions, whereas the media function of dioptrics tended toward prostheses for seeing and recognizing reality, essentially an instrumental function. With the aid of dioptric techniques, weaknesses of human vision could be compensated for or corrected, its function expanded, made more effective, even perfected. "Am making eyeglasses to see the Moon bigger," noted Leonardo da Vinci, 1512, in the *Codex Atlanticus.*[15]

The epistemological, even moral dichotomy that finds expression here has etched itself deeply into the history of Western Christian knowledge culture. One of the most influential and popular works of edifying literature is the writings of the Lutheran Johann Arnd(t), *Wahres Christenthum (True Christianity);* the first volume appeared in 1605, and it has been translated into most European languages and reprinted in many different versions up to and including the present day. The beginning of the first volume is devoted to questions of the appearance of God on Earth, problems of iconicity/visualization, indeed, even in the strict sense: techniques of the visual. In Arndt's work, to look through, *perspicere*, has entirely positive connotations even in the case of the destructive action of a burning mirror through which the sun as a divine "power from above" (chapter 5, "Wherein True Faith Consists") sets fire to wood on Earth. By contrast, the technology of the camera obscura is judged to be decidedly negative. The dark chamber produces a dark and false world from the bright exterior of the reality created by God, and it is a product of pride and arrogance. "This is to say that man, through the deplorable fall is entirely—regrettably!—darkened in his heart and reason; indeed, he has become a completely reversed image—from the image of God to the image of Satan."[16]

In his *Ars magna lucis et umbrae* (1645–46), Athanasius Kircher celebrates shadow, and therefore the projected image, as a profound and negative alternative model to divine light and formulates this within its own small metaphysics.[17] The master text of this story is well known and has been discussed at length with reference to media and apparatus theory, for example, in Jean-Louis Baudry's essays:[18] the allegory of the cave from Plato's *Republic*, the philosopher from Athens or its offshore island of light (Aegina). Common man is a captive in the darkened, twilight cave of non- or prephilosophical awareness. The production of the shadows that appear remains just as hidden to him as the carriers of the figures that are projected. He would have to turn around, or better, he gets turned around to achieve knowledge. Education is a process of perverting, of twisting, of turning attention in one direction to another.

There are many interpretations of Plato's master text. Here we will concentrate on pointing up the simple comparisons as found in *Republic* 514a–17a.

Here is a selection of oscillating dualisms from Plato's analogy (or allegory) of the cave:

> upper ... lower
> darkness ... light
> evil ... good
> false ... true
> not philosophical ... philosophical

> empirical ... intellectual
> sensible ... ideal
> dead ... alive
> bad (existing in reality) state ... ideal (conceivable) state

Etymologically, the derivation of the term *projection* is straightforward. The Latin *proicio* means that I put something forward (in a spatial sense), throw it out, throw it forward, but I also let it go in the sense of abandoning it, spurning it, rejecting it, relinquishing it; in principle, *proicio* describes an active, constructive process. *Proiectio* incorporates the meaning of throwing forward and devising in the sense of changing a form. The projector is not only an apparatus that throws images; it is also the designer.

The connotation of the term, however, is anything but glorified idealistically. In the Latin noun *proiectus* there also resonates perfidy, that which has been overthrown, that which is despised. In throwing something forward, it is also forced down, and from this something new can originate.

Transformation: Alchemy

In ancient Greece, *chymeia* signified work with materials that were poured and fused, technologically speaking, metallurgy and dyeing. To give form to and harden what was raw and fluid or to immerse surfaces in dyes and give them an attractive appearance are intentions that are inherent in projection as operations. The tradition of alchemical ideas and practices can be traced far back to the civilizations of ancient China, Egypt, India, and Byzance, to the Greco-Roman world in Europe and the Golden Age of Arab-Islamic science. The strong movements in the first centuries C.E. were picked up again in the eighth and ninth centuries by Arab-Islamic civilization and reformulated. In the Arab world, these experiments and treatises were called *al-kimyá*.

In the European Middle Ages, the Hermetic tradition of natural philosophy was brought to Spain, where it became established as the program of *alchimia*. During its great florescence in Christian-influenced Europe, alchemy was heavily appropriated by the dominant religion. Chemical transformation processes were analogized to the processes whereby unbelievers or doubters became reformed characters. The way from physical (sensual) experience to the metaphysical vision with its climax of mystical union was in its vanishing point very closely connected to the idea of projection.

FIGURE 17.2. Even into the Age of Enlightenment, the two subfields of optics were ascribed contrary moral and epistemological designations. Eyeglasses were good, because they facilitated a clear view of God's Word and creation for near- and farsighted people. This sanctioned clear-sightedness is found as a motif in many paintings from the fourteenth, fifteenth, and sixteenth centuries; the illustration shows a detail from the Wildung Altar of 1404 by Conrad von Soest, with one of the oldest depictions of eyeglasses. From *Zeiss Werkschrift* 27 (1958): 179.

FIGURE 17.3. In Johann Arndt's *Vier Büchern vom wahren Christenthum,* which was first published in 1610 and continues to be reprinted up to the present day, there are regular juxtapositions. Eyeglasses facilitate insights into God-given nature.

FIGURE 17.4. The projected image of the little man in the camera obscura, on the other hand, is bad, "occulted," "inverted," "alien"; "an image of Satan," as stated in Arnd(t)'s book. (Both illustrations are taken from an eighteenth-century edition; the latest edition, to my knowledge, is from 1996.) From Johann Arndt, *Sechs Bücher vom wahren Christentum nebst dessen Paradies-Gärtlein* (1610; repr., Bielefeld, Germany: Missionsverlag der Evang.-Luth. Gebetsgemeinschaften, 1996), chapter 40, v. 26.

FIGURE 17.5. In his didactic instructions concerning the visible world (1658), Johannes Amos Comenius drew "looking through" and "looking at" together in the figure of Prudence (*prudentia*/cleverness). In her left hand, the allegory carries a telescope for the pro-spective gaze into the future, and in the right, a mirror for retrospection. That is, only the sagacious are capable of remembrance and of the visionary look forward. From *Milliaria*, facsimile ed. (Osnabrück, Germany: Otto Zeller, 1964), 224.

FIGURE 17.6. The stages 35/36 and 39/40 in an alchemical treatise from the eighteenth century with the transmutation qualities of separation, conjunction, coagulation, and projection. MS 974 in the Bibliothèque de l'Arsenal, Paris (MS 974), cited in Stanislas Klossowski de Rola, *Alchemy, the Secret Art* (London: Thames and Hudson, 1973), 109.

The "vast diversity of things, that neither are nor can be, but are thought and believed"[19] flowered to such an incredible degree due to this wild and wonderful world of the imagination and nature experiments, perhaps their very last florescence, before modern civilization began its relentless labor of standardization and universalization.

Alchemy was not invented to hone concepts. The iridescent multifariousness of the meanings of its semantic fragments is one of its characteristics. Alchemy was processing the particular, the individual, including in the texts it generated. There are still alchemical tracts that have not been read and understood by more than a handful of adepts and later exegetes. In this sense, alchemy is an elitist theory and practice. It is "a dream that one listens to but can only stammer in its telling. When people ceased to dream by their fires and listen to themselves in material things, the dream of alchemy retreated into the night."[20]

The connotative vibration of the creation of something qualitatively new, which is bound up with projection, is expressed lucidly and in many and diverse ways in experimental alchemy. In the early modern era, alchemical theory and practice acted as a means of (self-)understanding a subject that was as yet unsettled, abeyant, tending toward the empirical, and possibly with a thoroughly provisional future, which sought to position itself in relation to the other, to that which was not, or not yet, understood—including nature. The cosmos hummed, and to listen to it was exhilarating. The cosmos projected itself acoustically into the souls of those adepts who had decided to participate in the world and not merely to observe it. Understood in this way, alchemy as a new model for experiencing and working on the world did not point back to the magical past but forward into a possible future.

In numerous alchemical treatises, the final stage of the *magnum opus*, when base matter is transmuted into the beautiful gleaming stuff of coveted gold, is *proiectio*. According to the model followed, this is either the seventh or twelfth stage in the transformation process. In experimental and practical terms, *proiectio* means a process of throwing. Precisely calculated amounts of pulverized *lapis philosophorum* (philosopher's stone) were coated with heated wax and thrown onto the molten base metal. If the *lapis* truly possessed the transformative power that was attributed to it, this last union of substances would lead to the desired transmutation.

In a late medieval treatise, which was translated in 1608 into challenging

German, Roger Bacon was at great pains to point out that the material used to project, the *lapis philosophorum*, must itself contain precious substances. To achieve the most beautiful appearance of the Moon (silver) and the Sun (gold), female and male, respectively, in Bacon's text as in most other European languages, but not German, the philosopher's stone must contain that which is sought to be achieved. "If we seek projection to the red or white work therefore we should take it from the gold."[21]

It was important for us to undertake this brief excursion because many of the natural philosophers who studied optics and the technologies of transfer, like Roger Bacon and Giambattista della Porta, were also practicing alchemists and were very well acquainted with the theoretical literature on the subject. Many experiments, for example, including Isaac Newton's, were inspired by alchemy.[22] His famous *experimentum crucis* demonstrating that a prism could decompose white sunlight into a spectrum of colors that could not be broken down further used the experimental setup of a special camera obscura. In Newton's experiment, sunlight was not used as an energy source to project objects outside of the dark chamber; light was itself the object being refracted by prisms inside the chamber. The generation of substantial color qualities is but a reading and written variant of alchemical transmutation practices. In Mark Rothko's (1903–70) painting, one can study variations, as well as in works by Yves Klein or Sigmar Polke.[23]

The semantic field of alchemy is very closely connected with a tradition from which there have been repeated attempts to extricate Vilém Flusser. (But why? To modernize him retrospectively from a Western perspective? Does he need that?) It is the tradition of magical natural philosophical thought, particularly its experimental varieties. By the architects of rigorous hierarchies, it is still defamed as epistemologically inferior. For such people, the porously encoded heuristics of Flusser are anyway unscientific, heretical, and therefore unusable. Other historians of science and technology, by contrast, are beginning to discover this dimension as an important source of productive knowledge. Thus we should not ignore them either for any spurious reasons. In the heart of Flusser's hometown of Prague is the Alchemists' Lane.

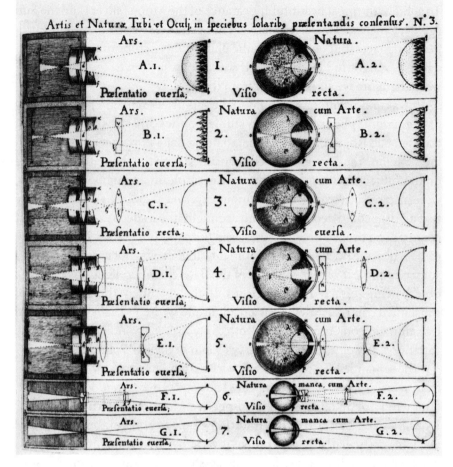

FIGURE 17.7. Christoph Scheiner's contrasting of the projection of sunlight in the artificial eye of the telescope's tube *(tubus)* with its diverse lenses (left: *ars*) and in the natural, human eye (right: *natura*). Striking is the differentiated use of vocabulary: regarding the natural process, he speaks of correct or inverse "beholding" *(visio recta e versa)*, and seeing technically is referred to as representing *(praesentatio)*. The variants 6 and 7 show combinations. From Christoph Scheiner, *Rosa ursina sive sol* (Bracciano, Italy: Andreas Phaeus, 1626–30), book 2, chapter 23, III.

Projections of the Impure

In astronomy and astrophysics, projection retains the meaning of depicting an illuminated celestial body through the eyepiece of a telescope on photographic film or white paper. Because he unwisely got embroiled in an argument over priority of discovery, the Bavarian-Swabian Jesuit, mathematician, and physicist Christoph Scheiner (1573–1650) rose to dubious prominence in the early history of European science. Galileo Galilei of all people, the most famous physicist of the visible and powerful thinker from Tuscany, Scheiner challenged around 1610, asserting that Galileo had not been the first to have observed something outrageous with the aid of a telescope: the sun is not the pure, brightly shining divine orb that the established church took it to be and praised incessantly in metaphors. On the contrary, in magnification, the star that provides the energy for our system exhibits strange spots. Although he interpreted them incorrectly at first, as passing clouds or satellite-like foreign bodies, Scheiner insisted that he had been the first to observe them. This provoked the great investigator of nature to a bitter feud and exaggerated polemics with the provincial teacher from Germany and led Scheiner to punish Galileo later. At the Roman Inquisition, tribunal about the correct world system—Copernican or Ptolemaic—Scheiner got one of his students to write the decisive expertise against the heliocentrist Galileo. In this Scheiner used his influence with Pope Urban VIII, to the disadvantage of the Tuscan master of the stars.[24]

The effects of the feud and dispute brought Scheiner the opposite of what he had wanted for his career. His actual achievements in the fields of invention and astronomy faded into the background and were all but forgotten, especially because in his main work, *Rosa ursina sive sol* (The ursine rose, or the sun), which he wrote from 1626 to 1630 and which is one of the most beautiful folios on astronomy, he could not resist including a lengthy account reiterating his claim to priority and criticizing his rival.

The second book (Liber II) of Scheiner's work about the sun introduces in detail the instruments he used to make his observations, how they are constructed, and what they are capable of—most unusual for a scientist in this period. Among the instruments described is a most intriguing artifact for the genealogy of projection, which Scheiner—also well known for his invention of a stork's beak to copy images—was probably the first to use. Already in 1612, in his correspondence with Galileo, the Jesuit had mentioned that using a camera obscura construction and various flat mirrors, he was able to project sunspots onto white paper to draw them; that is, he could reproduce them two-dimensionally and then copy them.[25] This was not new: others had done this before him, including Johannes Kepler. Covering the ocular of his telescope with colored transparent material to protect his eyes from the bright sunlight was not novel either—only the name of his instrument for observing the sun was: helioscope.

In *Rosa ursina sive sol*, however, a helioscopic apparatus appears as a projection device,[26] which in the third book is presented in a developed form as a "heliotropic telescope" and "telescopic heliotrope,"[27] respectively. Here Scheiner adapted one of the common Dutch telescopes of the time and fitted it with two convex lenses, which he ground to shape himself. One served as lens, the other as eyepiece. By positioning the eyepiece lens beyond the focal point of the rays refracted by the other lens, Scheiner produced the projection of a small image of the sun in front of the eyepiece. Normally, this would have appeared in the lens of the eyepiece. The telescopic heliotrope variant was also a so-called equatorial telescope. The telescope was on a mount and, aligned with the Earth's axis, could be turned easily to correspond with the position of the sun.

Matter of decisive import for an archaeology of seeing by technical means and for projection as a scopic strategy is found in the third book of *Rosa ursina sive sol* in variantological abundance. Seventy very fine full-page engravings meticulously map the forms and movements of the sunspots, which at first Scheiner had thought were passing by the fiery star. In the meantime, he assumed, like Galileo, that these formations, regarded as impurities, were a material part of the sun and its activity. The logical consequence of this was to conclude that the sun rotated on its axis, for otherwise the slowly moving spots and their continual reappearance could not be explained. The copper-plate engravings in the folio were made from the drawings done from projections. They were printed proof that the eye of God that looked down on the Earth and other planets of our solar system was not pure light but was formed. The *anima mundi*, which provided for all life on Earth, had therefore to be understood not only as an actively forming agent but as itself formed.[28]

To comply with the accepted view at least on an allegorical level, Scheiner chose for this formedness the image of the rose, a full-blooming *rosa ursina*, which in Christian mythology represents the most beautiful of all roses and symbolizes love, fertility, and the bearer of all life; in his treatise on the art of

FIGURE 17.8. Scheiner's walk-in heliotropic apparatus (bottom). From *Rosa ursina sive sol* (Bracciano, Italy: Andreas Phaeus, 1626–30), frontispiece.

combination, Gottfried Wilhelm Leibniz used the rose in place of Ramon Llull's "A" as a symbol for God.

Around sixty years later, an attempt was made to salvage the purity of the eye of God in the heavens. In his book on the artificial eye of the telescope (*Oculus artificialis teledioptricus sive telescopium*, 1685), Johannes Zahn (1631–1707) assigned this media technology the power to restore purity in reality. Through projection, the impurities in the heavens would be removed and brought down to Earth in the image. In an illustration showing a camera obscura with a helioscopic observation instrument, Zahn wrote on the aperture facing the sun, "Maculas etiam caelo deducit ab alto" (it even pulls down the spots from the heavens). For Zahn, projection was a strategy to restore lost purity, just as before it had contributed to understanding that light that is thrown first attains its fascinating beauty when it takes on form in the encounter with dark materials.

Scheiner's helioscopic observation instrument had another odd feature that is of general significance for the role of observation in physics. Around the instrument, he built a casing in which the projection was captured. One of the plates in *Rosa ursina sive sol*[29] shows Scheiner's assistant reaching into the casing to trace the projected image while Scheiner himself sits in the background in a position where he cannot see the projection. He is working at a desk on the observations, making calculations, and giving instructions to his assistant. In the case of the larger examples of this instrument—Scheiner built helioscopes up to twenty-two meters long—his assistants could have actually climbed into the casing and would then have been included in the observation space. In this way, the results would have emerged through communication between assistant as internal and astronomer as external observer. That this division of labor, which necessitated communication between the parties involved, could lead to considerable uncertainties in the results that were written down is highly probable.[30] However, there is no doubt that this model had paradigmatic character for modern scientific practice.

Projections of the Nonvisible

In the "physics of the visible" of the early modern age, the boundaries between the inside and outside of bodies were fairly secure. The methods for investigating nature involved optomechanical devices and chemical apparatuses. In the course of the Age of Enlightenment, electrification, and the emergence of a technology industry, which also took hold of the sciences, interest grew in crossing these boundaries. The inner cohesion and functioning of bio-, geo-, techno-, physio-, and psychological bodies aroused the burning curiosity of researchers from different disciplines. They began to investigate various aspects of things that up to that point had been invisible and attempted to visualize them: first and foremost the energy that holds everything together internally and at the same time can have such destructive effects.

The murderous climax of this development was the dropping of atomic bombs on Hiroshima and Nagasaki in the mid-twentieth century. After the ontological experience of Auschwitz, this was the point at which a hyper-cultivated technological rationale tipped over into its potential opposite. Nuclear research and its applications systematically went beyond anything that conformed to human dimensions. The investigation of the smallest internal particles of nature and relations of mass and energy between them led to the development of a weapon with the power for unimaginable destruction. At the moment when the bombs detonated, the two Japanese cities were lit up as though by a gigantic frenzy of flashbulbs. The dying bodies of people were projected as shadows on the stones of roads, on staircases, and on the walls of houses. The Japanese-born American cultural researcher Akira Mizuta Lippit has studied these phenomena and has begun to make them the starting point of an idiosyncratic theory of film.[31]

In 1945, Kodak discovered that when developed, its photographic paper was sprinkled with strange impurities. Much later, it was found that the spots originated from the radioactivity released by the first nuclear test, which had eaten into the packaging. The high sensitivity of photographic emulsions was discovered by Antoine Henri Becquerel (1852–1908) in 1896, who demonstrated that the radiation emitted by uranium salts can affect photochemical emulsions even when they are kept in darkness. For the cleaning-up operation after the nuclear catastrophe in Chernobyl, among those who were called up was the professional photographer Oleg Veklenko. The negatives of film that he exposed at the site show milky patches where the radiation has eaten into the film. Its contact with the photochemical material produced a projection of radioactivity that is not visible to the naked eye.[32]

The technological and cultural fields impinged on by Becquerel's discovery not only gave rise to the industrial form of cinema. Shortly before (1895), living bodies had been rendered transparent by X-rays; in the literal

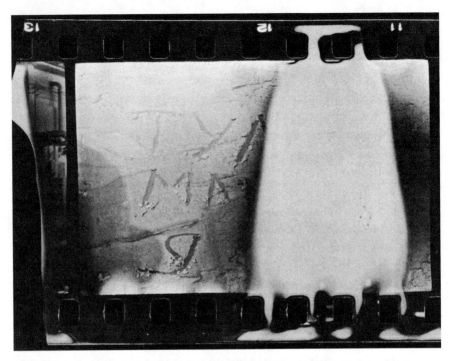

FIGURE 17.9. The Russian photographer Oleg Veklenko took pictures during the clean-up operation after the nuclear catastrophe at Chernobyl (Ukraine). Radioactivity ate into the film negatives, giving rise to milky streaks. From Charlotte Bigg and Jochen Hennig, "Spuren des Unsichtbaren: Fotografie macht Radioaktivität sichtbar," *Kultur & Technik* 31, no. 2 (2007): 22.

meaning of the expression, they were irradiated by the enlightenment. This brought the interior of bodies into the realm of the visible. Sensational phenomena, which beforehand had only been accessible to visual experience in corpses, were turned inside out: they were not only visible on a screen but also projected as images onto photosensitive material and in this way, as two-dimensional surfaces, in turn became objects exerting attraction. By this time, the transcription of neurotic bodies into visual notations of hysteria by the French neurologist Jean-Martin Charcot (1825–93) were well known. Around 1895, about the same time as Röntgen made his discoveries, Sigmund Freud began to describe his own inner phenomenology, which culminated in *The Interpretation of Dreams* (1900).

All these examples are relatively late implementations of the project to visualize the nonvisible. At bottom, the Aufklärung, the Enlightenment, is the historical period in which the project of visualization, of rendering visible

FIGURE 17.10. The Danish artist Jacob Kirkegaard spent a number of weeks in 2005 in Chernobyl, where for hours he filmed and made sound recordings of the radioactivity levels in abandoned, contaminated rooms in the forbidden zone. The result was his 2006 audiovisual installation *AION*.

what one could not see, was pursued involving huge effort and with fascinating experiments. At the center of interest was the discovery of electricity as something which could be produced artificially, energy at once destructive and useful that was wrested from nature and tamed so that it became available for humans to utilize.

For many people at the time—both with scientific and theological interests—the fascinating electrical phenomena were even a direct expression of the divine. Electricity was for them the medium of God's omnipotence. Following the strict Catholic and Jesuit traditions, at first they once more made the attempt to conceive of theology and natural philosophy as one. The discovery of electricity and magnetic and galvanic phenomena they linked with the notion of God's immediate presence in the world: a new conception of God that gave rise to a new understanding of "soul and corporeality, of mind and matter, of life and substance."[33] Taking over from and also complementing medieval metaphysics of light, it designed a new image of the divine which was intimately connected with the primeval fire: "Magnetism and electricity appeared as the most obvious portrayal of divine power in the world and in things; as the hidden power that creates the life, motion, and heat which pervades the entire universe."[34] What Athanasius Kircher had only hinted at in his studies of magnetism in the seventeenth century the "electrical theologians" put into effect with varying degrees of radicalism: "'God the magnet' becomes... the magnetic force of nature. The de-personalization of the idea of God in its extension leads inevitably to... practically equating the divined spirit, as the *vis magnetica dei*, with the all-ensouling power of nature."[35]

A precondition for constructing such extravagant edifices of thought and images was, paradoxically, the fact that we cannot directly perceive electricity. "We do not possess... an electrical sense," L. Graetz noted down about this banal circumstance in one of the standard nineteenth-century manuals.[36] Electricity cannot be smelled or felt or heard; moreover and particularly, it is not visible. Only when it comes into contact or runs through matter is it perceptible to the sense faculties, for example, as heat, motion, vibrations, phenomena of light or shadow. It is the ideal phenomenon of something that is distant in the present, provided that the voltage ratio is right, the current flows, and the archaic forces of attraction and repulsion can unfold. Here Walter Benjamin's concept of the "aura" finds a nice equivalent.[37]

A good half-century after the domination of Catholic-Jesuit thought, the methods for conceiving the relationship of God and nature both in scientific and empirical terms had become radically different, not least due to the influence of René Descartes and Isaac Newton. At the beginning of the eighteenth century, there was a veritable wave of experimental theology, which for the practice of science represented an attempt to apprehend God better and to teach knowledge about God in a more effective way. The umbrella term covering the various branches was "physicotheology." A characteristic of this theological approach was its many variants, which each engaged with a particularity of nature, for example, "astrotheology"

FIGURE 17.11. *Opus Tertiae Diei: "Et vocavit Deus aridam terram congregationesque aquarum appelavit maria."* The separation of the waters and dry land on the third day of Creation (right). Engraving from Jacob Scheuchzer's magnificent *Physica sacra*, which was published in Augsburg and Ulm between 1731 and 1735 in four volumes.

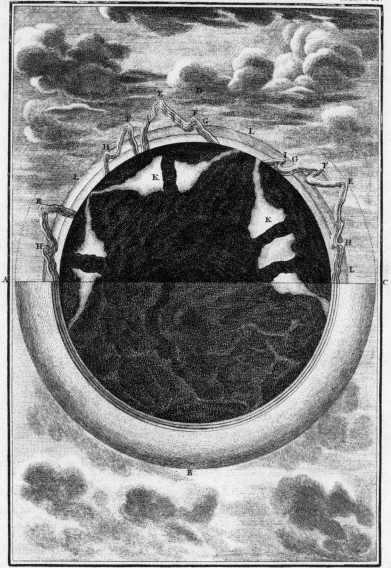

TAB. VII.

GENESIS Cap. I. v. 9. 10. I. Buch Mosis Cap. I. v. 9. 10.
Opus tertiæ Diei. Drittes Tagwerck.

I. G. Pintz sculp.

(focusing on celestial objects), "insectotheology" (relating to entymology with an especially bizarre subdiscipline of "acridotheology," which sought to deduce the wisdom and goodness of God from the characteristics of locusts), "brontotheology" (study of thunder and storms), "lithotheology" (geology), "ichthyotheology" (ichthyology), and "sismotheology" (deriving from the study of earthquakes).[38] Johann Jacob Scheuchzer's *Heilige Physik (Physica sacra)*, which was published in 1735 in Augsburg, can be read as an early bible of this movement, particularly from a pictorial perspective. In its many large-format engravings, the story of the Creation and conditions in nature unfolds with high drama (Figure 17.11).

Physicotheology had strong connections with the German Pietist movement, which became established from the late seventeenth century onward as a particularly Bible-studying and socially committed variant of Protestantism that shares much with the British Puritan movement.[39] The physicotheologians worked on a revised version of the concept of nature and its creator and mover or, to be more exact, on a substantiation of the presence of God in nature. "The world which is perceived by the senses... is astonishingly functional and purposefully arranged; it is therefore a reasonable argument for the existence of God as well as an opportunity to recognize his omnipotence/wisdom/providence and inducement to praise and love him," as Michel summarizes this idea.[40]

Let us look at how this functioned taking the example of an Englishman. John Freke (1688–1756) was a theologically well-versed English surgeon who worked at St. Bartholomew's Hospital in London. His special area was diseases of the eye, and he invented a number of medical instruments. Freke caused a stir with an early natural philosophical essay on electricity in which he argued theologically. In 1746, Freke sent his *Essay to Shew the Cause of Electricity* to the Royal Society in London, the powerful aristocratic custodian of all knowledge about nature and its technical applications in the United Kingdom.

In this essay, electricity is for Freke the primary phenomenon of the sublunary world, "the First Principle in Nature" as he wrote in the appendix to his treatise, significantly with an exceptionally long title.[41] With the early machines that were being used to produce frictional electricity in mind, Freke places particular emphasis on stating that the "electrical fire" does not originate in the "apparatuses" or in any of their physical or technical components. Instead, electricity resides in the air, which he refers to as "paebulum vitae," nourishment or food of life. Like many medical practitioners and doctors in the deep time of a developing relationship between science, art, and technology, Freke was a vitalist.

Thus for Freke, electricity is inextricably bound up with all things living, permeating the animal, the vegetable, and the mineral alike. It makes the blood red ("it is rubefying the blood") and is therefore "flamma vitalis," the very flame or fire of life,[42] as he expressed it in bombastic images of the visible. Analogous to Robert Fludd's designation for the sun (Fludd was also a practicing physician), the devout medic from London also liked to refer to electricity as "anima mundi," as the soul or mover of the visible world. Taking this as a starting point, one could attempt to write a genealogy of expanded animation[43] from a media-archaeological perspective.

In the technical section of his essay, Freke describes simple technical devices for generating weak artificial electricity and lists materials that are good and bad conductors yet again. And he gives an exact description of what happens when an electrified body meets a nonelectrified body, when the electricity jumps over from one body to another and the electrical fire discharges acoustically with a crack and optically with a spark: crack-and-spark, this is electrical audiovision in an archaic electrotechnical form.[44]

The possibility of projecting the crack-and-spark is already present in Freke's approach. One of the most striking and mysterious formulations, also with regard to a theology of electricity, is found in Freke's preface. With electricity, one becomes directly acquainted with the "Officer of God Almighty."[45] The term *officer* not only implies the execution of a function but also contains the notion of an indicator: electricity becomes the visual display for the actions of God Almighty.

Before Luigi Galvani (1737–98) in Bologna began to use amputated frogs' legs as oscillographs to prove the existence of natural, animal electricity by way of a scientific spectacle, Georg Christoph Lichtenberg (1742–99) was experimenting in Göttingen on imaging processes for this form of energy, the scientific principles of which were not understood at the time. In a lecture at the Göttingen Königliche Societät der Wissenschaften on February 21, 1778, Lichtenberg presented his experiments. In his introduction, Lichtenberg said that "the great physicists had concentrated on the spots that the electrical discharges of the Leyden Jar produce on polished bodies,"[46] which are judged to possess a strange beauty. When in 1777 he began to work with Alessandro Volta's (1745–1827) improved electrophorus, he made a chance discovery that amazed him. In the fine powder that Lichtenberg had produced by rubbing the resin "cake," the discharges of electrostatic electricity left strange patterns in the material, particularly marked on the convex side of the cake. Enthralled, Lichtenberg began to play around systematically with these phenomena using many variations and spread more powder on the plates to obtain more striking visualizations.

After he had thought over the first experiments and results in more detail, Lichtenberg concluded "that the figures either originate from the passage of the electrical matter from the positively charged lid via the resin cake into the lower lid, or from its flowing into the surface of the resin cake." These "flows" from the lid into the cake, which formed the thrown-up powder into stars, and which he had particularly observed at night, Lichtenberg called "projections."[47] Furthermore, he noticed that the star forms in the scattered material (we recall the *lapis philosophorum*) corresponded to a positive charge of the resin cake, whereas a negative charge brought forth more circular structures. Lichtenberg compared the structures to astronomical motifs: "Sometimes there were almost innumerable stars, Milky Ways, and larger suns."[48] The inspiration for this was obviously the idea of the projection of microstructural structures onto the macro-cosmos.

The importance of Lichtenberg figures for the history of science and for aesthetics;[49] the many adepts at his experiments and the countless modifications that the physicist made himself have been described many times elsewhere and are not of prime interest for our present investigation. The reason why they are included here is that the Lichtenberg figures represent an enrichment of the quest to discover the multifarious concepts and practices of projection. In this case, light is not necessary as a medium to produce an image, in the sense of a throwing-forward-and-onto. The material—in this case, powdered resin—was formed directly by the electrical charge that was generated between the metal lid and the charged resin underside of the electrophorus.[50] The image, the frozen state of a moment in time, was the expression of the world as a specific potential state.

In the interpretations of contemporary commentators on Lichtenberg's experiments, as well as on the experimental work of the physicist and "father of acoustics" Ernst Florens Friedrich Chladni (1756–1827) and the much later experiments of the physicist Jules Antoine Lissajous (1822–80), the ideas of the *theologi electrici* resonate. The reason why I have given them space to speak for themselves here has to do with Flusserian thought.

Vilém Flusser's idea of a new imagination, which he never tired of underlining and which made him so many young friends in the fields of design, architecture, and art, has strong metaphysical dimensions and, although not always compelling, interesting theological aspects. If, in place of electricity, one thinks of the organization of commands for the electrical circuits in a computer, the algorithms, or the code, and if one inserts these ubiquitous and seemingly all-powerful controlling entities in Flusser's anthropological argument about the necessary passage through the abstraction of the zero-dimension, we obtain a picture that corresponds in a special way with the notions of the presence of God in the world that we are familiar with from the *theologi electrici*. Flusser, however, surprises us with a completely different cast.

In a conversation with Miklós Peternák and László Béke in 1990 on "Religion, Remembrance, and Synthetic Images," the cultural philosopher from Prague began with the remark that he had the feeling that the older he got, the more Jewish he was becoming. At the very end, Flusser talked about his idea of the new imagination that is connected to the concept of the techno-imaginary. Here Flusser referred explicitly to Immanuel Kant's philosophy and concluded his argument with a surprising and harsh proposition:

> This is what Kant was referring to in *The Critique of Pure Reason*. It is a force that allows me to understand the abstract, and from there one goes back to—if one wants to—theology. Up to now we had [this]: God revealed his countenance on Mount Sinai, and in Auschwitz it was covered up again. In this way it could be said that the history of humankind lies between Mount Sinai and Auschwitz. And now, after Auschwitz, God is completely absent, there is no longer a God, we project ourselves in order to get back to him. If I may say so, perhaps it's completely un-Jewish—I don't know, I'm not very good at this, but it is my understanding of Jewishness—I would say that synthetic images are an answer to Auschwitz.[51]

According to the Old Testament, the veiling of God's countenance took place in the precise moment when it was unveiled on Mount Sinai (2 Exodus 33:18–23) in that the appearance of his glory (the indwelling of God in the world, which is expressed in the Kabbalah by the tenth Sephirot of Shekhinah) took place through a cloud.[52] The appearance of God was nubigenous. Flusser, however, is interested here above all in elucidating the possibility of a new design of the world as a projection of ourselves into something entirely

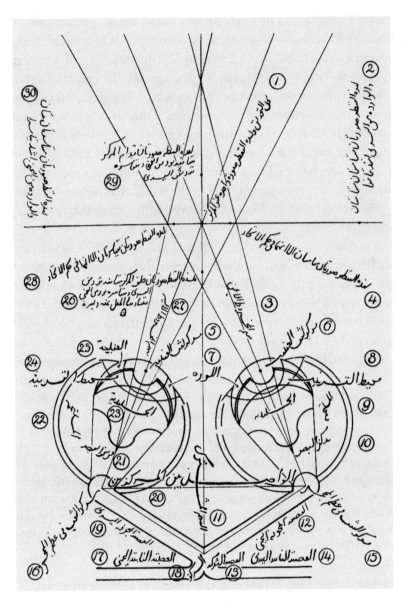

FIGURE 17.12. Depiction of binocular vision as a process of double projection from points in the outside world into the eye. From the eye, the information is sent to the crossing optic nerves in the brain. The brain assesses, evaluates, and orders the information from which an image is then generated. According to the Arab perspectivists around 1000 C.E., the correct image exists exclusively in the imagination. Transcription of a drawing by Ahmed ibn Muhammad ibn Ja'far from the year 1000; that is, before Ibn al-Haytham's superb *Book of Optics* of circa 1021, in Ruggero Pierantoni, *L'occhio e l'idea* (Turin, 1981), 13.

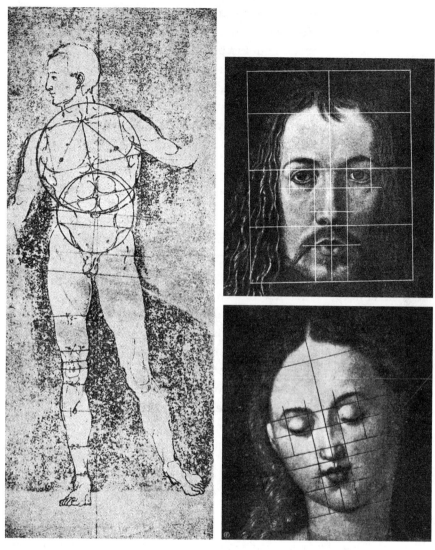

FIGURE 17.13. From "Ideal Figures of Portrait Painting," in Georg Wolff, *Mathematik und Malerei* (Leipzig, Germany: Teubner, 1916), 70–73.

other. Abstraction becomes the precondition of the hope of being able to reestablish a connection with God. Flusser does not think of the techno-imaginary as disconnection but as a possibility of connecting. By choosing the reference of Moses on Mount Sinai, Flusser glides alongside the ideas of God of the *theologi electrici*. Cloud and pillar of fire, in which the appearance of God takes place, are images that operate between absence and presence, between materializing and presence of mind, which are comparable to the mysterious nature of electricity.

At the End an Extraordinary Encounter: Flusser with Heidegger

Flusser's concept of projection as an activity and a way of thinking which is bound up with the gesture of designing is indebted on one side to traditions that are profoundly magical. If one attempts to locate them in the development of philosophy, one has to go back to the Presocratics. On the other side, the concept is indebted to a twentieth-century thinker who readily proclaimed that his postphilosophical thought drew on the poetic philosophers of the epoch before Socrates, namely, Martin Heidegger.

In his Bochum Lectures, Flusser, the cultural philosopher who fled Prague because of the Nazis, engaged often and in parts profoundly with the work of the German phenomenologist. "Heidegger is not a good philosopher, but he is an important one," he said at the end of a section that was actually about Husserl's phenomenology.[53] In this argumentative passage, Flusser circles around the epistemological importance of design in Heidegger's philosophy:

> In Heidegger's opinion ... we are determined, in narrowness, in *angst*, in angustia. ... Suddenly something bursts open, and we open up, we decide. With this decision, around us opens up a clearing in the thicket, so to speak, the clearing of being. ... Suddenly one realizes one is not falling to one's death. One realizes that death is not a problem, because where I am, death is not; and where death is, I am not. In a way one has got death over and done with. ... One is no longer an existence in decline, but is beginning to design oneself in the direction leading away from death. Heidegger calls this design for being. In this situation says Heidegger, angst changes suddenly; it becomes care. I take care that, I care for, ... I procure something for myself. What Heidegger means to say here is that thought only then begins to be design. I am not of the same opinion as Heidegger, I believe that when we

begin to design, we are no longer subjects of objects, but projects for objects, designs for things and no longer the subjugated of things.[54]

In conclusion, let us engage with the word games of these two so very different thinkers and expand them with one of our own. According to Flusser, projection is connected with a movement that reaches into the future. He understands future as "realizing one's potential" in contrast to the "having become unreal" of the past (this pairing in the argument is also connected with the ontological experience of Auschwitz). However, if I understand the deep layers of the relations between the arts, sciences, and technologies as something that is not complete, that which was in the past and will be in the future can always be opened up anew, then a further concept of Heidegger's insinuates itself in the train of thought: revealing (entbergen) in the sense of insightful recognition. Heidegger's well-known idea of technology is a particular "way of revealing," of "bringing-forth," or of "unconcealing" of that which in an Aristotelian sense cannot show itself or generate itself of its own accord: technology as a specific way of making, as a way of production, as *poiesis*, as Bernard Stiegler demonstrates in his Heidegger exegesis. "As production *(poiesis)* technics is a 'way of revealing.' Like poiesis, it brings into being what is not"[55]—but could be and may have been, I would like to add.

Turned genealogically and placed by the side of the Flusserian concept of projection, the juxtapositioning of the two results in an exciting situatedness for the individual, for us, the possibility of an in-between with potential (Figure 17.14).

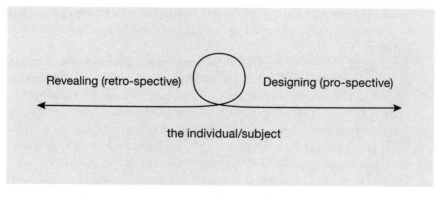

FIGURE 17.14. Schema. By S. Zielinski.

To be on the move too much in only one direction (or too little in the other) results in never-ending suffering, if one gives credence to the findings of the Berlin philosopher Michael Theunissen on the pathology of time.[56] He sees the pathological forms of the awareness of time as spanning the poles of extreme melancholy and paranoia—of being too much within time as a persistent state of not-being-able-to-forget, on one hand, and of being too little within time as a continual flight from that which has been. Vilém Flusser, however, did not exist or think in the in-between. He could be both deeply melancholy and at the same time forcefully and rapidly seeking to fly from remembering.

Translated by Gloria Custance

Notes

1 This is the term *manchfaltig,* which I use in the sense defined in Lorenz Oken's *Lehrbuch der Naturphilosophie* (Zurich, Germany: Schultheß, 1843), which to my mind perfectly matches particularly colorful phenomena from a genealogical perspective.
2 Kazimir Malevich, *Suprematismus: Die gegenstandslose Welt* (Cologne, Germany: DuMont 1962), 200.
3 A kindred spirit, e.g., is Ludovico Antonio Muratori in his *Della forza della fantasia umana* (Venice, 1745); the text is replete with media metaphors of burning, imprinting, and so on, information on the matrix of the mind.
4 Cf. the essays by Hans Belting, Siegfried Zielinski, and Franziska Latell in *Variantology 4: On Deep Time Relations of Arts, Sciences and Technologies in the Arabic-Islamic World and Beyond,* ed. Siegfried Zielinski and Eckhard Fürlus (Cologne, Germany: Walther König, 2010), and especially Belting's monograph *Florenz und Bagdad: Eine westöstliche Geschichte des Blicks* (Munich, Germany: C. H. Beck, 2008).
5 Ernst Kapp, *Philosophie der Technik* (Braunschweig, Germany: Westermann, 1877), 30–31. Published in English as *Elements of a Philosophy of Technology,* edited by Jeffrey West Kirkwood and Leif Weatherby, translated by Lauren K. Wolfe (Minneapolis: University of Minnesota Press, 2018).
6 The interview titled "On Religion, Memory, and Synthetic Image" is on the DVD *We Shall Survive in the Memory of Others,* released by C3 in Budapest in collaboration with the Vilém Flusser Archive (Cologne, Germany, 2010).
7 Quoted from the typescript "Eine neue Einbildungskraft," the manuscript used for Flusser's essay in Volker Bohn's compilation *Bildlichkeit* (1990). Manuscript in the Vilém Flusser Archive at the Berlin University of the Arts.

8 Eugene S. Ferguson, *Engineering and the Mind's Eye* (Cambridge, Mass.: MIT Press, 1992).
9 Ludwig Wittgenstein, *Tractatus logico-philosophicus: With an Introduction by Bertrand Russell* (London: Paul, Trench, Trubner, 1922), §3.11, http://www.kfs.org/~jonathan/witt/t311en.html.
10 Henry Miller, "The Red Herring and the Diamond-Backed Terrapin," in *Art and Cinema*, San Francisco Museum of Art (1947). Contributors to the symposium and proceedings included Luis Buñuel, Maya Deren, Hans Richter, and the Whitney Brothers.
11 Miller, 4.
12 Benjamin, "The Work of Art in the Age of Mechanical Reproduction," in *Illuminations: Essays and Reflections* (New York: Random House, 1988), 226–27.
13 I thank Christian Posthofen for drawing my attention to Donald W. Winnicott (1896–1971). I should also like to take this opportunity to thank him for his continuing support of our work.
14 Alexander Galloway, *Außer Betrieb: Das müßige Interface*, translated by Niklas Schrape (Cologne, Germany: Walther König, 2010).
15 Quoted in Heinz Herbert Mann, "Optische Instrumente," in *Erkenntnis, Erfindung, Konstruktion*, ed. H. Holländer (Berlin: Gebrüder Mann, 2000), 362, who lists the precise references of the notes that contain Leonardo da Vinci's remark.
16 Quotation from the 1996 reprinted edition of *Sechs Bücher zum wahren Christentum nebst dessen Paradiesgärtlein* (Bielefeld, Germany: Missionsverlag), 1:72ff. The engravings/sketches are obviously new. In the picture showing the camera obscura, the projected figure is dressed.
17 We discuss this briefly in our introduction to *Variantology 3* by way of contrast to the positive role of shadow in Chinese traditions: Zielinski and Fürlus, "Ars brevis lucis & umbrae," in *Variantology 3: On Deep Time Relations of Arts, Sciences and Technologies in China and Elsewhere*, ed. Siegfried Zielinski and Eckhard Fürlus, in cooperation with Nadine Minkwitz (Cologne, Germany: Walther König, 2008).
18 See, e.g., Jean-Louis Baudry, "Ideologische Effekte—erzeugt vom Basisapparat," *Eikon* 5 (1993): 34–43.
19 Charles Mackay, *Zeichen und Wunder: Aus den Annalen des Wahns* (Frankfurt am Main, Germany: Eichborn, 1992), 8. Originally *Memoirs of Extraordinary Popular Delusions and the Madness of Crowds* (London: Office of the National Illustrated Library, 1841).
20 Hans Eduard Fierz-David, *Die Entwicklungsgeschichte der Chemie: Eine Studie* (Basel, Switzerland: Birkhäuser, 1952), 132.
21 Roger Bacon, *Vom Stein der Weisen und von den vornembsten Tinkturen des Goldes* (Leipzig, Germany: Jacob Apel 1608), 57. I found this bizarre work

in 1992, in the University of Salzburg library, bound together in one volume with a version of *Monas Hieroglyphica* by the English mathematician, physician, cartographer, and alchemist John Dee (1527–1608).

22 See the thought-provoking biography by Michael White, *Isaac Newton: The Last Sorcerer* (Reading, Mass.: Addison-Wesley, 1998), which investigates Newton's relationship to alchemy at length, a facet of Newton's life and work that had previously been ignored.

23 See the last chapter of Michel Butor's *Die Alchemie und ihre Sprache* (Frankfurt am Main, Germany: Fischer, 1990). On Klein and Polke in this context, see the detailed study by Ulli Seeger, *Alchemie des Sehens* (Cologne, Germany: Walther König, 2003), 51, who refers, for example, to "the philosopher's stone" as the "powder of projection."

24 The polemic, unworthy both with regard to its subject matter and its form, is portrayed in William R. Shea, *Galileo, Scheiner, and the Interpretation of the Sunspots* (1970), and in Anton von Braunmühl's Scheiner biography, more from the perspective of the German Jesuit: Anton von Braunmühl, *Christoph Scheiner als Mathematiker, Physiker und Astronom* (Bamberg, Germany: Buchner, 1891).

25 Braunmühl, *Christoph Scheiner*, 17.

26 Christoph Scheiner, *Rosa ursina sive sol, Liber II* (1626–30), 105.

27 Scheiner, 349.

28 Astrophysical experiments have shown that the spots on the solar photosphere are the site of strong magnetic fields.

29 Scheiner, *Rosa ursina sive sol, Liber II*, 150.

30 On the history, including cultural, of quantification in astronomical observations, see Simon Schaffer, "Astronomers Mark Time," *Science in Context* 2, no. 1 (1988): 115–45.

31 Akira Mizuta Lippit, *Atomic Light (Shadow Optics)* (Minneapolis: University of Minnesota Press, 2005).

32 See Charlotte Bigg and Jochen Henning, *Spuren des Unsichtbaren* (2007), 23.

33 Quoted by Ernst Benz, *Theologie der Elektrizität: Zur Begegnung und Auseinandersetzung von Theologie und Naturwissenschaften im 17. und 18. Jahrhundert* (Wiesbaden, Germany: Franz Steiner, 1971), 6.

34 Benz, 7.

35 Benz, 14.

36 Leo Graetz, *Die Elektricität und ihre Anwendungen zur Beleuchtung, Kraftübertragung, Metallurgie, Telephonie und Telegraphie* (Stuttgart, Germany: Engelhorn, 1883), xi.

37 Cf. Walter Benjamin's famous essay "The Work of Art in the Age of Technical Reproducibility." Originally *Das Kunstwerk im Zeitalter seiner technischen Reproduzierbarkeit* (1936), various editions.

38 See Myles W. Jackson, "'Elektrisierte' Theologie: Johann Heinrich Winkler und die Elektrizität in Leipzig in der Mitte des 18. Jahrhunderts," in *Musik, Kunst und Wissenschaft im Zeitalter Johann Sebastian Bachs* (Hildesheim, Germany: Olms, 2005), 61; for a recent study, see Paul Michel, *Physikotheologie: Ursprünge, Leistung und Niedergang einer Denkform* (Zurich, Germany: Beer, 2008), 3–4, which seeks to define and disambiguate terms.
39 Such a crazy genealogy has, unsurprisingly, not yet been written. But it is likely that one can trace a lineage from the English Puritans to the cybernetic hippies of the U.S. West Coast.
40 Michel, *Physikotheologie*, 3.
41 John Freke, *An Essay to Shew the Cause of Electricity, and Why Some Things Are Non-electricable* (London: Innis, 1746), 59.
42 Freke, 3ff. and esp. 4 (condensed summary).
43 I tried to do this in two lectures given at two differents places: at the EGS in Saas Fee 2007–8 and in 2007 at Tate Modern in London, at the symposium Pervasive Animation. See "Expanded Animation: A Short Genealogy in Words and Images," in *Pervasive Animation*, edited by Suzanne Buchan, 25–51 (London: Routledge, 2013).
44 Cf. Freke, *Essay to Shew the Cause of Electricity*, 36–37.
45 Freke, v.
46 Georg Christoph Lichtenberg, *Über eine neue Methode, die Natur und die Bewegung der elektrischen Materie zu erforschen* (Leipzig, Germany: Geest & Portig, 1956), 20.
47 Quoted from Lichtenberg, 22–23.
48 Lichtenberg, 21.
49 See, e.g., Dieter Kliche, "Lichtenbergsche Figuren: Physik und Ästhetik," *Trajekte* 6 (2003): 35–37.
50 In this connection, it would be interesting to investigate the genesis of the psychoanalytical concept of projection, which for example was understood by C. G. Jung—who was well grounded in the world of alchemical ideas—as transmission, "transference."
51 The conversation is on the DVD *We Shall Survive in the Memory of Others*.
52 Giorgio Agamben discusses the last of the ten Sephirot in connection with his version of Guy Debord's analysis of the *Society of the Spectacle* in *The Coming Community* (Minneapolis: University of Minnesota Press, 1993). Agamben's focus is the idea of "isolation" (80.1) and the disconnection of this Sephirot of appearance/imaging from the entire system of divine attributes (the exaggerated appearance as the Fall). My thanks to Florian Hadler for the interpretation of this clue.
53 Vilém Flusser, *Kommunikologie weiter denken: Die Bochumer Vorlesungen*, ed. Siegfried Zielinski and Silvia Wagnermaier (Frankfurt am Main, Germany: Fischer, 2009), 83.

54 Brief summary quoted in Flusser, 180–81.
55 Cf. Heidegger, "Die Frage nach der Technik" (lecture, 1953), in *Vorträge und Aufsätze,* 3rd ed. (Pfullingen, Germany: Neske, 1954). The quotation by Bernard Stiegler is from his *Technics and Time: The Fault of Epimetheus* (Stanford, Calif.: Stanford University Press, 1998), 9.
56 Michael Theunissen, *Negative Theologie der Zeit* (Frankfurt am Main, Germany: Suhrkamp, 1991), esp. chapter 3, "Verdunkelde Zukunft—Zur Pathologie des Daseins in der Zeit," 197–281.

18   Allah's Automata

Where Ancient Oriental Learning Intersects with Early Modern Europe. A Media-Archaeological Miniature by Way of Introduction

The deeper I venture in my imaginary time machine into the present of times past, into the tension-filled relationships existing there among the arts, sciences, and technologies, the more urgent seems the necessity to begin to unfold these diverse cultural relationships in a horizontal fashion. The exhibition on Allah's Automata *that I was able to organize in 2015 at the ZKM Center for Art and Media Karlsruhe, together with Peter Weibel, was one of the most thrilling moments in my imaginary excursions into the Near East. It was followed by the exhibition "DIA-LOGOS": Ramon Llull and the Ars Combinatoria (Karlsruhe, 2018; Lausanne, 2018–19). The following programmatic essay discursively opened up my thinking of a technoculture that has its basis in the Golden Age of Arab-Islamic intellectual culture.*

This essay was originally published in 2015.

⁂

In our present lives, the only place free of media is paradise—in other words, a place that does not exist. For the three religions all based on one God and on one book, Christianity, Islam, and Judaism, this utopian place is highly sacred and thus self-referential. God is the only signifier capable of signifying himself. God is *lux*. He sees himself "as the self-beholding light." This is the place and at the same time the event of *self-contemplation*. No medium is required for this *noēsis noēseōs*.[1] It is perfect, media-free, and thus cannot be substantiated by us.

Every other form of contemplation is dependent on reflection, on light as lumen (with varying technical values), and is inconceivable without a medium. When a religious institution wishes to articulate its faith to its believers, it needs to use a medium to illustrate and convey the meaning of what it wishes to communicate. For example, God uses the engineer who believes in him to construct an automaton that, in turn, is used as a medium for praising him. "The . . . instrument is finished with the power and strength of Allah."[2] This is the concluding sentence in the manual by the Banū Mūsā, describing their music automaton in the ninth century. Staging and perception are all realized as a material and intellectual sensation about and through a third party. The Middle Ages, an epoch in which God's absence from earthly existence was celebrated particularly intensely and with dolorous passion and devotion, was the ideal media age. The production of simulacra and other representations in the form of icons, objects, sculptures, rhythms, chants, and poetry was in full swing. All artifacts had to signify God and to praise him.

At the end of this epoch, or to be precise, in the thirteenth century, the Catalan philosopher Ramon Llull, whom Charles Lohr called a "christianus arabicus,"[3] attempted to sum up the media-based foundations of religious communications. In his theological-philosophical masterpiece, the *Ars magna*,[4] this thinker and missionary who lived on the Balearics constructed symbolic machines, initially on paper and later in reality. His idea was that followers of the religions of the book and the word could use these devices to communicate with one another about those axioms of the divine that were binding for all: charity, goodness, absolution, glory, and so on. Today, Llull's rotating discs with their symbols for the axioms, their predicates, and their basic modes of function can be seen as early algorithmic artifacts, which were established at the interface between the old and new worlds, between the Orient and the Occident, and between worldviews and religions. With the aid of media machines, Llull attempted to accomplish something that disciples of the monotheistic religions, which all only recognize one reality, could not succeed in achieving.

The algorithm as a mathematical procedure is named for a Persian-born mathematician, Abū Ja'far Muḥammad ibn Mūsā al-Khwārizmī. This extremely devout Muslim spent two decades, from 810 to 830, in Baghdad. There, in the House of Wisdom (Bayt al-ḥikma), that laboratory of knowledge and transformation in the early Golden Age of Islam, he wrote his most important texts on algebra, music, mechanics, the calculation of intervals, and the measurement of time. Among other things, al-Khwārizmī compiled special "prayer tables for each latitude."[5] To determine the correct position vis-à-vis the holy city of Mecca, muezzins either had to be able to operate astrolabes themselves or they needed to enlist the services of a professional expert on the medium, a *muwaqqit* (astronomer). It must be assumed that also in Islam, cosmology and the practice of religion were very closely interrelated. Ayhan Ayteş, media archeologist from Turkey, has contributed an excellent article on this relationship.⁶

In the Christian tradition, sounding clocks became one of the most prominent sacred automata, before lending their rhythms to the movements of the European modern age and accelerating them to the point where "chronocracy" reigned.⁷ Clocks became the master machines of the modern world, as Lewis Mumford, who published his first book on literary utopias in 1922, frequently emphasized.⁸ "Sacred automata are automatic instruments whose movements express the sacred, and are an exhortation to be pious," writes Jörg Jochen Berns in the opening lines of his text *Uhr als Himmelsmaschine* (The clock as heavenly machine), where he describes the work of engineers (in this case, engineers with a Christian background) in transforming "heavenly work into work of the soul."⁹

In the Alexandrian, Chinese, Byzantine, Persian, and Arab traditions, the situation was not dissimilar, albeit with the difference that long before the invention of the mechanical clockwork in Europe, the Oriental, Near Eastern, and Arab cultures experimented with and, more to the point, constructed mechanical automata. These were intended as luxury items for the palaces of the rich and of the caliphs, as gifts for guests, to impress foreign emperors, as pieces to be looked at and listened to on public squares, in praise of God the Almighty, or to demonstrate to Him how close His brilliant creatures could get to Him in their inventiveness and creative abilities.

The techno-souls that drove these machines were filigree structures consisting of levers, hydraulic and pump mechanisms, which kept the mechanical hearts in permanent and regular motion. In principle, the hydraulics in the water clocks were based on the simple timekeepers familiar from ancient Egypt with regularly spaced markings on the inside of the bowl—Arab historians of technology have traced the construction to around 1500 B.C.E. These archaic devices functioned analogously to the hourglass, invented even earlier; however, in the case of the clepsydra (water thief, as the Greeks called it), water flowed from one vessel into another. The amount of the displaced liquid indicated the time. It is thought that the clepsydrae were constructed for measuring predetermined times, such as how long speeches at court proceedings or political debates were meant to take. Via China, where there were similar water clocks, and from Hellenistic Alexandria, Byzantium, and the Persian Empire, specialist knowledge was refined and spread to the Arabian Peninsula and from there to Andalusia in Europe. It is reported that, in the eleventh century, there were still two impressive water clocks on public squares in Toledo. Among other uses, these clocks sounded loudly at prescribed intervals to call the faithful to prayer.

Possibly the most impressive horologium in terms of sound, which was powered by water, was built at the end of the eleventh century by a Chinese engineer, Su Song, for the palace gardens in Kaifeng, around the same time as Shen Kuo's experiments with the projection of flying birds in the camera obscura. Su Song's sounding installation, audible from a great distance, was ten meters high and driven by a waterwheel with a diameter of around four meters that already utilized a stop-and-go mechanism,[10] a principle essential for the later mechanical clock technology in Europe of the late Middle Ages (and, later still, in the film camera). In Su Song's automaton, kinetic figures with cymbals, trumpets, and gongs are reported to have marked the passage of time.

In her essay, Nadia Ambrosetti cites this context with reference to an essay by science historian Salim T. S. al-Hassani: the entirety of the individual elements in the legendary elephant clock by the engineer al-Jazarī (1136–circa 1206), who came from present-day Anatolia, represented everything that the Near Eastern world knew at that time of mondial knowledge.[11] The animal carrying the clock is a reference to India, whence Arab natural philosophers had imported a large portion of their advanced knowledge of mathematics and astronomy (Figure 18.1). The elephant is the bearer of the clockwork and acts as its black box; it contains the hydraulic system that was developed in the Greek tradition. On its back is a magnificent Persian rug, a tribute to their Iranian neighbors from whom the Arabs had not only taken over the stories from *The 1001 Nights,* which contain descriptions of a rich culture of artifacts and automata. A large number of the philosophers of nature, mathematicians, and engineers who worked in ninth-century Baghdad and Basra came from present-day Iran or commuted between the Arabian and Iranian cities. The rider or mahout sitting behind the elephant's head and beating time rhythmically wears a turban and is, in most of the sketches, black, which could be interpreted as a reference to the neighbors in India. And the scribe at the center of the arrangement, who is writing the time down on a disc, wears clothes that identify him as a Muslim. The dragons that are the mechanical interface of the movement echo Chinese mythology. The two falcons above them, from whose mouths the iron balls drop into the dragons' mouths, invoke Arab cultures where falcons represent male

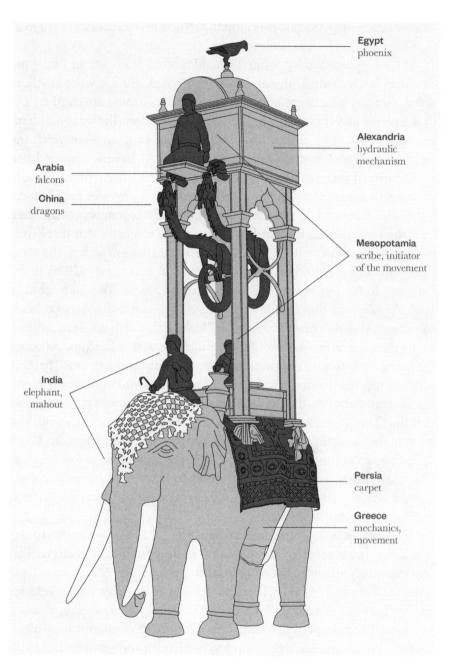

FIGURE 18.1. Al-Jazarī's elephant clock, full view, showing all its mondial references. Illustration by Clemens Jahn, 2015.

vitality. And finally, the phoenix atop the structure references the origin of the water-driven clock in ancient Egypt.

The interpretation by Salim T. S. al-Hassani, from whom I take my cue here, is speculative. The inventor of the clock did not write anything about its multicultural connotations. Nevertheless, these are contained in its imagery—and anyway, it is a fascinating speculation. The eight-hundred-year-old elephant by al-Jazarī, who hailed from a region after which the most famous modern-day Arab media network, Al Jazeera, is named, can be interpreted as the role model for an early Islamic notion of the *Globale*, the current art event at ZKM in Karlsruhe, which provides the platform for *Allah's Automata*. Even before the House of Wisdom was established by Caliph al-Ma'mūn, the Muslim realm was so extensive that there were important points of contact, as well as cross-fertilization, with all the other developed cultures that existed in the world at that time—China in the Far East, India, Persia, and the Greco-Roman world. The Arab scholars positively absorbed these cultures of knowledge and technology (al-Jazarī sometimes even described himself as an "Indian"[12]), and thus they functioned as the interface between the different traditions, at that juncture in history when the ancient world began to transform into the modern one. The Arab scholars created the first renaissance,[13] and to this end used all the media the ancient authors had provided them. For example, it was via Samarkand that the Chinese methods of papermaking reached the Islamic world. Vast amounts of translation work permitted Persian, Latin, and particularly Greek sources to enter the contemporary reality of Baghdad's House of Wisdom. The texts rendered in Arabic, which often only survived because of these translations, in many cases formed the basis of disciplines that were essential for the European modern age.

The Banū Mūsā—Muḥammed, Aḥmed, and al-Ḥasan—were the sons of Mūsā ibn Shākir, who became rich as the leader of a band of desert robbers and eventually also established a reputation as a scholar and astronomer. A friend of Caliph al-Ma'mūn, he enabled his sons to study and do research at the House of Wisdom, which consisted mainly in organizing, making, checking, and interpreting translations. Their areas of interest were highly diverse—mathematics and geometry, mechanics and pneumatics, music and astronomy. One example: the only reason why European readers have access to books V–VII of the famous work of mathematics and geometry on conical sections, written in Greek in Alexandria in the third century by

Apollonius Pergaeus (or Apollonius of Perga), is because of the translations by the Banū Mūsā. The Greek original is irretrievably lost.[14] The same applies to much of Heron of Alexandria's writings on mechanics, without which the automata of both the Arab-Islamic world and early modern Europe would have been inconceivable. Heron's work has mostly come down to us via its translation into Arabic, which was kept in Constantinople's palaces housing manuscripts and books, where they were copied and translated by Europeans. Today, young engineers at the Technical University of Berlin can still read this classic of Greek-Alexandrian mechanics in the 1900 bilingual edition in Arabic and German.[15] Just how important the Banū Mūsā were for education in mathematics and geometry in the nineteenth century in Germany, with an educational system aware of deep-time dimensions, can be seen from a document published in 1885. The *Liber trium fratrum de geometria* by those clever sons of the educated robber was translated and published by Maximilian Curtze, a schoolmaster at the Royal Grammar School in Thorn (present-day Toruń, Poland) and was commissioned by the German Academy of Natural Scientists for teaching purposes.[16]

The exhibition *Allah's Automata* is a brief excursion through the Golden Age of Arab-Islamic cultures. Focusing on exceptional examples, it explores the rich and fascinating world of the automated mechanical devices developed by Muslim engineers in the period from the early ninth to the early thirteenth centuries, which were described in manuals and some of which were actually built. We focus on texts that have been acclaimed in the specialist literature as outstanding testimonies to early automaton construction. In this way, the exhibition offers an operational canon especially relevant for the archaeology of the media and the arts. In the history of science and technology, isolated references to this subject crop up occasionally, mainly as marginalia. However, the entire complex is now in the process of being rediscovered and researched for a materialist history of arts and the media, which can be developed within the context of the arts, sciences, and technology.

Four master manuscripts on automaton construction from Baghdad, present-day Turkey, and Andalusia are on show together and in their entirety for the first time and are placed in their genealogical contexts. Two of the most famous automata have been built especially for this exhibition and are presented as working models.

The *Kitāb al-ḥiyal* (*The Book of Knowledge of Ingenious Mechanical Devices*) is the best-known manuscript by the Banū Mūsā ibn Shākir. The brothers wrote it in the first half of the ninth century (around 830) in Baghdad's House of Wisdom. The compendium is full of sketches and precise instructions of how to build one hundred different mechanical models *(shakl)*. These devices include a diversity of artifacts, kinetic sculptures, and automata—the latter being literally "devices that move of their own accord": hydraulically and mechanically operated drink dispensers, pneumatically activated animals that make noises, oil lamps that automatically lengthen their wicks and refill their oil and also adjust their wind guards in such a way that the flame is protected and the lamps burn indefinitely. They are a kind of sacred *perpetuum mobile* comparable to the Sacred Light familiar from Catholic churches as the symbolic light that shines constantly in the house of the Christian God. And in fact, the Muslim constructors named this artifact "Sirāj Allāh" (the Lamp of God).[17]

In the ninth century, the mathematician al-Khwārizmī made an interesting classification of the Banū Mūsā's book. He first divided it up according to the disciplines of philosophy, dignity, logic, medicine, arithmetic, geometry, astronomy, and music. He then subdivided it into two sections, familiar from the early European modern era in the theory of mechanics: weight and movement, and self-propelled machines—in other words, the art of imaginative inventions.[18]

Only three almost complete manuscripts of the Banū Mūsā's *Kitāb al-ḥiyal* have survived: one is in the library of the Topkapı Palace in Istanbul (Manuscript Ahmet III 3474); a second copy (circa 1250–1300) is in the Vatican Library's collection of manuscripts in Rome; and the third—a copy dating from 1209—is shared between the Oriental manuscript collections of the Universitäts- und Forschungsbibliothek Erfurt/Gotha, Thuringia, Germany (seventy folios), and the Staatsbibliothek in Berlin (seventy-five folios). We are pleased to be able to exhibit in this exhibition the copy from the Vatican Apostolic Library and the manuscript divided between the Oriental collections in Berlin and Gotha. The latter is presented for the first time as a complete document, united in the virtual reality of a digitized manuscript.[19]

The second manuscript by the Banū Mūsā ibn Shākir is even more spectacular. The treatise, titled *Kitāb al-urghanun* by Arab historians of science, dates from around the mid-ninth century and was presumably written slightly later than the book with the one hundred inventions by Aḥmad, who is considered the best engineer of the three brothers.[20] *The Instrument Which Plays by Itself* is the literal translation of the manuscript's title, which George

Farmer describes as a "solitary exemplar" in his famous work on the early history of the organ.[21] Today, this second treatise is believed lost. The only known copy was, for a long time, in the Three Moons College of the Greek Orthodox Church in Beirut, Lebanon. With the help of George Saliba, an expert on the history of science and a native of Beirut who teaches at Columbia University in New York, we were able to gain access to photographic negatives of the manuscript that are held in the Université Saint-Joseph's Bibliothèque Orientale in Beirut. Louis Cheikho, the founder of a magazine for Arabic studies, *al-Mashriq* (The East), described this instrument of the Banū Mūsā in a 1906 issue of the journal, using these photographs.[22] This journal article was the basis for George Farmer's English translation. The manuscript *The Instrument Which Plays by Itself* from Baghdad, which dates to around 850, was well worth the "almost criminological research" that George Saliba describes in his contribution to the exhibition catalog. Mona Sanjakdar Chaarani has deciphered the fascinating contents of the text for us.[23] Indeed, one could say from the perspective of an arts and media archaeologist that the manuscript describes not only a flautist who plays continuously but at the same time an archaic programmable universal musical instrument. The title itself proclaims the generalizable import that the Banū Mūsā ascribed to the technology they outline. They obviously intended their sounding artifact to be understood independently of any specific performance by a sornay player. They gave exhaustive explanations of a modern music automaton that could not only vary its rhythm but was also technically capable of being programmed for any melodies. At the very beginning of the text, the idea of a universal instrument is clearly outlined: "We wish to explain how an instrument ... is made which plays by itself continuously, whatever melody ... we wish, sometimes in a slow rhythm ... and sometimes in a quick rhythm ... and also that we may change from melody to melody when we so desire."[24]

For an archaeologist of the arts and media interested in physical artifacts and their logic, the idea of a programmable music automaton inspires amazing thoughts. By producing such an automaton, present-day generations of media artists can learn something about the skills and abilities that, almost twelve hundred years ago, culminated in the construction of controllable objects—nowadays, this is known as physical computing. According to conventional wisdom, the origins of such machines would be located in the early European modern age, or even the Enlightenment. Thus we at the

Berlin University of the Arts decided to model the principal functional elements of the Banū Mūsā's automaton. Because of the valuable manuscripts in its immediate vicinity at the exhibition, we could not produce a model powered by hydraulics and pneumatics. The universal music automaton dating from the ninth century contains a twenty-first-century pneumatic and electric soul; the hydraulic mechanism is a 2D animation.

A reconstruction, dating from 2002, of al-Jazarī's *al-Jāmi' bayn al-'ilm wa-'l-'amal an-nāfi' fī ṣinā'at al-ḥiyal* (1206), his legendary *Compendium on the Theory and Practice of the Mechanical Arts*, was given to me as a present by the book's editor, Fuat Sezgin. Together with the English translation by British historian of science and technology Donald Routledge Hill, published as the *Book of Knowledge of Ingenious Mechanical Devices* (1974), it formed the basis of our work in researching the Arab-Islamic automata and proved so invaluable for this work that we present it here, together with the most precious unique items and the digitized images. Fuat Sezgin's reproduction features one of the oldest copies of al-Jazarī's manuscript, the original of which, held in the Süleymaniye Library, we were unable to secure for this exhibition.

The masterpieces among the automata described by al-Jazarī were the clocks designed to be looked at and, most importantly, to be heard from a great distance. These horological machines can be described in today's terminology as ancient audiovisual installations. They were complex structures that not only indicated the time so that it was rendered visible but also made time sound so that people could hear it, thus objectifying it. These artifacts counted the hours or smaller units of time and intoned the temporal framework within which people moved—a function comparable to that of the muezzin, who called people to prayer from their towers at regular intervals, providing a temporal religious structure in urban and rural environments. Etymologically, the word *minaret* is derived from "lighthouse" *(manāra)*, although this metaphor is misleading because the minaret is primarily an acoustic medium.

Two of al-Jazarī's outstanding masterpieces are presented in detail in the book *Allah's Automaton* by Ulrich Alertz and Ayhan Ayteş: the horologium taken to Aachen and given to Charlemagne as a gift by Hārūn al-Rashīd at the beginning of the ninth century and the legendary elephant clock which in the meantime is the technical superstar of engineering prowess from the Golden Age of Islamic culture. A number of years ago, a giant replica of the elephant clock was installed in a shopping mall in Dubai. The costs involved in loans of such models are astronomical. So we traveled to Bursa, the former capital of the Ottoman Empire, where people feel such a close affinity with "their engineer" al-Jazarī and his technical skills, that they built for us a new model especially for this exhibition. We are particularly pleased with this version of the elephant clock because it is not hidden under decorative splendor.

Each one of the automata designed and/or actually built[25] by Ibn al-Razzāz al-Jazarī from the el-Jezireh region he made in praise of Allah and in honor of the latter's powerful representatives on Earth, who provided the engineer with his commissions and rewarded him regally. He designed and described them as if they were mechanical objectifications of the cosmic spirit and thus they are truly Allah's automata. Indeed, al-Jazarī repeatedly emphasized this fact in his texts. Hārūn al-Rashīd's horologium featured a group of artificial musicians, micro-automata within the macro-automaton, so to speak. At the end of the twelve hours of night, the twelve windows positioned in a row above the musicians' heads, which to begin with are all closed in the dark blue of night, are open and shine brightly. This is a strong image for God's automata: the divine light which causes the day to break; the Almighty is all-knowing and thus He is also the constructor of this artifact. And al-Jazarī closes this chapter of his manual[26] with the words "Allah is all-knowing."

Beyond their religious and ludic significance, a large number of these devices also possessed useful functions, something that competed with their sacred connotations. The most trivial of these were the hydraulic and pneumatic automata designed to get participants in drinking parties or other forms of revelry drunk as quickly as possible. However, the larger mechanical devices were intended to intervene in the given natural environment and to change it drastically for the benefit of its inhabitants, to transform extremely arid tracts of land into fields of thriving vegetation. To do this, water for irrigation had to be raised from wells deep in the ground. Hoists, pumps, wheelworks, and other mechanisms actually represented instances of animation in the most direct sense of the word—ensoulment in the sense of bringing inanimate material to life.

The concluding section of al-Jazarī's famous book on the theory and practice of the mechanical arts is devoted to a magnificent little mechanical box, more precisely, to the technical question of how to lock a container for valuables in such a way that knowledge of a relatively complex code is necessary to reopen it. As a solution, al-Jazarī proposes a combination of four locks located next to one another. The mechanism that opens the box only works when the three levels of key, which every lock possesses independently, are set correctly. This is easy to achieve for people who know the code. For people without this knowledge advanced cryptologic skills are necessary to succeed in opening the casket.

A text several decades older than al-Jazarī's manuscript is the *Kitāb al-asrār fī natā'ij al-afkār (The Book of Secrets in the Result of Ideas)*, the work of a (presumably) Andalusian engineer, Aḥmed ibn Khalaf al-Murādī. It was started in the eleventh century. The only known manuscript (1266) to have survived is in the Biblioteca Medicea Laurenziana in Florence and has been restored by a group based in Milan, Leonardo3, on behalf of the Museum for Islamic Art in Qatar. Massimiliano Lisa and Mario Taddei described al-Murādī's book for us and have made their interactive interpretation of the manuscript available for the exhibition.[27]

The existence of this even earlier manuscript that describes with great precision the technology required for hydraulically and pneumatically driven automata demonstrates three things:

- Between the ninth and twelfth centuries, the knowledge of physics, mathematics, and geometry necessary to construct these kinds of artifacts was not restricted to a few isolated instances in Mesopotamia but should rather—at least as far as the elite is concerned—be considered an established form of cultural technology.
- The pioneers in the production of media-based attractions in the form of autonomously moving devices and of other highly effective artifacts made no secret of the way their artificial paradise machines worked. The idea of concealing the mechanisms of these automata, transforming them into black boxes, did not occur until the early modern Christian period and became even more hermetic in the Enlightenment and the Age of Industrialization.
- The mammoth translation task undertaken by Muslim scholars, a task in which many Christian clerics assisted, allowed Arab-Islamic culture to gather the pertinent knowledge from the ancient world and, in many instances, to develop it further. Between 800 and 1200, a culture of experimentation sprang up, in combination with the Chinese, Indian,

Persian, and Arabian knowledge traditions; a culture that was, among other things, full of surprises. Over and beyond established knowledge, Allah's automata gave rise to spectacular innovations not found in Europe until the early modern era and modern times: a draft of mondial modernity before the idea assumed global proportions.

I would like to suggest how, in the Golden Age of the Arab-Islamic culture of experimentation, the provenances and futures were interlinked. Our investigations of the deep time of specific constellations in the relations between arts, sciences, and technologies always assume that the arrows of time, which must be described here, point in both directions.

Ctesibius of Alexandria (285–222 B.C.E.), a contemporary of Archimedes of Syracuse (circa 287–212 B.C.E.), is considered the initiator of research on pneumatics as a science to produce movement artificially. The son of a hairdresser who is said to have worked as a barber as a young man, Ctesibius is an early example of a culture of experimentation which saw itself as a "generator of surprises." Among other things, he devoted many experiments to the water clock and is also considered the inventor of the hydraulically and pneumatically driven organ.[28]

In his studies of pneumatics, Philo of Byzantium, who lived in the third century B.C.E., drew extensively on Ctesibius and expanded on the latter's experiments by constructing a large number of artifacts. In his summary, Aage G. Drachmann lists as many as sixty-five mechanical models, of which we find a number of almost identical copies in the Banū Mūsā's *Book of Ingenious Devices*, particularly the many automatic drinking and serving devices.[29] In addition, Philo attempted to explain the theory behind his physical models. One noteworthy fact is his understanding—in the Presocratic tradition—of air as a body composed of minute particles. As he saw it, the interaction between water and this body in manipulated vessels generated the energy that could set other objects in motion.

In the first century B.C.E., an Alexandrian and a Roman shared the privilege of being hailed as the greatest engineer and inventor in our field—Heron of Alexandria and Marius Vitruvius Pollio (Vitruvius) paid their respects to Ctesibius and Philo. At the same time, Heron and Vitruvius both independently developed their own impressive theaters of automata. These included water clocks and organs whose pressure was generated automatically by means of hydraulic systems (Figure 18.2).

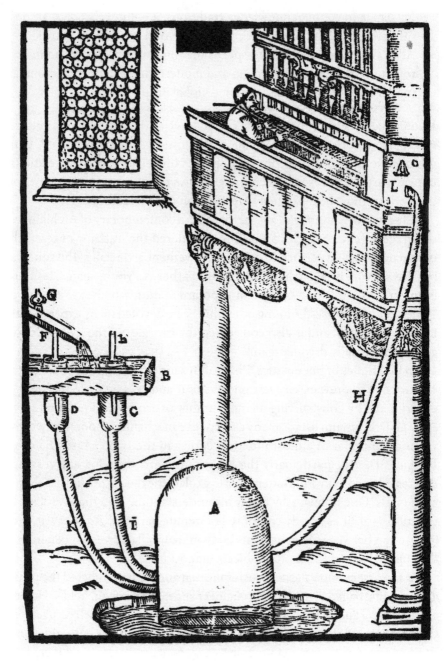

FIGURE 18.2. Vitruvius's water organ, according to Giovanni Branca in *Le machine* (Rome: Jacobo Manuci & Jacomo Mascardi, 1629), part 3, plate 19 (21).

FIGURE 18.3. Flute player and water organ, statuette from Hellenistic Alexandria, Musée du Louvre, Paris, Inv. CA 426.

However, for the Banū Mūsā and also for al-Jazarī and al-Murādī, Heron of Alexandria's *Pneumatica et Automata*[30] was the most important source of inspiration. Closely relating and interweaving theoretical and physics-related aspects with practical considerations, Heron elaborated all manner of technical elements and structures that the Arabs took up and developed for use in their automata (Figures 18.3–18.5). Following Philo, Heron also began by considering the air as something just as material as the other three Empedoclean elements (fire, water, earth) and accordingly constructed his dramatic world of artifacts, instruments, and automata from different combinations of the four. Various kinds of levers, pipes (channels), cogwheels, pulleys, flap valves (of the kind used in the Banū Mūsā's music automata), and weight regulators created Heron's world, where magic amphorae could alternately serve wine or water or even a mixture of the two, where tin birds twittered and monks whistled and twirled. Heron's theater of automata made

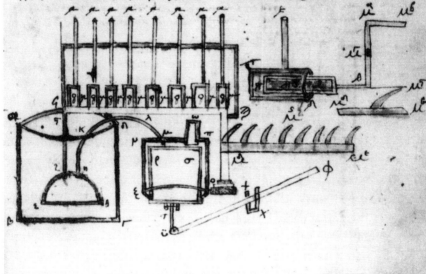

FIGURE 18.4. Heron of Alexandria, water organ, *Pneumatica*, Biblioteca Nationale Marciana, Cod. Gr. Zan. 516, folio 184r.

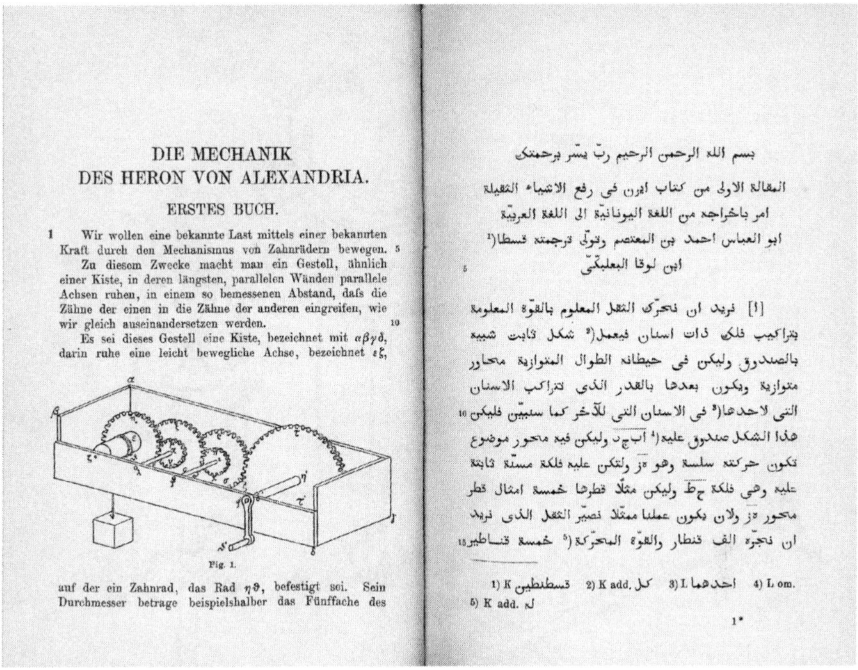

DIE MECHANIK
DES HERON VON ALEXANDRIA.

ERSTES BUCH.

1 Wir wollen eine bekannte Last mittels einer bekannten Kraft durch den Mechanismus von Zahnrädern bewegen. Zu diesem Zwecke macht man ein Gestell, ähnlich einer Kiste, in deren längsten, parallelen Wänden parallele Achsen ruhen, in einem so bemessenen Abstand, dafs die Zähne der einen in die Zähne der anderen eingreifen, wie wir gleich auseinandersetzen werden.
Es sei dieses Gestell eine Kiste, bezeichnet mit $\alpha\beta\gamma\delta$, darin ruhe eine leicht bewegliche Achse, bezeichnet $\varepsilon\zeta$,

Fig. 1.

auf der ein Zahnrad, das Rad $\eta\vartheta$, befestigt sei. Sein Durchmesser betrage beispielshalber das Fünffache des

بسم الله الرحمن الرحيم ربّ يسّر برحمتك

المقالة الاولى من كتاب ايرن فى رفع الاشياء الثقيلة
امر باخراجه من اللغة اليونانيّة الى اللغة العربيّة
ابو العباس احمد بن المعتصم وتولّى ترجمته قسطا(1
ابن لوقا البعلبكّى

[ا] نريد ان نحرّك الثقل المعلوم بالقوّة المعلومة بتراكيب فلك ذات اسنان فيعمل(2 شكل ثابت شبيه بالصندوق وليكن فى حيطانه الطوال المتوازية محاور متوازية ويكون بعدها بالقدر الذى تتراكب الاسنان التى لاحدها(3 فى الاسنان التى للآخر كما سنبيّن فليكن هذا الشكل صندوق عليه(4 ابجد وليكن فيه محور موضوع تكون حركته سلسة وهو هز ولتكن عليه فلكة مسنّنة ثابتة عليه وهى فلكة حط وليكن مثلًا قطرها خمسة امثال قطر محور هز ولان يكون عملنا ممثّلا نصيّر الثقل الذى نريد ان نجرّه الف قنطار والقوّة المحرّكة(5 خمسة قناطير

1) K قسطنطين 2) K add. كل 3) L احدهما 4) L om.
5) K add. لا

1*

FIGURE 18.5. Heron of Alexandria, in *Mechanics,* vol. 1 (Leipzig, Germany: Teubner, 1900).

people appear on the stage or disappear from it in a decidedly strange manner, opened doors as if by magic, and moved heavy objects in predefined geometric patterns. Even at that time, Heron had a conception of what programs are. He organized the commands for moving physical objects by winding the end of the strings connected to the objects around wooden slats in a specific order. The process of unwinding then determined the direction, speed, and duration of the movement. Unquestionably, the Banū Mūsā's programming idea using the rotating cylinders with pins is more advanced, but in most cases their technical models simply copy Heron's.

If we point the arrow of time from this interface between the ancient and the new worlds in the direction of the present, it flies through the period that has long been considered the era when the production of divine automata first took place. The early modern era with its mechanical cabinets of wonder and arsenal of instruments, machines, and automata was long considered the period when *the* (only) modern age was invented, in other words, the European one. Natural philosophers such as the Neapolitan Giambattista della Porta, the English physician and philosopher Robert Fludd, and most

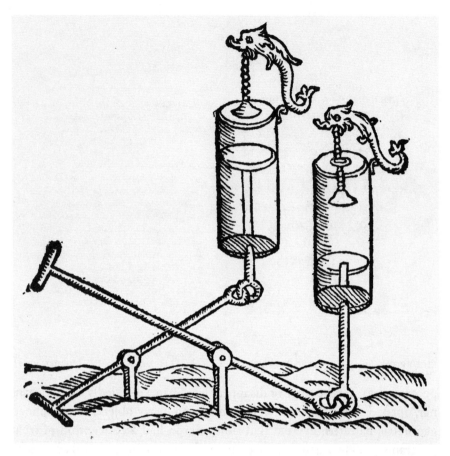

FIGURE 18.6. Floats. From Giambattista della Porta, *Pneumaticorum libri tres* (Naples: G. G. Carlino, 1601), 65.

prominently, the German Jesuit and long-standing head of the Collegio Romano in Rome Athanasius Kircher became the much-admired protagonists of a theater of machines and automata whose origins—if traced back to deep time at all—were attributed to the Greeks and, in particular, to the Hellenes.[31]

The most interesting object in della Porta's slim volume *Pneumaticorum libri tres* (1601), which has only seventy pages and contains a large number of inventions from Heron's *Pneumatica*, is an organ whose pipes are opened and closed by the pins of a rotating cylinder and whose power supply comes from a rotating waterwheel. In the following section, della Porta gives a rough sketch of Vitruvius's water organ, which is technically incorrect as far as the permanent supply of air is concerned.[32] There is no word about the

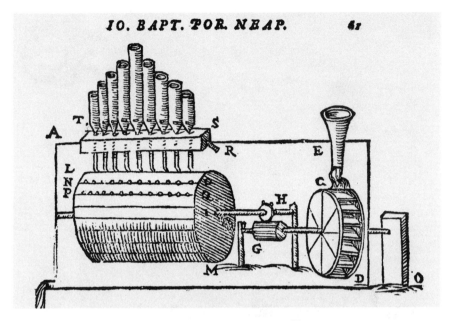

FIGURE 18.7. Water organ. From della Porta, *Pneumaticorum libri tres,* 61.

inventor(s) of the programmable cylinder, even though access to Arabic literature was not a problem in sixteenth-century Naples (Figures 18.6 and 18.7).

A little less than two decades after Giambattista della Porta, Robert Fludd simply copied the instrument he described, whom Fludd held in high esteem. However, there is one detail of Fludd's water-powered organ cylinder that places the automaton of God in a different light. The energy, the water, cascades from a pipe attached to a cloud in the sky onto the technical appliance: *Lord aux chiottes,* God relieves himself, as philosopher Georges Bataille put it when, in 1928, he published the first version of his *Histoire de l'oeil (Story of the Eye)* (Figure 18.8).

One of the main universal music theories of the Baroque was Athanasius Kircher's *Musurgia universalis* (1650). The book includes a wealth of elements familiar from the Greek-Alexandrian, Byzantine, Roman, and Arabic traditions which are combined with the knowledge that went into the late medieval glockenspiel of the fourteenth and fifteenth centuries. Kircher invested a great deal of effort in the practicalities of how the cylinder with pins could be programmed with as many tunes as possible. And, on a double page in the center of the book he presents a magnificent illustration of his own automata theater, which confirms my theory that Kircher is the legitimate

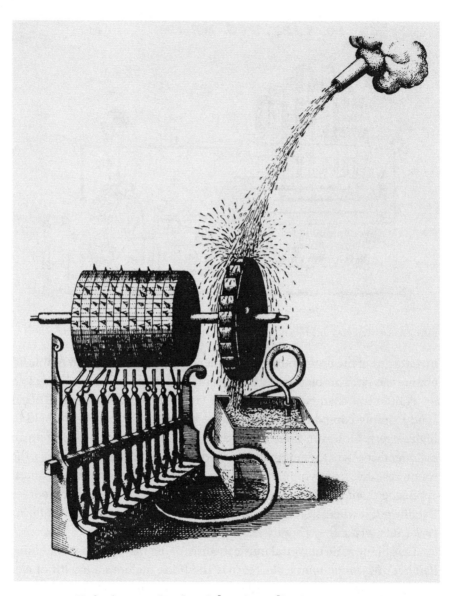

FIGURE 18.8. Hydraulic organ: "Lord Auch [aux chiottes]"—this was the pseudonym under which Georges Bataille published his *Histoire de l'œil (Story of the Eye)* in 1928. Here God seems to drive the pipe organ by ejaculating directly onto the water wheel connected to it. Etching from the works of Robert Fludd, early seventeenth century (based on a drawing by Giambattista della Porta, late sixteenth century).

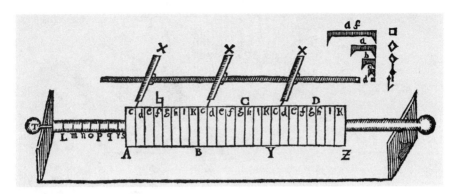

FIGURE 18.9. Instructions for arranging pins on a cylinder so that it can play several musical pieces. The depiction is two-dimensional. The cylinder can be shifted sideways so that the pins programming the various music pieces (indicated by letters on the cylinder) do not touch it, but instead the levers, which are marked by crosses in the diagram. At the top right are "bridges" for various note values. From Athanasius Kircher, *Musurgia universalis*.

founding father of the Hollywood dream factory (Figures 18.9 and 18.10). The mighty cylinder, which occupies almost half of the engraving, moves not only the keys of the central organ but also the figurines of a strange quartet of mechanical figures in the upper left-hand corner of the illustration. This is obviously a tribute to Pythagoras and his story of the blacksmiths striking their anvils at different intervals. In the upper middle of the picture a mobile angel figure directs the entire scene, and in the upper right-hand segment of the engraving, a mechanical group of dancers revolves around a kinetic skeleton. This danse macabre on the revolving stage is driven by a second channel into which water is also fed.[33]

In the twenty-first century, a narrative positing the originary nature of European modernity is not useful. By inventing the modern age, European culture of the early modern era and, to an even greater extent, that of the Enlightenment declared itself the measure of all things. Everything in the arts, the sciences, and technology that is considered stale, primitive, out-of-date, or innovative, complex, and advanced is determined from the perspective of the self-appointed avant-garde in Rome, Paris, London, and, temporarily, Saint Petersburg. Modernity and the modern mutated into hegemonic postulations in the clash of cultures, religions, and nations, which, at the beginning of the third millennium, once again turned out to be primarily a battle for economic, political, and religious dominance on this finite planet.

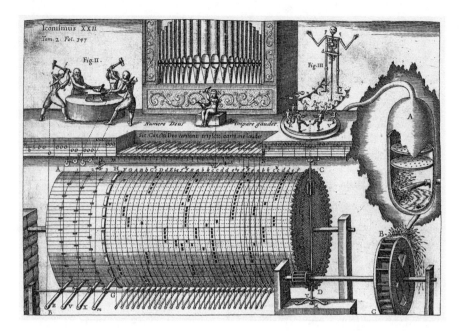

FIGURE 18.10. The so-called Pythagorean organ conceptualized by Athanasius Kircher in his treatise on the theory and praxis of music, *Musurgia universalis*. The programmable cam cylinder is working mechanically and hydraulic pneumatically in the same way as the Banū Mūsā's automaton did eight hundred years before. The scene with the cyclopes hammering at a forge refers to the famous legend of the physical origin of musical tuning by the Greek philosopher Pythagoras, in Athanasius Kircher, *Musurgia universalis,* etching plate between pp. 345 and 346.

From the perspective of deep-time relations between the arts, sciences, and technologies, what I consider to be modern is, more than anything else, a particular attitude toward the world. This attitude is characterized by a courageous and experimental approach to the things that surround individuals and among which they live; it is not an attitude governed by testing, by appropriation, and certainly not by exploitation. The world we live in is seen as something mutable, in exactly the same way as individuals are considered mutable from the perspective of those around them. Ideally, it is both the individual and the composite whole that change, each to its own advantage.

In many ways, the three sons of Mūsā ibn Shākir in ninth-century Baghdad formed the kind of community in terms of their thinking and work that, in the modern world, would be referred to as a team. They engaged with all kinds of intellectual disciplines and traditions, came together to create new entities and, at the same time, came up with surprising ideas of their own.

The notion of a universally programmable mechanical heart for controlling all manner of musical instruments was one such brazenly new idea. It has influenced the way that mechanical music automata are constructed right up to the present day. The projection of the program, the arrangement of the pins on a music automaton's cylinder, onto a two-dimensional surface we recognize as the same patterns that people encountered on nineteenth-century punch cards used to operate industrial machines.[34]

Apparently, al-Murādī and al-Jazarī, the Andalusian and the Anatolian engineer, respectively, were not familiar with the text that described the mechanism of the rotating cylinder with pins. Nevertheless, they astonished audiences with the diversity and the enormous complexity and precision of their timekeeping automata. Their technical knowledge and their aesthetic confidence in depicting their models and automata in manuals, which can still be used today to construct them, make Lewis Mumford's remark about the clock as the master machine of the modern European industrial age pale by comparison. In the way that it interweaves the divine and the profane, the Golden Age of Arab-Islamic technologies brought forth outstanding examples of the interplay between the sciences and the arts in technical objects, which simply did not require the concept of media art installations.

Allah's Automata may at first sound antiquated and conservative, but from the perspective of media archaeology it is the discussion of a hot spot where a variety of different discourses from art, science, and technology vigorously intersect. The time machine, which offers our miniature of a first exhibition on this subject in Germany, will also traverse future present days in which these past ones will play an important role.

Notes

1 Stephan Meier-Oeser, "Medienphilosophische Konzeptionen in der Erkenntnis-und Zeichentheorie des Mittelalters," *Das Mittelalter* vol. 15, no. 2 (2010): 245.
2 Farmer, *The Organ of the Ancients* (London: William Reeves, 1931), 114.
3 Charles Lohr, "Ramon Llull: 'Christianus Arabicus,'" *Randa* 19 (1986): 7–34.
4 See Amador Vega, "Ramon Llull: A Logic of Invention," in *Variantology 2*, ed. Siegfried Zielinski and David Link, 45–64 (Cologne, Germany: Walther König, 2006).
5 Thomas F. Glick, "Islamic Technology," in Jan Kyrre Berg Olsen Friis, ed., *A Companion to the Philosophy of Technology* (Chichester, U.K.: Wiley-Blackwell, 2009), 34.
6 Ayhan Ayteş, "Divine Water Clock: Reading al-Jazarī in the Light of al-Ghazālī's Mechanistic Universe Argument," in *Allah's Automata*, edited by Siegfried Zielinski and Peter Weibel (Ostfildern, Germany: Hatje & Cantz, 2015).
7 See Peter Weibel, *Die Beschleunigung der Bilder in der Chronokratie* (Bern, Switzerland: Benteli, 1987).
8 Lewis Mumford, *The Story of Utopias* (New York: Boni and Liveright, 1922).
9 Jörg Jochen Berns, *Die Jagd auf die Nymphe Echo: Zur Technisierung der Wahrnehmung in der Frühen Neuzeit* (Bremen, Germany: Edition Lumière, 2011), 441.
10 See Ahmad Y. al-Hasan and Donald R. Hill, *Islamic Technology: An Illustrated History* (Cambridge: Cambrigde University Press, 1986), 55–56.
11 See Nadia Ambrosetti's essay "Wavering between the True and the False: A Short Excursion through Greek and Arab Automata," in *Allah's Automata*, esp. 46.
12 See Claus-Peter Haase, "Modest Variations: Theoretical Tradition and Practical Innovation in the Mechanical Arts from Antiquity to the Arab Middle Ages," in Zielinski and Fürlus, *Variantology 4*, 195–213.
13 On this, see the seminal study by George Saliba, *Islamic Science and the Making of the European Renaissance* (Cambridge, Mass.: MIT Press, 2007).
14 Pergaeus Apollonius, *Conics, Books V to VII: The Arabic Translation of the Lost Greek Original in the Version of the Banū Mūsā*, ed. Gerald J. Toomer (New York: Springer, 1990). This text machine also begins—as is usual in the manuscripts of the Muslim philosophers—with an invocation of Allah: "In the name of God, the merciful, the forgiving. I have no success, except through God" (620).
15 Ludwid Nix, ed., *Herons von Alexandria Mechanik und Katoptrik* (Leipzig, Germany: Teubner, 1900).
16 *Verba Filiorum Moysi, filii Sekir, id est Maumeti, Hameti et Hasan: Liber*

trium fratrum de geometria (Halle, Germany: Kaiserlich-Leopoldinisch-Carolinische Deutsche Akademie der Naturforscher, 1885).

17 In addition to Nadia Ambrosetti's article, see Fuat Sezgin and Eckhard Neubauer, *Wissenschaft und Technik im Islam: Vol. 5. Katalog der Instrumentensammlung des Institutes für Geschichte der arabisch-islamischen Wissenschaften* (Frankfurt am Main, Germany: Goethe University Frankfurt, 2003), 46–47. A description of the Eternal Flame is also found in Donald R. Hill, *A History of Engineering in Classical and Medieval Times* (London: Croom Helm, 1984), 359ff.

18 Banū Mūsā Ibn Shākir, *Kitāb al-hiyal [The Book of Ingenious Devices]*, ed. Ahmad Y. al-Hasan (Aleppo: Institute for the History of Arabic Science, University of Aleppo, 1981), 15.

19 The Turkish Ministry of Culture declined to lend the Istanbul copy to us because it is considered one of Turkey's most significant items of cultural heritage and they do not allow it to leave the country.

20 See, e.g., Farmer, *Organ of the Ancients*, 87.

21 Farmer.

22 An early but incomplete translation into German was undertaken by the archaeologist Eilhard Wiedemann based on the article in *al-Mashriq* 9 (1906). This essay, "Über Musikautomaten bei den Arabern," which first drew my attention to the work of the Banū Mūsā, was published in an Italian periodical, *Centenario della nascita di Michele Amari* 2 (1910): 164–85, and also circulated separately. See the reprint in the first volumen of the *Gesammelte Schriften zur arabisch-islamischen Wissenschaftsgeschichte (Veröffentlichungen des Instituts für Geschichte der Arabisch-Islamischen Wissenschaftsgeschichte)*, ed. Dorothea Girke (Frankfurt am Main, Germany: Goethe University Frankfurt, 1984). Imad Samir, who collaborated with the Max Planck Institute for the History of Science in Berlin, has produced a new translation for our project, commissioned by ZKM, Karlsruhe. Unlike the early translation, Samir's is complete and more accurate in many details. Together with the 1931 English translation by Henry George Farmer it is reprinted in *Allah's Automata*, 69–86.

23 George Saliba, "The Mysterious Provenance of Banū Mūsā's Treatise on Music," in *Allah's Automata*, 58–64. An early French version of Chaarani's text, *L'orgue hydraulique des Banu Mûsa*, has long been available online: http://www.muslimheritage.com/article/l'orgue-hydraulique-des-banu-mûsa-hydraulic-organbanu-musa.

24 Quoted from Farmer, *Organ of the Ancients*, 88.

25 For a detailed discussion of al-Jazarī's world of automata, see Zielinski and Fürlus, *Variantology 4*.

26 The Institute for the History of Arabic Science at the University of Aleppo hosted a major research project to reconstruct this valuable knowledge in

which al-Jazarī's manuscript plays a central part. Ahmad Y. al-Hasan has reedited it in Arabic under the title *al-Jāmi' bayn al-'ilm wa-'l-'amal an-nāfi' fī ṣinā'at al-ḥiyal (A Compendium on the Theory and Practice of the Mechanical Arts)* (Aleppo: Institute for the History of Arabic Science, University of Aleppo, 1979).

27 Massimiliano Lisa, Edoardo Zanon, and Mario Taddei, "The Manuscript by al-Murādī from Andalusia," in *Allah's Automata*, 123–29.

28 Aage Gerhardt Drachmann, *Ktesibios, Philon and Heron: A Study in Ancient Pneumatics* (Copenhagen: Munksgaard, 1948), 3.

29 Drachmann, 67–68.

30 Here the two-volume Leipzig edition *Herons von Alexandria Druckwerke und Automatentheater: Pneumatica et Automata,* procured by Wilhelm Schmidt and published in 1899, was used and also contains an appendix on Vitruvius's water clocks.

31 On the aforementioned protagonists of an early mechanical art, see Siegfried Zielinski, *Deep Time of the Media* (Boston, MA: MIT Press 2006).

32 Giambattista della Porta, *Pneumaticorum libri tres* (Naples, 1601), 60–66.

33 Athanasius Kircher, *Musurgia universalis sive ars magna consoni et dissoni,* 342–343.

34 For *Soundart*, the publication edited by Peter Weibel to accompany the exhibition of the same name at ZKM, Karlsruhe, I have assembled, under the title "Lüologie, Techno-Seelen, Künstliche Paradise," fragments of an an-archaeology of sound art. The essay is included in this volume as chapter 16.

Publication History

"History as Entertainment and Provocation: The TV Series *Holocaust* in West Germany." First published as "History as Entertainment and Provocation," *New German Critique* 19 (1980). Slightly revised in *Germans and Jews Since the Holocaust: The Changing Situation in West Germany*, edited by Anson Rabinbach and Jack Zipes (New York: Holmes & Meier, 1986).

"Media Archaeology: Searching for Different Orders of Envisioning." First published as "Medienarchäologie. In der Suchbewegung nach den unterschiedlichen Ordnungen des Visionierens," *Eikon* 9 (1994).

"Seven Items on the Net." First published as "Seven Items on the Net," *C Theory*, May 31, 1995. Available online at http://ctheory.net/ctheory_wp/seven-items-on-the-net/.

"Toward a Dramaturgy of Differences." First published as "Towards a Dramaturgy of Differences," in *Interfacing Realities*, edited by Joke Brouwer and Carla Hoekendijk (Rotterdam, Netherlands: V2 Organisatie, 1997).

"From Territories to Intervals: Some Preliminary Thoughts on the Economy of Time/the Time." First published as "From Territories to Intervals: Some Preliminary Thoughts on the Economy of Time/the Time," in *net.condition*, edited by Peter Weibel and Timothy Druckrey (Cambridge, Mass.: MIT Press, 2001).

"On the Difficulty to Think Twofold in One." First published as "On the Difficulty to Think Twofold in One," in *Sciences of the Interface*, edited by Hans Diebner et al. (Tübingen, Germany: Genista, 2001), written together with Nils Röller.

"The Art of Design: (Manifesto) On the State of Affairs and Their Agility." First published as "The Art of Design—[Manifesto] on the State of Affairs and Their Agility/Kunst der Gestaltung—[Manifest] zum Stand der Dinge und ihrer Bewegungsmöglichkeiten," in *Analog Mensch Digital—Design an der Schnittstelle*, bilingual exhibition catalog, edited by Shutterstock (Berlin: Shutterstock, 2014).

"'Too Many Images!—We Have to React': Theses toward an Apparatical Prosthesis for Seeing—in the Context of Godard's *Histoire(s) du Cinéma*."

First published as "'Zu viele Bilder—wir müssen reagieren!' Thesen zu einer apparativen Sehprothese im Kontext von Godards 'Histoire(s) du cinéma,'" in *Strategien des Scheins. Kunst, Computer, Neue Medien*, edited by Florian Rötzer and Peter Weibel (Munich, Germany: Boer, 1991).

"The Audiovisual Time Machine: Concluding Theses on the Cultural Technique of the Video Recorder." First published as "Audiovisuelle Zeitmaschine: Schlußthesen zur Kulturtechnik des Videorecorders," in *Zur Geschichte des Videorecorders*, by Siegfried Zielinski (Berlin: Volker Spiess, 1985).

"War and Media: Marginalia of a Genealogy, in Legends and Images." First published as "Krieg & Medien. Marginalien einer Genealogie in Legenden & Bildern," in *Bilderschlachten. 2000 Jahre Nachrichten aus dem Krieg. Technik-Medien-Kunst*, edited by Thomas Schneider et al. (Göttingen, Germany: Vandenhoeck und Ruprecht, 2009).

"*Theologi electrici*: A Few Passages." First published as "Theologici electrici. Einige Passagen," in *Theologie und Politik. Walter Benjamin und ein Paradigma der Moderne*, edited by Bernd Witte and Mauro Ponzi (Berlin: ESV, 2005).

"Historic Modes of the Audiovisual Apparatus." First published as "Historic Modes of the Audiovisual Apparatus," *Iris* 17 (1994). Special issue: "Spectateurs et publics de cinéma" ("Spectators and Audiences of Cinema").

"'To All!' The Struggle of the German Workers Radio Movement, 1918–1933." First published as "'An Alle!,'" in *Wem gehört die Welt—Kunst und Gesellschaft in der Weimarer Republik*, edited by Neue Gesellschaft für Bildende Kunst Berlin (Berlin: NGBK, 1977), in the original, much longer version written together with Erwin Reiss and Thomas Radevagen. This text was based on a more theoretical essay written together with Erwin Reiss in 1975 and published in 1976 as "Internationaler Medienzusammenhang— am Beispiel der Entstehung des Rundfunks in England, Frankreich und Deutschland," in *Das Argument*, Sonderband AS 10 (Karlsruhe, Germany: Massen/Medien/Politik, 1976).

"Urban Music Box, Urban Hearing: Avraamov's *Symphony of Sirens* in Baku and Moscow, 1922–1923—A Media-Archaeological Miniature." First published as "Stadt als Musicbox: Die Hupensymphonie von Avraamov in Baku und Moskau 1922/23. Eine medienarchäologische Miniatur," in *Programm und Programmatik. Kultur- und medienwissenschaftliche*

Analysen, edited by Ludwig Fischer (Konstanz, Germany: UVK, 2005). Translated into Portuguese for *Ghrebh: Journal for Communication* 9 (2007).

"How One Sees." First published as "How One Sees," in *Variantology 4: On Deep Time Relations of Arts, Sciences, and Technologies in the Arabic-Islamic World and Elsewhere,* edited by Siegfried Zielinski and Eckhard Fürlus (Cologne, Germany: Walther König, 2010), written together with Franziska Latell.

"Lüology, Techno-souls, Artificial Paradises: Fragments of an An-archaeology of Sound Arts." To be published in Peter Weibel, ed., *Sound Art: Sound as a Medium of Art* (Cambridge, Mass.: MIT Press, forthcoming).

"Designing and Revealing: Some Aspects of a Genealogy of Projection." First published as *Entwerfen und Entbergen. Aspekte einer Genealogie der Projektion,* International Flusser Lectures (Cologne, Germany: Walther König, 2010).

"Allah's Automata: Where Ancient Oriental Learning Intersects with Early Modern Europe. A Media-Archaeological Miniature by Way of Introduction." First published as "Allah's Automata: Where Ancient Oriental Learning Intersects with Early Modern Europe; A Media-Archaeological Miniature by Way of Introduction," in *Allah's Automata: Artifacts of the Arab-Islamic Renaissance (800–1200),* edited by Siegfried Zielinski and Peter Weibel (Berlin: Hatje Cantz, 2015).

Index

Aachen, 306, 386
Abenteuer des Werner Holt, Die (film), 9
Abou al-Hassan ibn Ali Ahmed Tlemsani, 321
abstract art, 339
abstraction, 39, 88, 166, 332, 340, 367, 370
academia, xiii, 47
Academia Secretorum Naturae, 158
Academy of Media Arts, Cologne, xix, 38, 42, 334
acoustics, 68, 200, 313, 366; mechanico-, 265; topographical, 265
Adenauer, Konrad, 215
Adorno, Theodor W., xxii, 6–8, 171, 183, 309
advertising, 4, 6, 7, 30, 133, 240
AEG (Telefunken), 212, 233
Aegina, 345
Aeon, 71
aesthetics, xii, 9, 38, 80–81, 95, 113, 120, 132, 183, 204, 253, 305, 366; commodity, 7–8, 20, 27; practice of, xix, 40–42, 46–47, 92; techno-, 52, 59, 130, 199, 207
aether, 182
Affaire Blum (film), 9
Africa, 147, 207, 272
agency (social), 86, 192, 340
agitation, 13, 16–17, 35, 236, 239, 241
Agrigento, 272
airplane, 254, 262–63
alchemy, 93, 339, 346, 352–53, 374
Alertz, Ulrich, 386
algebra, 378
algorithm(s), 48, 80, 120, 367, 378; military, 160; and organization, 102, 342
alienation of work, 266
Allah, 318, 323, 336, 378, 387

Allah's Automata (exhibition), 337, 377, 382–83, 389, 399
allegory, 180; of the cave (Plato), 345; drum (Athanasius Kircher), 190–92; theory of, 179
All-Russian Association of the Time League, 65
alphabet, 89
Alphaville (film), 100
Alptraum als Lebenslauf (film), 3
Althusser, Louis, xiv, 188
Ampex, 129, 130, 142
analog, 99, 120, 128, 135, 142
Anaxagoras, xiv
Anaximander, xiv
Andalusia, 321, 379, 383
Anders, Günther, xxii, 183
Angerer, Marie-Luise, xxiii
Angriff der Gegenwart auf die übrige Zeit, Der (film), 59
anima, 315
anima mundi (world soul), 309, 317, 356, 365
animation, 102, 319, 386; expanded, 306, 315, 365
animus, 315
Anschütz, Ottomar, 141
antifascism, 9, 25, 225, 227
anti-Semitism, 6–7, 9–10, 13–14, 27
Apocalypse Now (film), 28
Apollonius Pergaeus (Apollonius of Perga), 325–26, 328, 337, 383
apparatus: media, xviii, 65, 141, 169, 320; military, 211; musical, 265; optical, xi, 106; organizational/state, 35, 211; theory/heuristics (Jean-Louis Baudry), 54, 78, 100, 117, 345; visual/audiovisual, 40, 71, 112, 117, 152, 183, 187, 189–93, 195, 198–99, 320, 344, 346

408 | INDEX

Arabic-Islamic culture, 155, 157, 382–83
Arbeiterfunk—Der Neue Rundfunk (periodical), 217, 220, 222, 224, 234–35, 237, 239–42, 246, 249–51
Arbeiterfunktag (Workers Radio Day), 249
Arbeiter-Kurzwellen-Amateur (periodical), 240
Arbeiter-Radio-Bewegung (German Workers Radio Movement), xviii, 207, 210, 231, 215, 217–18, 222–24, 227, 233–48, 255
Arbeiter-Radio-Bund (ARB, Workers Radio League), 218, 222–24, 235–36, 238–40, 242
Arbeiter Radio-Klub (Workers Radio Club), 215–16, 230, 233, 236, 242
Arbeiter-Sender (periodical), 225–28, 236–37, 239–42, 244, 247–48, 250–51, 259
archaeology (Michel Foucault), xv, xvii; (an)archaeology, xviii, 35, 305; of cinema (Jean-Luc Godard), 99, 121; of media, 37, 127, 312, 356, 383, 399; of music, 320; of optics, 276
Archimedes of Syracuse, 149, 320, 335, 337
architecture, 53, 177, 179, 339, 367
archive, xii
ARD (German public broadcasting system), 4
Aristotle, xiv, 51, 269, 299, 300
Aristoxenus, 309, 327
Arndt, Johann (Johan Arnd), 345, 348–49
Arnheim, Rudolf, xxii
ars combinatoria, 49
Ars Electronica (festival, Linz), 204
Arsenal Berlin (cinema), 26, 143
ars memoriae, 51
Artaud, Antonin, 45–46
art history, 193, 383
artifact, xvi, xix, 37, 46, 52–53, 103, 118, 122, 130, 141, 156, 179, 305, 315, 325, 340, 342, 344, 378, 380, 384, 385, 388, 391

artificial paradise, 162, 176, 305, 388
Art in the Cinema (festival and symposium), 341
artist, xvi, xix, 40, 46, 48, 51, 59, 72–74, 91, 99, 139, 218, 239, 341
art of combination. See *ars combinatoria*
Ashby, W. Ross, xxii
astrolabe, 378
astronomy, 156, 269–70, 300, 355, 366, 378, 384
astrophysics, ixii, 34, 300, 355
atom, xiv, 272, 273
Atomists, 174, 309
audience: radio, 223, 226, 233, 243; research, 12, 31, 145; theater, 252; TV, 3–4, 6–7, 14–15, 26, 129–30, 132–33, 136–38, 162
Audion-Versuchserlaubnis (radio operating licenses), 213
audiovision, xix, 37, 317, 365
Aufhäuser, Siegfried, 246
Augsburg, 362, 364
Augstein, Rudolf, 18
Auschwitz, 14, 18, 20, 21, 24, 359, 367, 371
Aus einem deutschen Leben (film), 10
Ausschuß für Rundfunkstörungen (Committee on Broadcast Interference), 234
authority (of state), 19, 223
automation of music, 329
automaton, 81, 154, 155, 351–53, 380, 383; divine, 218, 323, 377, 395; meta-automaton, 218; music, 350, 378, 385, 386–87, 399; and time, 306
avant-garde, xviii, xxi, 46, 63, 65, 80, 99, 113, 131, 182, 198, 285, 308, 320, 341, 397
aviation. See airplane
Avraamov, Arseny M. (Arseny M. Krasnokutsky), 261–62, 264–65
Axel Springer (publishing house), 12
Ayteş, Ayhan, 379, 386
Azerbaijan, 262

INDEX | 409

Baacke, Dieter, xxii
Baake, Curt, 223
Bächlin, Peter, xxii
Bacon, Francis, 330
Bacon, Roger, 295, 299, 303, 306, 316, 377
Baden-Baden, 251, 276–77
Badiou, Alain, xvi
Baghdad, 293, 323, 325, 330, 350, 402, 404, 406, 408
Baku, 285–87, 289
Balestrini, Nanni, xxii
Baltrušaitis, Jurgis, xvii
Banū Mūsā brothers (Muḥammad, Aḥmad, Al-Ḥasan), 325–26, 328–29, 331, 378, 382–86, 389, 391, 393, 398
Barck, Karlheinz, 53
Barthes, Roland, xxii, 61, 100, 198
Baruchello, Gianfranco, xxii
Basel, 285
Basra, 269, 323, 380
Bastlermeister (periodical supplement), 240
Bastler-Zeitung (periodical), 240
Bataille, Georges, xx, 39, 45–46, 50, 59, 73, 180, 395–96
Baudrillard, Jean, xiv, xxii
Baudry, Jean-Louis, xxii, 54, 100, 188–90, 198, 201, 345
Bauer (publishing house), 13
Bayer, Konrad, xxii
Bayt al-ḥikma (House of Wisdom, Baghdad), 323, 378, 382, 384
Bazin, André, xxii
BBC (British Broadcasting Corporation), 12
Becquerel, Antoine H., 359
Béke, László, 340, 367
Bekhterev, Vladimir, 68
Belting, Hans, xxii, 270, 275
Benjamin, Walter, xi, xviii, xxii, 169–70, 174, 178–86, 195, 255, 341, 362
Bense, Max, xxii
Bentham, Jeremy, 35
Bergen-Belsen, 10

Berger, Hans, 181
Berger, John, xxii
Berger, René, xxii
Bergson, Henri, xxii
Berlin, xvi, xviii, 5, 9, 11, 19, 22–23, 25–26, 28, 32, 91, 93, 143, 162, 178, 209, 215, 220, 222, 226–27, 233, 235–38, 242–43, 249, 251, 306, 384
Berlin Academy of the Arts, 306
Berliner Welle (radio broadcaster), 238
Berlin Musical Instruments Museum of the State, Institute for Music Research, 306
Berlin University of the Arts, 386
Berns, Jörg J., 379
Berz, Peter, xxiii
Bexte, Peter, xxii
Bible, 312, 364
Biblioteca Medicea Laurenziana, Florence, 388
Bibliothèque Orientale, Beirut, 326, 385
binary: code, 47, 68, 100, 264; question, 85
Binswanger, Ludwig, 190
biology, xx, 52
biomechanics, 68, 264
bioscope, 195
Birkenau. *See* Auschwitz
Birmingham School of Cultural Studies, xix
black box, 320, 380, 388
Black Sea, 262
blasphemy, 174
Block, René, 306
Bloody May, 220
Blumler, Jay G., xxii
Blumröder, Christoph von, 306
Blut und Ehre: Jugend unter Hitler (TV series), 30
body: human, 60–61, 68, 73, 148, 176, 179, 192, 199, 205, 271, 285, 293, 309, 315, 341; physical, 39, 174, 177, 317, 327, 365, 389; and soul, 340
Bódy, Gábor, xxii
Bokanowski, Patrick, 99

Bologna, 285, 365
bomb, 9, 13, 23, 116, 162, 164; atomic, 359
bourgeoisie, 210, 225, 237, 243, 250
brain, 68, 70, 77, 105, 182, 193, 274, 278, 283–85, 293, 300, 368
Brandt, Willy, 19
Braun, Alfred, 251
Braun, Christina von, xxiii
Brecht, Bertolt, xi, xviii, xxii, 8, 54, 57, 74, 79–81, 251–55
Bredow, Hans, 207, 211, 236, 243
Brief an die Mutter, Ein (film), 26
Brno, 261
broadcast, 11, 15, 19, 99, 105–6, 129–30, 132–38, 140, 143, 146, 161–62, 181, 209–12, 215–18, 226–27, 229–30, 233–39, 243–46, 249–52
Brussels, 292
Buber, Martin, 89, 184
Budapest, 340
Bühler, Karl, xxii
Burch, Noel, 187, 206
Byzantium, 147, 315, 319–20, 326, 379, 389

Cahun, Claude, xxii
Cairo, 269–70, 300
camera, film and video, 65, 106, 112, 128, 131, 162, 166, 181, 189, 199, 264
camera obscura, 192, 193, 269, 270, 301, 321, 344–45, 349, 353, 356, 358, 373, 380
Cameron, James, 70
CanalPlus, 119, 122
capitalism, 7, 24, 212, 218, 220, 226, 243, 251
Carstens, Karl, 19–20
Cartesian: ego, 84–85, 88; doubt, 83; dream, 82; geometry, 189, 202
Cassirer, Ernst, xxii, 86–87
Castells, Manuel, xxiii
catoptrics, 342
catoptric theater *(theatrum catoptricum)*, 344

CBS, 21, 30
celestial bodies/objects, 271, 355, 364
centralization, 207
Cesi, Federico, 158
Chaarani, Mona S., 385, 401
Chad, 166–67
chaos, xv, 46, 314; pilot, 73; theory, 77
Chaplin, Charlie, 111, 113
Chapuis, Alfred, 321–22
Charcot, Jean-Martin, 360
Charlemagne, 306, 333, 386
Cheikho, Louis, 326, 385
Chekiang (Hangchow), 270
Cheng-Yih, Chen, 313
Chernobyl, 359, 360, 361
Children in the Concentration Camps (exhibition), 25
Chladni, Ernst F. F., 68, 178, 366
Chladni figures, 176, 180
Christianity, 13, 299, 302, 318, 344, 346, 356, 377, 379, 388
chronocracy (Peter Weibel), 58, 71, 379
chronophotography, 37
Chronos, 71
church: Christian, 8, 13, 170, 247, 249, 320, 325, 355, 384, 385
cinema, xii, xix, xxi, 10, 26, 28, 37, 39–40, 60–61, 63, 72, 74, 78–79, 99, 102–3, 105–14, 117–21, 129, 137–40, 156, 184, 187–88, 195, 197–202, 264, 318, 339, 341, 359
cinematography. *See* cinema
circuit: electrical, 210, 367; closed, 44
class struggle, 215, 217–18, 220, 223, 225, 227, 250, 254
Clement IV (pope), 282
clepsydra (water clock), 320–21, 379
clinamen: in Lucretius, xiv
clock, 61, 71, 140, 157, 319, 322, 326, 380, 399; elephant, 306, 380–82, 386–87; water, 306, 320, 321, 323, 335, 379, 389, 402
Cocteau, Jean, 113
code: binary, 68, 100, 264; Morse, 159, 230, 236, 265

INDEX | 411

Cohen, Leonard, 119
Cohen, Marcel, 47
Cohen-Séat, Gilbert, xxii
collaboration, xx, 95, 247
Collège de France, Paris, 61
Collegio Romano, Rome, 394
Cologne, xix–xx, 38, 42
Columbia University, 87, 385
Coming Home (film), 28
Comintern Radio, 235, 239
commodity, 8, 28, 105, 132; aesthetics, 7, 8, 20; cultural, 7, 113
communication, xiv, 54, 59, 68, 82, 99, 102, 107, 117, 122, 139, 147–49, 212, 252, 358, 378; coded, 113; electronic, ix, 49, 52, 82, 103; mass, 8, 27, 129, 130, 133, 135–36, 138, 141, 251
communism, 25, 161, 212, 216, 218, 220, 224, 235–38, 241, 250, 264–65
community, 82, 137, 216, 398
Comolli, Jean-Louis, xxii, 100, 117, 190, 198
compass, 171
complexity, 37, 52, 74, 119, 327, 399
computer, 44, 60, 77–78, 82, 89, 100, 164, 193, 201, 339, 367; game, 344; supercomputer, 71
concentration camp, 4, 10, 14, 18–20, 25, 27
conformity, 92
Constantinople, 155, 383
constructivism, 63
consumer, 8, 132, 139, 240
continuity, 46–47, 82, 188–89, 199, 272
control, xiii, 59–60, 72, 77–78, 81, 95, 131, 135, 139, 144–45, 162, 164, 183, 192–93, 212, 234, 328–29, 331–32, 367–85, 399
cosmology, 378
courier. *See* messenger
Crary, Jonathan, 193–94
creativity, 59, 73, 112
criptologia. *See* cryptology
crisis, 24, 87; economic, 226, 234; identity, 10

critical engineering, 207
critical thought, xiii, xv, xvi, 7–8
critique: cultural, 46, 59, 70, 79, 80, 147, 170–71, 179, 182–84, 188, 207, 235
cryptology, 152, 388
Ctesibius of Alexandria, 320, 326–27, 389
cultural industry, 7, 17, 21, 30, 79, 202, 204
cultural technique, xix, 93, 127, 129, 139, 141, 158, 307, 312, 314, 333
culture: hegemonic, 35, 193, 397; and technology, 190, 340, 397
Curtze, Maximilian, 383
cybernetics, 44, 95, 99–100, 375
cyberpunk, 100
cylinder: programmable, 395, 397, 399; rotating, 157, 191, 328–32, 393–95
Czechoslovakia, 10, 224

Dachau, 18, 20, 24
Dadaism, 182
Damascus, 301, 326
Daratt (film), 166–67
dark chamber. *See under* projection
data: glove, 50, 103; helmet, 50; processing, 80–81
David (film), 10
Davies, Douglas, 131
Day After, The (film), 30
Debord, Guy, xxii, 43, 73, 82, 375
Debray, Régis, xxiii
De Bry, Theodor, 309
decentralization, 135, 141, 199, 205
Dee, John, 171, 374
deep time, xiv, xix–xix, xxv, 57, 155, 157, 305, 308, 313, 332, 339, 365, 383, 389, 394, 398
Deer Hunter, The (film), 28
DEFA (GDR state-owned film studio), 9–10
Deir El-Balamand, 326
DeLanda, Manuel, xxiii
Deleuze, Gilles, xv, xxii, 37–38, 46, 119, 204, 315
della Porta, Giambattista, 158–60, 167–68, 331, 353, 393–96

Democritus, xiv, 272, 299, 309
Deren, Maya, xxii, 373
Descartes, René, 82–85, 89, 192–93, 272, 297–98, 300, 302–3, 344, 362
design, xiii, 37, 82, 91–95, 100, 136, 218, 264, 265, 326, 339–40, 362, 367, 370–71, 387
designer, 40, 44, 120, 139, 152, 156, 346
desire, 17, 39, 43, 45, 50, 109–10, 198, 318
determinism, 84, 170
Deutscher Radio-Club (German Radio Club), 212
Deutsche Volkspartei, 215
Deutsche Welle (German public broadcaster), 217–18
Deutschland, erwache! (film), 9
diagram, 66, 188, 276, 278, 283, 289, 292–93, 302, 306–7, 397
dialectic, 6, 30, 34, 198, 314
dialogue, xv, 43, 49, 116, 183–84
Diary of Anne Frank, The (film), 10
difference engine, xiii
digital, 89, 113, 120; design, 82, 99–100; fascism, 51
Digital Domain, 70
dioptrics, 342, 344, 358
discontinuity, 53
discourse, xvii, xxii, 3–4, 39, 46–47, 88, 99, 101–3, 106, 111
display, 111, 162, 170, 317, 344, 365
dispositif, 117, 188, 197–98, 205
disruption, 47, 91
dissonance, 91
diversity, xx, 37, 85, 352
Diviš, Prokop, 173, 184
Doane, Mary A., xxii
Dotzler, Bernhard, xxiii
dpd (German news agency), 22
Drachmann, Aage G., 389
Drahtloser Amateur Sendedienst (DASD, Wireless Amateur Transmission Service), 237
drama, 7, 13, 19, 20, 27, 31, 80, 180, 251
dream factory. *See* Hollywood

Dröge, Franz, xxii
Droz, Edmond, 321, 30
Druckrey, Timothy, xxiii
dualism, 40, 54, 80, 110, 184–85, 342, 345
Dubai, 387
Duchamp, Marcel, 45–46, 318
Duguet, Anne-Marie, xxii
Dulac, Germain, xxii
Dupré, Wilhelm, 127
Dynasty (TV series), 136

Eastern Front (World War II), 18, 209
Ebeling, Knut, xxiii
Eco, Umberto, xxii
ecological thinking, xv
economy, 68, 72–74, 92, 180, 232, 264; abolition of the (Georges Bataille), 59; of friendship, xxv; of information, 60; political, 59, 207
Edison, Thomas A., 141, 197
editing table (film post production), 110–11
Eggeling, Viking, 182
Egypt, 270, 319, 321, 333, 346, 379, 382
Ehe im Schatten (film), 9
Ehmer, Hermann K., xxii
eidola (idols), 273
Eidophor, 180
Einstein, Albert, xi, 87, 243
Einstürzende Neubauten (band), 261
Eisenstein, Sergei, xxii, 103, 116, 264
electrical sense, 173, 362
electricity, 65, 169–86, 317, 361, 362, 364–67, 370
electroencephalogram, 181
electromagnetism, xix, 127–31, 135–37, 178–79, 265
electronic reproduction, 105, 130, 141
electronics, 72, 181, 235, 317
electrotechnology, xviii
Elektrischer Schnellseher, 195
elite, 3, 388; cultural, 57; political, 20
Elsaesser, Thomas, xxii
Empedocles, xiv, 174, 272, 299
Engell, Lorenz, xxiii

Engels, Friedrich, 70
engineering, xii, 38, 78, 207, 323, 331, 386
enlightenment, 6–8, 331–32, 344–45, 359–60, 385, 388, 397
entertainment, 3, 6, 11, 27–28, 30, 80, 161–62, 202, 211, 212, 215, 231, 319
envisioning, 35, 38, 40, 42
Enzensberger, Hans Magnus, xxii
epic theatre (Bertolt Brecht), 255
Epicurus, xiv
epistemology, 86–87, 102, 187, 190, 192, 299, 344, 347, 353, 370
Erich Maria Remarque Peace Center, Osnabrück, 147
Ernst, Wolfgang, xxiii
Erwerbslosenfunk (Radio for the Unemployed), 226
eternal flame, 325
ethics, 20, 30, 45, 82, 88–89, 166, 184, 314
etymology, 78, 83, 86, 182, 320, 346, 386
Euclid, 271–72, 299, 302, 340
Euclidean geometry, 80
European High Middle Ages. *See* Middle Ages, European
European Media Art Festival Osnabrück, 147
exclusion, xviii, xxi, 220
expanded cinema, 101, 122, 198
expanded hermeneutics, xii, xix, 127
experiment: *cultura experimentalis*, xvii; experimental practice, xiii, xix–xx, 41, 131, 152, 169, 179, 182–83, 192, 352, 353, 362; thought, xxi, 101, 103
exploitation, 92, 220, 243, 398
eye: anatomy of, 268–300; artificial (*oculus artificialis*), xi, 344, 354, 358

"Factory Whistles" (poem), 266
falcon (symbol), 323, 380
Falkenberg, Alfred, 223–24
Falklands War, 166
Fallujah, 147
Fārābī, Abū Naṣr al-, 270–71, 327
Far East, 270, 306, 382
Farmer, George, 325, 385

Farocki, Harun, 57
fascism, xviii, 3–5, 9–18, 22–31, 51, 162, 198, 222, 224–27, 341
Fassbinder, Rainer Werner, 10
Fatih library, Istanbul, 276
Ferguson, Eugene S., 340
Fichte, Hubert, 88
film: biographical, 28; feature, 3–4, 9, 14, 28, 132, 136–40, 166
Filmkritik (periodical), 114
filmstrip, 141
fire-writing (Johann W. Ritter), 169, 178, 180
First World War (World War I), 57, 63
Flathaus, Alfred, 223
Flight of the Lindberghs, The (radio play), 251–55
Fludd, Robert, 160, 172, 293–95, 309–10, 317, 331, 365, 393, 395–96
Flusser, Vilém, xii–xiii, xx, 89, 183, 339–44, 353, 366–67, 370–72
Fluxus, 131
FM Einheit (musician), 261
Fontana, Giovanni, 152–55
form: geometrical, 159; meta-, 87
Foucault, Michel, xv–xvii, xx, 35–38, 188, 205–6, 272
Frank, Rudolf, 251
Frankfurt/Main, 10, 26
Franklin, Benjamin, 170
Freier Radio-Bund Deutschlands (FRBD, Free Radio League of Germany), 224–28, 238–39, 241–42
Freke, John, 185, 317, 364–65
Freud, Sigmund, 46, 117, 188, 340, 360
Freund, Gisèle, xxii
Fricker, Johann L., 173, 184
Fulda, 172
Fuller, Matthew, xxiii
Fülöp-Miller, René, xxii, 61, 184
Funkstunde (periodical), 251
futurism: Russian, 68, 264

Gabriel, Ulrike, xxiii
Gai Savoir, Le (film), 106, 110, 122

414 | INDEX

Galen of Pergamon, 268, 272, 299, 302
Galileo Galilei, 192, 344, 355–56
Galloway, Alexander, xxiii, 344
Galvani, Luigi, 365
galvanism, 169, 173, 176, 183, 361, 365
Gao, Shiming, xxiii
garden, xvi, xxiv, 35, 321, 380; Garden of Eden, 47
gas chamber, 11, 16, 18
Gastev, Aleksei K., xxii, 63, 66, 68, 261, 263–67
gaze, 38–39, 106–7, 110, 190, 195, 270–71, 299, 305, 340, 346; history of the, 42
gear wheel, 157
Geissler, Heiner, 21
gender, 172, 315, 323
genealogy (Friedrich Nietzsche, Michel Foucault), xv–xvi, xviii, xxi, 38, 152, 156, 187, 272, 276, 299, 315, 320, 335, 339, 341, 355, 365, 375
generators of surprise (Mahlon Hoagland), xvi–xvii
genocide, 10–12, 15, 18, 25
geometry, 189, 270, 293, 297, 309, 323, 325, 382–84, 388; non-Euclidean, 80
German Communist Party (KPD), 212, 218, 220, 222, 224, 237–38
Germans in WWII, The (TV series), 136
Germany: East (German Democratic Republic), 9; West (German Federal Republic), 3–22, 24, 26–31, 162, 207
Germany, Year Zero (film), 120–21
Giedion, Siegfried, xxii
Gilbert, William, 171
globalization, 92
glockenspiel, 331, 395
Gmelin, Otto F., xxii
Godard, Jean-Luc, xxii, 72, 98–101, 104–22, 198, 205
Goethe, Johann W. von, 170
Golden Age of Arab-Islamic culture/science, 307, 323, 346, 377–78, 383, 386, 389, 399
Göttingen Königliche Societät der Wissenschaften (Göttingen Royal Society of Sciences), 365
Gould, Stephen J., 54
Grafe, Frieda, xxii
Graham, Dan, 131
Graz, xii, 57, 59
Greece, ancient, xiiii, 71, 83, 86, 150, 271, 302, 305, 307, 309, 319–26, 334, 346, 379–85, 394–95, 398
Greek orthodox church, 325, 385
Green, Gerald, 3
Griffiths, Keith, 116
Groß-Berliner Einheitsausschuß werktätiger Rundfunkhörer (Unity Board for the Working-Class Listeners of Greater Berlin), 227
Gruner & Jahr (publishing house), 13
Guattari, Félix, xv, xxii, 37–38, 46, 183, 204, 315
Gudi flutes, 313–14, 334
guerrilla television, xviii
Gumbrecht, Hans U., xxii

Habermas, Jürgen, xviii, xxii
hacker, xviii
Haken, Bruno N., 251
Hall, Stuart, xxii
Hamburg, 237
Hansen, Miriam, xxiii
Haraway, Donna, xxiii
hardware, 133, 138, 145, 176, 197, 331
Harlan, Veit, 27, 30
harmony, 109, 171, 180, 309; cosmic/divine, 308–9
Haroun, Mahamat-Saleh, 166–67
Hārūn al-Rashīd (fifth Abbasid Caliph), 306, 320, 386–87
Hasan al-Rammah, 155–56
Hauptmann von Köln, Der (film), 9
Hausen, Karin, xvii
Hayles, N. Katherine, xxiii
Heath, Stephen, xxii, 134
hegemony, 57, 187, 199, 308–9; cultural, 8; European, 333
Heidegger, Martin, xv, 370–71

Heider, Fritz, xxii
Heinrich Hertz Institute, Berlin, 235
Heisenberg uncertainty principle, 103, 116
Heißenbüttel, Helmut, xxii
helioscope, 356, 358
Hellenism, 305, 309, 315, 319
Helvíkovice, 173
Henschel, 162
Herbert, Nick, 116
Hero of Alexandria (Heron of Alexandria), 155, 325–26, 383, 389, 391–93
Herrmann, Hans-Christian von, xxiii
Heterodoxies (Johann W. Ritter), 179, 183
heterogeneity, xii, xvii–xxi, 14, 37, 46, 51, 74, 77, 195, 201, 205, 302, 312, 342
heterology (Georges Bataille), xx
heterotopia (Michel Foucault), xx
heuristics, 189, 195, 198, 353
Hickethier, Knut, xxii
hierarchy, xiii, xviii, 45, 53, 353
Hill, Donald R., 336, 386
Hillis, Danny, 71
Hindenburg, Paul von, 238
Hiroshima, 79, 359
Hirschberg, Julius, 268, 289
Histoire(s) du cinéma (film), 99, 101, 104–6, 109, 110–22, 198, 205
history: chronological, 37; discontinuity of, 193; oral, 26
history of science, 150, 271–72, 276, 297, 325, 340, 355, 366, 383, 385
Hoagland, Mahlon B., xvi
Hocke, Gustav R., xvii, 42
Hoffmann Group (neo-Nazi group), 23
Hoggart, Richard, xxii
Holl, Ute, xxiii
Höllerer, Walter F., xxii
Hollywood, 397
Holocaust, the, xviii, 3–32
Holocaust (TV series), xviii, 3–32
Holy Inquisition. *See* Roman Inquisition
Holzer, Horst, xxii
Homer, 148

homo artefactus, 94
homogeneous, 54
Hood, Stuart, xxii
Horkheimer, Max, xxii, 7–8, 72, 171, 183
Hörl, Erich, xxiii
horologium. *See* clock
horror, 11, 27, 164, 166
Howkins, John, 137
Howlings in Favour of Sade (film), 43
Hübner, Heinz, 12
Huhtamo, Erkki, xxiii
Hui, Yuk, xxiii
Huillet, Danièle, xxii
humanities, xvii–xviii; German-speaking, xiii
Humboldt, Wilhelm von, 83
Hunayn ibn Isḥāq al-ʿIbādī, Abū Zayd, 268, 274
Husserl, Edmund G. A., 88, 188, 370
hydraulics, 149, 319–29, 332, 379–80, 386–89, 396, 398

Ibn al-Haytham (Abū ʿAlī al-Ḥasan ibn al-Ḥasan), 269–81, 283, 289, 293, 299–302, 340, 368
Ich war neunzehn (film), 9
icon, 37, 50, 51, 105, 112, 272, 345, 378
iconic turn, 99
iconoclasm, xi
iconography, 80, 100, 192
iconoscope, 181
identity, 20, 51, 59, 73, 195, 205
ideology, xviii, 6–7, 14, 20, 24, 27, 73, 79, 132, 146, 183, 188, 252, 255, 308
illusion, 46, 51, 60, 103, 105–7, 110, 117, 130, 188–90, 195, 199, 204, 246, 341–42, 344
image: computer-generated, 39; technical, 100, 102
imaginary–real–symbolic (Jacques Lacan), xiv, 102, 200, 204
imagination, xx, 40, 52–55, 59, 73–74, 92, 94, 107, 117, 152, 193, 204, 265, 340, 352, 368; Flusser on, 339–40, 367
IMAX theater, 201

immigrant, 24
incandescence, 176
industrialization, 83, 141, 320, 388
Industrial Light and Magic, 70
infinity, 71, 88, 339
informatics. *See* computer
information, 27, 40, 52, 59–60, 72, 85, 89, 149, 204, 332, 367, 372; theory, 182
In jenen Tagen (film), 9
Innis, Harold, xxii
installation (art), 131, 204–5, 305, 316, 399
Institut für Sprache im technischen Zeitalter (Institute for Language in the Technical Age, Technical University Berlin), 127
instrument: measuring, 60–61, 379; musical, 180, 262, 265, 309, 313, 320, 328, 332, 385, 399; radio-musical, 265; string, 262, 329
interactivity, 44, 59, 73, 100, 116, 120, 122, 145, 205, 388
interdiscursivity, xiii, xvii–xviii, 37, 50, 188, 339
interface, 44, 49–50, 54, 72–73, 77–89, 116, 122, 202, 279, 344, 378, 380, 382, 393
internet, xix, 43, 49–51, 89, 169, 339
interobjectivity, xv
Iris (periodical), 187–88
Islam, 150, 155–57, 272, 274, 307, 315, 322–24, 332, 346, 377–78, 382–83, 386, 388, 399
Istanbul, 147, 155, 276, 384
Ives, Charles, 264

Jabès, Edmond, 47, 54
Jacob's ladder, 312–13
Jakob der Lügner (film), 9
Janssen, Jules, 156
Jarry, Alfred, 71
Jazarī, Ibn al-Razzāz al-, 306, 318, 320–21, 323–25, 336, 380–82, 386–88, 391, 399, 402
Jena, 176
Jerusalem, 147

Jesuits, 170, 173, 361, 362
Jewish culture, 3, 8–20, 24–30, 367, 377
Jiahu, 313
Jidl mit der Fiedel (film), 26
Joseph Süss Oppenheimer (film), 27
journalism, xiii, 15, 19, 139, 164
Jiddische Kino, Das (film), 26
Jud Süss (film), 27, 30

Kabbalah, 47, 367
Kaifeng, 321, 380
Kairos, 71, 80, 312; poet, 70, 73
Kaiserhofstrasse 12 (film), 28
Kamāl al-Dīn al-Fārisī, 279, 281
Kamper, Dietmar, xvi, xxiii–xxiv, 107, 122
Kant, Immanuel, 367
Kapp, Ernst C., xi, xxii, 340
Katz, Elihu, xxii
Kazan, 65, 70
Kepler, Johannes, 192, 271–72, 274, 293, 296–97, 300, 344, 356
Khwārizmī, Abū Jaʿfar Muḥammad ibn Mūsā al-, 378, 384
Kinder aus Nr. 67 oder, Die (film), 28
Kindī, Yaʿqūb al-, 271, 273
kinetics, 40, 117, 152, 315, 321, 325, 380, 384, 397
King (TV series), 28
Kircher, Athanasius, 169–74, 180, 190–93, 204, 275, 285, 331–32, 345, 362, 394–98
Kirkegaard, Jacob, 361
kitsch, 255
Kittler, Friedrich A., xii, xxiii, 80, 309
Klein, Yves, 353
Kluge, Alexander E., 59
Knilli, Friedrich, xii, xvi–xxii, 3, 15, 31–32, 86, 162, 259
Knödler-Bunte, Eberhard, xxii
Knowbotic Research (group of artists), 46, 48, 204
knowledge: fragmentation of, 83; objective, xiii; organization of, xviii, 46
Kodak, 359
Kohl, Helmut, 27
Kommunales Kino Frankfurt (cinema), 26

Königlich Bayerische Akademie der Wissenschaften (Royal Bavarian Academy of Science), Munich, 179
Kracauer, Siegfried, xxii
Kramer, Harry, xxii
Krämer, Sybille, xxiii
Kremlin, 265
Krieg, Peter, 201
Kriwet, Ferdinand, xxii
Kroker, Arthur, xxiii
Kroker, Marilouise, xxiii
Krupp und Krause (TV series), 9
Kuhle Wampe (film), 57, 251
Kuleshov, Lev, xxii, 81, 182
Künzel, Werner, xxiii
Kyeser, Konrad, 156–58, 168
Kymographion, 61

labor, xvii, 26–27, 65, 70, 79, 81, 83, 140, 145, 161, 241, 246, 265–66, 352, 358
laboratory, xvi, 41–42, 66, 68, 77, 93, 143, 152, 176, 378
Lacan, Jacques, xii, xiv, 46, 50, 188, 197–98, 200, 206
La Mettrie, Julien Offray de, 60
Lang, Fritz, 106, 116
language, xi–xii, 44, 47, 83, 103, 107, 189, 192, 205, 264; poetic, 68; of propaganda, xviii; universal, 61
lapis philosophorum, 352–53, 366
Lascaux (cave paintings), 100
Lask, Berta, 251
Lasswell, Harold D., xxii
L'Atalante (film), 113
Laurel, Brenda, xxiii
Lauretis, Teresa de, xxii
Led Zeppelin, 313
Lehrstück (didactic drama, Bertolt Brecht), 4, 80, 252, 255
Leibniz, Gottfried W., 87, 358
Leiris, Michel, 45
Lenin, Vladimir, 63, 116, 265
lens: camera, 39, 40, 106, 166, 269, 344; eye, 276, 278–79, 285, 293, 342; telescope, 149, 354, 356
Leonardo da Vinci, 156, 344, 373

Leonardo3 (research group), 388
Lessing, Gotthold E., 131
letterpress, 83
Levin, Thomas Y., xxiii
Lévi-Strauss, Claude, 99
Lévy, Pierre, xxiii
Lichey, Georg, 251
Lichtenberg, Georg C., 180–81, 365–66
light (*lux*), 78, 86, 117, 131, 149, 152, 189, 193, 269–75, 289, 293, 299, 342, 353–56, 358, 362, 366, 386; divine/sacred, 85, 323, 345, 377, 384, 387; electric, 170–71, 173–76, 178–82
light-figure (Johann W. Ritter), 178, 180–81
Lightning over the Water (film), 111
Lilienthal, Peter, 4, 10
Lindberg, David C., 271–73
Lindbergh, Charles, 252–53
Link, David, xxiii
Lippit, Akira M., 359
Lisa, Massimiliano, 388
Lischka, Gerhard J., xxii
Lissajous, Jules A., 366
Llull, Ramon, 358, 377–78
logic, 52, 80, 103, 120, 193, 384
logos, xx
Lohr, Charles, 378
London, 93, 147, 273, 308, 317, 364–65, 375, 397
Lo Sardo, Eugenio, 253, 171
loudspeaker, xii, 77, 236
Lovink, Geert, xxiii
Lucas, George W., Jr., 70
Lucretius, xiv–xv, 179, 299
Luhmann, Niklas, 88
lumen, 275, 377
luminocentrism, 274
lüology, 308, 313, 402
Lyotard, Jean-François, xxiii, 71

Mach, Ernst W. J. W., 271
machine: blind, 102; fantasy (René Fülöp Miller), 61, 184; human, 63, 85, 264; human–machine interface, 72, 85, 88, 116, 202; illusion-, 60; image-,

60; media-, 38, 50–51, 54, 73, 116; mediation-, xiii; military, 60; paradise, 85, 89, 388; programmable, 84–85, 89, 329, 385, 399; regulating-, xiii; sight, 187, 344; wish-, 188–90. *See also* time machine
magnetic head (video recorder), 113, 128
magnetism, 113, 171–76, 317, 361–62, 374
Maidanek, 20
Maletzke, Gerhard, xxii
Malevich, Kazimir S., xxii, 339
Mama, ich lebe (film), 9
Ma'mūn, Caliph al-, 323, 382
manchfaltig, 372
manifesto, xix, 91, 220, 222, 262, 339
Manovich, Lev, xxiii
Man Ray, 182
Marek, Kurt W., xvii
Marey, Étienne-Jules, 40, 58, 60–65, 141, 156
Mariátegui, José C., 89
Marinetti, Filippo T., xxii
marketing, 12, 21, 28, 50, 145, 202
Marquis de Sade, Donatien A. F., 39, 43, 171
Marx, Karl, 70
Marx, Wilhelm, 215
Marxism, 99, 169, 226, 238, 250
Mashriq, al-, (periodical), 326, 385
Masina, Giulietta, 107
mass culture, 7, 26, 30
mass production, 7, 117
Massumi, Brian, xxiii
master model, 269, 276, 293, 299
materialism, xiv, 63, 81, 299, 314, 383
materiality: medial, x, 201; techno-aesthetic, 130
materiology, xvi
mathematics, xvii, 48, 87–88, 309, 323, 325, 380, 382–83, 388
matrix, 181–82, 372
Maurolico, Francesco, 289, 292, 300
McLuhan, H. Marshall, xi, xxii, 83, 100, 199

McQuail, Denis, xxii
Mecca, 378
mechanical device, 315, 318, 320, 323, 325, 358, 379, 383–89
mechanics, 152, 183, 254, 325, 378, 382–84
media: binary code of, 100; critical, xvi; critique, 59, 182, 183; design, 100; linear, 135; mass, 6, 8, 13, 16, 22, 161, 171, 182–83, 210, 212, 250, 252, 255, 265; media-human, 50–51, 54, 73, 116; media-machine, 50–51, 54, 73, 166
media archaeology. *See* archaeology
media art, xix, 72–74, 77, 170, 176
media studies, xiii, 86–88
media theory, xiii, 77, 88–89, 156, 190, 345
media thinking, xii–xiii, xvi–xvii, xxi, 207, 305; genealogy of, xiv, xxi
Mediterranean Sea, 306, 312
Mein Kampf (film), 9, 11
Mekas, Jonas, 99
melancholy, 70, 95, 244, 372
Merleau-Ponty, Maurice, 107
Mersch, Dieter, xxiii
Mersenne, Marin, 309
Mertens, Herbert, xvii
Mesopotamia, 269–70, 323, 388
message, 68, 129, 132, 135, 147–50, 158–59, 167, 209–11, 265; metaphysical, xii
messenger, 147–48, 167
Messina, 172, 289
metaphor, 6, 20–21, 51–53, 134, 187, 355, 372, 386
metaphysics, xii, 88, 160, 170, 173, 183, 275, 294, 300, 318, 335, 345, 346, 362, 367
Metz, Christian, xxii, 102, 198
Meyer-Eppler, Werner, xxii
microscope, 344
Middle Ages, European, 270, 289, 305, 321, 331, 346, 378, 380
Miéville, Anne-Marie, 99
Milan, 388
militancy, 22

military, 60–61, 131, 143, 147–48, 156–57, 160, 164, 166, 207, 211–13, 220, 247, 250
Miller, Henry, 341
Mills, Mara, xxiii
minaret, 320, 386
minority, social, 20
mirror, xv, 149, 163, 190–92, 198, 341–45, 350, 356
Mitrofanova, Alla, xxiii
modernity, xxi, 171, 308, 318, 332, 333, 389, 397; southern, 269
Modern Times (film), 111–12
Mohists, 269–70
Moles, Abraham A., xxii, 40
Mon, Franz, xxii
money, 72, 162, 256
monochord, 310
moon, 34, 79, 136, 147, 180–81, 275, 344, 353
Mörder sind unter uns, Die (film), 9
Moscow, 63, 67, 161, 235, 262, 264
Mount Sinai, 367, 370
multiplicity, xx, 38, 171
Mulvey, Laura, xxii
Mumford, Lewis, 379, 399
Murādī, Aḥmed ibn Khalaf al-, 388, 391, 399
Muristos, 150–51
Museum for Islamic Art, Qatar, 388
music, xx–xxi, 4, 7, 21, 44, 68, 78, 104, 107, 112–13, 119, 160, 172, 177, 180, 201, 204, 215, 236, 249, 250, 252, 261–67, 305–15, 320–34, 378, 382, 384–87, 391, 395, 397–99
musicology, 61, 313
Mussolini, Benito A. A., 225
Mutoscope, 195
Muybridge, Eadweard, 40
Myograph, 61

Nackt unter Wölfen (film), 9
Nagasaki, 359
Nake, Frieder, xxii
Nancy, Jean-Luc, xv

Nannen, Henri, 18
Naples, xviii, 172, 395
NASA, 202
National Socialism (Nazism), xviii, 3, 8–28, 136, 161–62, 181, 225, 228, 230, 238, 370
natural history, 61, 178
naturalism, 180, 303
natural philosophy, 160, 169, 171, 173, 271, 300, 312, 315, 323, 361
Nazi Germany. *See* National Socialism
NBC, 3, 6, 28
Needham, Joseph, 312, 334
Nees, Georg, xxii
Negt, Oskar R., 140
Nelson, Theodor (Ted) H., xxii, 169
Neoplatonism, 309
net. *See* internet
net culture, 48
network, electronic, xi, xii, xx, 52–53, 78, 82, 92, 95, 181, 183, 233, 382
Neuchâtel, 330
Neue Rundfunk, Der (periodical), 217, 242
Neukrantz, Klaus, 242, 250
Neumann, John von, xxii
New Crowned Hope (festival), 166
newspaper, 11, 13, 16, 22, 181, 211, 228, 239–42, 339
Newton, Isaac, 344, 353, 362
New York Times, 133
Nielsen Corporation, 133
Nietzsche, Friedrich W., xi, xv–xvi, 272
Nikritin, Solomon, xxii
North by Northwest (film), 116
nouveau roman, 100
Novalis (Georg Philipp Friedrich von Hardenberg), 35, 41, 78
November Revolution, 209, 212
nuclear: bomb, 22; catastrophe, 359–60; research, 359; war, 30
numerus sonorous, 309

objectification, 53, 318, 387
Occident, xxi, 155, 243, 378
October Revolution, 213

oculocentrism, 274
Oetinger, Friedrich C., 169, 174–75, 184
Old Testament, 367
OMNIMAX theater/cinema, 201
ontology, xv–xvi, xx, 6, 11, 21, 40, 80, 342, 359, 371
ophthalmology, 268, 274
optic nerve, 276, 289, 292–93, 342
optics (discipline), 172, 269, 270–83, 289, 292–93, 297, 299–303, 342, 344, 347, 353
organ (musical instrument, automaton), 320, 326–31, 385, 391, 395, 397–98
organism, 179, 318; cinematic, 103
organ projection (Ernst Kapp), xi, 340
Orient, xxi, 155, 378, 379
origin(s), xiii–xiv, 14, 52, 71, 92, 112, 129, 155, 157, 170, 319, 339, 340–41, 382, 385, 394, 398
Ørsted, Hans C., 178
oscillation, 180, 309
Ottoman Empire, 387

Padua, 152, 155, 285
Paech, Joachim, xxiii
Paik, Nam June, xxii, 103, 127, 131
painting, 100, 119, 131, 177, 179, 201, 339, 347, 353, 369
Palestine, 21
pamphlet, 58–59, 238, 262
Panopticon (Jeremy Bentham), 35
pantheism, 186
Papen, Franz von, 223
paradigm shift, 99
parents' generation, 17
Parikka, Jussi, xxiii
Paris, 61, 68, 308, 397
Pasolini, Pier Paolo, xiii, xxii, 91
Pasquinelli, Matteo, xxiii
Passion (film), 110
pathology, 39, 45, 190, 372
Peckham, John, 275, 300
pedagogy, 24
Peirce, Charles S., xxii
perception, xv, xxi, 40–41, 72, 84, 86, 105–7, 113, 127, 142, 169, 187, 205, 269, 271–75, 283, 293, 299–300, 340, 344
performance: artistic, 46, 131, 140, 252, 254, 261–65; sound, 321, 328, 385
perpetuum mobile, 325, 384
Persia, 83, 147, 274, 307, 315, 318, 328, 335, 378–82, 389
perspective (visual), 106, 130–31
Peternák, Miklós, xxiii, 340, 367
Petrograd, 65, 68, 75, 264–65, 267, 308, 397
Petrushka (composition), 264
Petsans, Johannes, 292
Petzold, Hartmut, xvii, xxiii
Pezold, Friederike, xxii
phenomenology, xv, 105, 116, 360, 370
Philips, 233
philology, xii, xviii–xix, xxi, 127
Philo of Byzantium, 320, 326, 389
philosophy, 46, 87–88, 107, 174, 188, 339, 367, 370, 384; dialogical, 82, 89; French, xv; history of, 83; of science, 87–88. *See also* natural philosophy
phoenix, 382
photography, 39–40, 60–63, 65, 105, 138, 141, 156, 166, 181, 183, 199, 224, 264, 326, 359–60, 385
physical computing, 385
Physical Technical Institute, St. Petersburg, 265
physicotheology, 173, 317, 362, 364
physiology, 60–61, 65, 68, 106, 197, 274, 275, 312
Pias, Claus, xxiii
Pijet, Georg W., 236, 250
pipe, 152, 313, 314, 328, 329, 391, 394–96
Pisa, 285
Pisters, Patricia, xxiii
Plato, xiv, 299–300, 345
Pleynet, Marcelin, xxii
plurality, 171, 195
pluriverse, 91
pneuma, 315
pneumatics, 151, 312, 315–16, 319, 321,

326, 329, 382, 384, 386–89, 391–92,
  394–95, 398
poet, xiv, 59, 73, 91, 94, 176, 261, 272
poetics, 68, 92, 170, 370; of relation, xx
poetry, xiii, 53, 70, 74, 100, 120, 131, 240,
  263–65, 378
political education, 12, 17
politics, xiii, xxi, 19, 26, 59, 161, 169–70,
  183, 212, 215, 224, 235, 239, 246
Polke, Sigmar, 353, 374
Polybius, 149–50
Polygraph, 61
polymath, 77, 169, 171, 269, 273
Popitz, Heinrich, 140
Posner, Hans, xxii
postwar generation, xxii, 17, 26
potential space (Donald W. Winnicott),
  xvi, 342
Prague, 340, 353, 367, 370
Praxinoscope, 195, 196
Presocratics, 42, 174, 272, 299, 370, 389
Prigogine, Ilya R., 54
prism, 270, 353
Professor Mamlock (film), 9
programmer, 95, 132; omnipotent, 84–85
project (Vilém Flusser), xii, 340, 342, 344,
  353, 366–67, 370–71
projection: apparatus, 188–89, 191–92,
  195, 199, 339, 344, 346; dark chamber,
  xi, 191–92, 269, 299, 353; screen, 102,
  111, 189, 201, 341–42; 70mm, 201;
  35mm, 201
projector, 110, 189, 192, 198, 346
proletariat, 63, 218, 220, 224, 227, 236,
  240–41, 244, 249–50, 257, 264
propaganda, xviii, 10, 161, 216, 228
prosthesis, 50, 105, 117
protest, 16, 20–21, 25, 228
Prussia, 224, 236, 249
psychoanalysis, 50, 99, 110, 184, 188, 339,
  342, 375
psychology, 7, 10, 60, 68, 162, 188, 199,
  299, 340, 342, 359
public debate. *See* discourse
Puissance de la parole (film), 112

pulsator mundi. *See* magnetism
pump, 262, 319, 387; air, 329; water, 379
punch card, 331–32, 399
Purkyně, Johannes E., 279
pyrotechnics, 317
Pythagoras, 305, 308–9, 314, 327, 397,
  398

Qi, 312
quantum mechanics, 87
Queneau, Raymond, xxii

radio: amateur, 212–16; antenna, 265;
  detector, 230–33; drama, 180, 251;
  military, 212–13; receiver, 210, 216,
  230–33, 252; shortwave transmitter,
  236–37; tube, 209, 213, 220, 231, 233,
  240
Radio Moscow, 161, 235
Radio-Sende-Spiel, 161
Ramsbott, Wolfgang, xxii
Rancière, Jacques, xv
Rat der Götter, Der (film), 9
rationality, one-dimensional, 45
Ray, Nicholas, 106, 111, 113
Rear Window (film), 106
Reasons for Knocking at an Empty House
  (video artwork), 131
Reck, Hans-Ulrich, xxiii
recording, 65, 113, 116, 128, 138,
  144, 156, 189, 236, 361; apparatus/
  device, 61, 112; cross-track, 128;
  electromagnetic, 127, 129–35
Red Army, 213, 261
Red Square, 262
reflexology, 68
Reichardt, Jasia, xxii
Reich Broadcasting Corporation, 207, 235
Reich Ministry of Public Enlightenment
  and Propaganda, 207, 235
Reich Postal Service, 211
Reichskristallnacht, 12
Reichs-Rundfunk-Gesellschaft (Reich
  Radio Society), 216
Reiss, Erwin, xxii, 162

422 | INDEX

religion, 170–71, 183, 249, 308, 335, 346, 372, 377–78, 397; monotheistic, 378
renaissance: Arabic-Islamic, 307, 315, 382; European, 269, 273, 305, 331
representation, xv, 43, 100, 182, 188, 270–71, 300, 378; antirepresentation, 102
resistance movement, 18–19, 25–26
Resistance: Not Conformism (exhibition), 25
Resnais, Alain, 99
revealing (entbergen), 371
revolver photographique, 156
Rheinberger, Hans-Jörg, xvi
rhizome, 37, 52
rhythm, 57, 63, 68, 73, 79, 89, 104, 112–13, 119, 119, 128–29, 135, 141, 146, 200, 261–62, 322, 326, 378–80, 385
Riga, 264–67, 287
Risner, Friedrich, 271, 273, 276, 283, 289
Ritter, Johann W., 169–70, 174–86, 333, 341
Roads, Curtis, 305
rock music, 4, 21, 313
Rogoff, Irit, xxiii
Röller, Nils, xxiii, 122
Roman Inquisition, 216, 355
romanticism, 78, 178
Rome, 147, 172, 190, 308, 384, 394, 397
Ronell, Avital, xxiii
Roots (TV series), 28
Ropohl, Günther, xix
Rosenbach, Ulrike, 131
Rosen für den Staatsanwalt (film), 9
Rösler, Gottlieb F., 174
Rossellini, Roberto, 120
Rössler, Otto E., 77, 82, 84–85, 88–89
Rostov, 262
Rotation (film), 9
Rote Fahne (periodical), 237
Rote Welle (radio broadcaster), 236
Rothko, Mark, 353
Rötzer, Florian, xxiii
Royal Society, London, 317, 364
Rühm, Gerhard, xxii

Rund um den Rundfunk—Rundschau des Volksfunk-Arbeiterfunk (periodical), 223–24
Runze, Ottokar, 4
Russolo, Luigi, xxii

Sabra, Abdelhamid I., 273–78, 280–81, 301
Sabra and Chatila massacre, 17–18
Saint Joseph University, Beirut, 326, 385
Saliba, George, 325–26, 385
San Francisco Museum of Modern Art, 341
Sans Soleil (film), 44
São Paolo, 261
Sartre, Jean-Paul C. A., 99
Satan, 345, 349
Saussure, Ferdinand de, xxii
Schallspiel, totales, xii
Schein, 195, 199
Scheiner, Christoph, 192, 285, 354–58
Schelling, Wilhelm J. S., 88, 184
Scherchen, Hermann, xxii
Scheuchzer, Johann J., 362, 364
Schiller, J. C. Friedrich von, 170
Schlaf der Gerechten, Der (film), 10
Schmidgen, Henning, xxiii, 335
Schmidt, Helmut H. W., 19
school, xvi, 8, 14, 24–25, 246, 252, 308–9, 314, 383; textbook, 12
Schramm, Matthias, 273, 279
Schramm, Wilbur, xxii
science, modern, 83, 93, 174, 193, 272, 297, 300
scientia subordinate, 309
scopic regime, 40
sculpture, 131, 177, 179, 378; kinetic, 325, 384
Second World War (World War II), xxi–xxii, 136, 161
self-consuming, 73
self-organization, 204, 230
semiotics, 100, 104, 134–35, 149
sensation, xiv, xxi, 39–40, 46, 54, 60, 71–71, 91, 137, 252, 261, 312, 360, 378

Serres, Michel, xxii
Sezgin, Fuat, 155, 336, 386
shadow, 48, 74, 152, 177, 269, 345, 359, 362, 373
Shakespeare, William, 49
Shannon, Claude E., xxii
Shen Kuo, 270, 380
shock, 16, 182
Sidon, 306
Siegert, Bernhard, xxiii
Siemens, 233
Sie sind frei, Dr. Korczak (film), 10
signal, xiv, xix, 60, 68, 112, 128, 148–49, 164, 230, 262–63
sign system, 50
Simon, Gérard, 273
Simondon, Gilbert, xxii
simulation, 82, 120, 204
singing, 312
"Sirāj Allāh" (the Lamp of God), 384
Situationist International, 43, 46, 73, 82
Six fois deux (film), 119
Skirrow, Gillian, 134
Smirnov, Andrey (Andrej Smirnov), xxiii, 267
Smith, A. Mark, 274
Social Democrats (SPD, German political party), 218, 222, 236, 238, 246, 250
Socrates, xiv, 370
Soest, Conrad von, 347
software, 139, 145, 197
sonimages, 99, 113
Sony, 145
soul, 35, 65, 173–74, 243–45, 306, 309, 315, 317–19, 340, 352, 361–62, 365, 379, 386–87
sound art, 48, 72, 105, 177, 179, 265, 305–9, 312, 315, 320, 332–33
Soviet occupied zone (in Germany), 9
Soviet Union, 10, 21, 63, 75, 262, 264
space: half-space (Johann W. Ritter), 177, 341. *See also* potential space
spectator, 107, 187–92, 195, 197–98, 341
speleology, 38, 42
Sphymograph, 61

Spiegel, Der (periodical), 4, 11, 18
spirit, 173–74, 178, 180, 244, 252, 264, 276, 285, 315, 318, 362, 372, 387
spiritus mundi, 174
Spraos, John, 198
Staatsbibliothek zu Berlin (Berlin State Library), 384
Stalin, Joseph V., 68, 239, 264
standardization, xx, 38, 79, 314, 352
Stärker als die Nacht (film), 9
Steirischer Herbst (festival), 57
Sterne (film), 9
Stern ohne Himmel (film), 28
Steyerl, Hito, xxiii
Stinnes, Hugo, 215
Stjernfelt, Frederik, xxiii
St. Petersburg. *See* Petrograd
Straub, Jean-Marie, 99
Stravinsky, Igor F., 264
structuralism, 99
structure: economic, 128; power, xiii
Stuckenschmidt, Hans H., xxii
Stück Himmel, Ein (TV series), 28
subject: ciné-subject, 187–90, 195, 197–205; subject-effect, 188, 201, 204; TV-subject, 200
subversion, 47, 82, 183
Süleymaniye Library, Istanbul, 386
Super 8 film, 11
surface, medial, xii, 169
surrealism, 45
surveillance, 131
Survival and Resistance (exhibition), 25
Su Song, 321, 380
symbol, xii, 39, 43, 50–52, 104–5, 112, 174, 200, 309, 323, 340, 356, 358, 384
symbolic: action, 38; form (Ernst Cassirer), 86–87; interaction, 204; machine, 378; turn, 199
symbolic (Jacques Lacan), xiv, 102, 204–5
Symphony of Factory Sirens (composition), 261–63
syntax, 39, 102–3, 119
synthesis, 83, 86
synthesizer, audiovisual, 112

system: complex, 94; computer-centered, 193; of knowledge, xviii; technical, xi, xix, 52–53, 60, 72, 107, 148, 162, 195, 236, 326, 339, 344
systems theory, 88

taboo, 10, 18, 21, 27
tactility, 182
Taddei, Mario, 388
Tag, Ein. Bericht aus einem deutschen Konzentrationslager. Januar 1939 (film), 10
tape: (electro)magnetic, xix, 135–38; recorder/machine, xix, 112, 117; speed, 128; video-, 105, 116–22, 127–33, 201. *See also* video
Taylor, Frederick W., 63
Taylorism, 63, 69; Russian, 66
technē, 305
technical a priori (Friedrich Kittler), xii
technical reproducibility (Walter Benjamin), 183, 195
Technical University, Petrograd, 68
Technical University of Berlin, xvii, 4, 15, 383
technics, 371
techniques, 94–95, 120, 129–30, 139, 193, 201, 204, 341, 344, 345. *See also* cultural technique
telecommunication, 43, 51, 53, 89, 149, 265
Telefunken, 162–63, 207, 209–12. *See also* AEG (Telefunken)
telegraph, 235; torch-, 147–49, 159
telegraphone, Poulsen, 112
telematics, xix, 38, 43, 49–53, 95, 102–3, 149, 169
teleology, xv
telephone, 16, 183
telepresence, 51, 202
telescope, xi, 34, 66, 344, 350, 354–56, 358
television (TV), 3–6, 9–16, 19–22, 27–28, 79, 99, 102–3, 112, 122, 127–41, 162–66, 181, 198–201, 339; Guerrilla, xviii; high-definition (HD), 199; music, 200; series, 6–16, 20–21, 26–30, 136; signal, 128, 164
territory, 57–58, 147, 149, 215
Tewje, der Milchmann (film), 26
Thacker, Eugene, xxiii
Thatcher, Margaret H., 166
theater, 54, 68, 78–80, 155, 162, 201, 251, 344; Heron of Alexandria on, 389, 391, 394, 395
theologi electrici, xxi, 169–86, 317, 364, 366–67, 370
theology, 160, 169–86, 275, 317–18, 361–62, 364–67, 370, 378
theremin, 265
Theremin, Lev S., 265
"Thereminvox," 265
Theunissen, Michael, 372
thinking without a banister (Hannah Arendt), xiii
Third Reich, 8–9, 216. *See also* National Socialism
Tholen, Georg C., xxiii
Thompson, Kristin, xxii
Thorn (Toruń), 383
Three Moons Monastery, Beirut, 325
time: audiovisual now-, 100; pathology of (Michael Theunissen), 372; policy, 73; temporal coordinates, xix, xxi, 127–40, 178, 320, 386; time-manipulation techniques (video), 129; visualization of, 103. *See also* deep time
time machine, xix, 71, 74, 112, 116–17, 128–33, 138, 141–42, 147, 377, 399
timeshifting (video recording), 133
Tiqqun, xxiii
Toledo, 321, 380
Toller, Ernst, 241, 251
Ton Steine Scherben (rock band), 119
Topkapı Palace, Istanbul, 336, 384
total war (German fascism), 162
Traber, Zacharias, 342
transcendence, 71, 84–85, 88, 197, 204
transdisciplinarity, xviii
tree structure, 52

Treusch-Dieter, Gerburg, xxiii
Trotz alledem! (film), 9
Turing, Alan M., xxii, 79
Tuscany, 355
TV. *See* television
TV Dé-coll/age (Wolf Vostell), 166
typewriter, xi, 111–12

unconscious, xii, 46, 110
universalism, 38, 169–70, 183, 332, 352
Universitäts- und Forschungsbibliothek Erfurt/Gotha, 284
University of Messina, 289
University of Padua, 152
University of Salzburg, xix, 374
Urban VIII (pope), 355
use-value, 8, 129, 134, 138–39, 144, 200–202
utopia, xv, xviii, xxiv, 41, 52, 120, 166, 169–70, 182, 266, 377, 379

VALIE EXPORT, xxii
variantology, xix, xxi, xxv, 35, 308
variations/variant, xiii–xv, xix, 82, 293, 353
Vatican, 171–72, 332, 384
Vaucanson, Jacques de, 330–31
Veklenko, Oleg, 359–60
Verhoeven, Michael, 4
Vernet, Marc, xxii
Versandhaus Arbeiterkult (Workers Culture Mail-Order Business), 236
Vertov, Dziga, xxii, 75, 116, 182, 264
Vesalius, Andreas (Andries van Wesel), 285–87, 292
vibration, 65, 155, 170, 173, 309, 362
victim, 7, 9, 12, 16, 18–21, 26–27, 30, 264
video: art, 131–32; recorder, xix, 100, 112, 117, 127–45
videodisc, 139, 145, 201
videotape. *See* tape
Vienna, 105, 122, 147, 166
Vietnam War, 28, 166
Viola, Bill, 131, 143
violence, 16, 20–21, 44, 110–11, 140

Virilio, Paul, xxii, 206
virtuality, 50, 51–52, 323
virtual reality (VR), 50, 51, 54, 116, 172, 202–3, 384
vision, theory of. *See* optics
visual studies, xiii
vitalism, 365
Vitruvius (Marius Vitruvius Pollio), 389–90, 394
volcano, 172
Volksempfänger VE 301 (radio receiver model), 161, 231
Volksfunk—Illustrierte Wochenschrift für Berlin-Magdeburg-Stettin (periodical), 223
Volta, Alessandro, 176, 365
voltage, 170, 173, 180, 362
Von Richtern und anderen Sympathisanten (film), 3–4
Vostell, Wolf, xxii, 165–66
V2_Institute for the Unstable Media, 49

Wagner, Herbert, 162
Wajdowicz, Roman, xxii
Walkman, 103, 146
Warszawianka (*Whirlwinds of Danger*, composition), 262
Watzlawick, Paul, xxii
WDR (West German public broadcaster), 15, 143
weapon, 23, 158, 164, 167, 207, 359. *See also* nuclear
Weibel, Peter, xxii, 57–58, 71, 88, 105, 122, 377
Weimar Republic, 207, 223, 250
Weiß, Helmut, 244
Welt in jenem Sommer, Die (film), 28
Wenders, Wim, 111
Western Allied occupied zones (in Germany), 9
Western Front, 209
Wetzel, Michael, xxiii
Weyl, Gerda, 249
Weyl, Hermann, xxii, 86–88
Wheel of Life, 195

Whitney Brothers, xxii, 373
Wiedemann, Eilhard E. G., 150, 270–73, 327–28, 336, 401
Wie ein Hirschberger Dänisch lernte (film), 10
Wiener, Norbert, xxii, 100
Wiener, Oswald, 54
Wiener, Philip, 275
Wiesel, Elie, 11, 19, 21
Wiking Jugend (neo-Nazi group), 22
Williams, Raymond, xxii, 134, 141
Winckel, Fritz, xxii
Windahl, Sven, xxii
Winkler, Hartmut, xxiii
Winnicott, Donald W., 342
Witelo, Erazmus Ciolek, 273, 275, 283–85, 289, 293, 300–301
Wittgenstein, Ludwig J. J., 54, 341
Wollen, Peter, xxii
Workers' Radio Movement. *See* Arbeiter-Radio-Bewegung
working class. *See* proletariat
world soul. See *anima mundi*
world spirit. See *spiritus mundi*
World War I. *See* First World War
World War II. *See* Second World War
World Wide Web, xix, 5. *See also* internet
writing, 83

Wrocław laboratory, 312
Wundt, Wilhelm M., 60
WZSPS (Soviet trade union radio broadcaster), 235

Xenophon, 149
X-ray, 359

Yiddish culture, 26
Yin dynasty, 313
Youngblood, Gene, 102, 120

Zahn, Johannes, xi, 358
ZDF (German public broadcaster), 4, 26–28
Zentralfunkstelle (Central Broadcasting Headquarters), 209
Zentralny Institut Truda (ZIT), Moscow, 63
Zentrums-Partei (Center Party), 215
zero-dimension (Vilém Flusser), 340, 367
Zhou dynasty, 313
ZKM | Center for Art and Media Karlsruhe, 337, 377, 382, 401–2
zodiac, 320
Zoopraxiscope, 195
Zylinska, Joanna, xxiii

(continuted from page ii)

40   OF SHEEP, ORANGES, AND YEAST: A MULTISPECIES IMPRESSION
Julian Yates

39   FUEL: A SPECULATIVE DICTIONARY
Karen Pinkus

38   WHAT WOULD ANIMALS SAY IF WE ASKED THE RIGHT QUESTIONS?
Vinciane Despret

37   MANIFESTLY HARAWAY
Donna J. Haraway

36   NEOFINALISM
Raymond Ruyer

35   INANIMATION: THEORIES OF INORGANIC LIFE
David Wills

34   ALL THOUGHTS ARE EQUAL: LARUELLE AND NONHUMAN PHILOSOPHY
John Ó Maoilearca

33   NECROMEDIA
Marcel O'Gorman

32   THE INTELLECTIVE SPACE: THINKING BEYOND COGNITION
Laurent Dubreuil

31   LARUELLE: AGAINST THE DIGITAL
Alexander R. Galloway

30   THE UNIVERSE OF THINGS: ON SPECULATIVE REALISM
Steven Shaviro

29   NEOCYBERNETICS AND NARRATIVE
Bruce Clarke

28   CINDERS
Jacques Derrida

27   HYPEROBJECTS: PHILOSOPHY AND ECOLOGY AFTER THE END OF THE WORLD
Timothy Morton

26   HUMANESIS: SOUND AND TECHNOLOGICAL POSTHUMANISM
David Cecchetto

25 ARTIST ANIMAL
Steve Baker

24 WITHOUT OFFENDING HUMANS: A CRITIQUE OF ANIMAL RIGHTS
Élisabeth de Fontenay

23 VAMPYROTEUTHIS INFERNALIS: A TREATISE, WITH A REPORT BY THE INSTITUT SCIENTIFIQUE DE RECHERCHE PARANATURALISTE
Vilém Flusser and Louis Bec

22 BODY DRIFT: BUTLER, HAYLES, HARAWAY
Arthur Kroker

21 HUMANIMAL: RACE, LAW, LANGUAGE
Kalpana Rahita Seshadri

20 ALIEN PHENOMENOLOGY, OR WHAT IT'S LIKE TO BE A THING
Ian Bogost

19 CIFERAE: A BESTIARY IN FIVE FINGERS
Tom Tyler

18 IMPROPER LIFE: TECHNOLOGY AND BIOPOLITICS FROM HEIDEGGER TO AGAMBEN
Timothy C. Campbell

17 SURFACE ENCOUNTERS: THINKING WITH ANIMALS AND ART
Ron Broglio

16 AGAINST ECOLOGICAL SOVEREIGNTY: ETHICS, BIOPOLITICS, AND SAVING THE NATURAL WORLD
Mick Smith

15 ANIMAL STORIES: NARRATING ACROSS SPECIES LINES
Susan McHugh

14 HUMAN ERROR: SPECIES-BEING AND MEDIA MACHINES
Dominic Pettman

13 JUNKWARE
Thierry Bardini

12 A FORAY INTO THE WORLDS OF ANIMALS AND HUMANS, *WITH* A THEORY OF MEANING
Jakob von Uexküll

11 INSECT MEDIA: AN ARCHAEOLOGY OF ANIMALS AND TECHNOLOGY
Jussi Parikka

10 COSMOPOLITICS II
   Isabelle Stengers

9 COSMOPOLITICS I
   Isabelle Stengers

8 WHAT IS POSTHUMANISM?
   Cary Wolfe

7 POLITICAL AFFECT: CONNECTING THE SOCIAL AND THE SOMATIC
   John Protevi

6 ANIMAL CAPITAL: RENDERING LIFE IN BIOPOLITICAL TIMES
   Nicole Shukin

5 DORSALITY: THINKING BACK THROUGH TECHNOLOGY AND POLITICS
   David Wills

4 BÍOS: BIOPOLITICS AND PHILOSOPHY
   Roberto Esposito

3 WHEN SPECIES MEET
   Donna J. Haraway

2 THE POETICS OF DNA
   Judith Roof

1 THE PARASITE
   Michel Serres

Siegfried Zielinski is Michel Foucault Professor of Media Archaeology and Techno-Culture at the European Graduate School in Saas-Fee, Switzerland; honorary doctor and professor of the Budapest University of Arts; until 2016, chair of media theory/archaeology and variantology of media at Berlin University of the Arts; and director of the Vilém Flusser Archive. He was founding rector (1994–2000) of the Academy of Media Arts Cologne and rector (2016–18) of the Karlsruhe University of Arts and Design. He is the author of *[. . . After the Media]: News from the Slow-Fading Twentieth Century* (Minnesota/Univocal, 2013) and coedited *Flusseriana* and *DIA-LOGOS: Ramon Llull's Method of Thought and Artistic Practice* (Minnesota, 2018).